APPRAISING LAKATOS

VIENNA CIRCLE INSTITUTE LIBRARY [2002]

1

Institut Wiener Kreis

Society for the Advancement of the Scientific World Conception

Series-Editor:

Friedrich Stadler

Director, Institut Wiener Kreis

and University of Vienna, Austria

Editorial Address:

Institut Wiener Kreis
Museumstrasse 5/2/19, A-1070 Wien, Austria
Tel.: +431/5261005 (international)
or 01/5261005 (national)
Fax.: +431/5248859 (international)
or 01/5248859 (national)
email: i_v_c@ping.at
homepage: http://ivc.philo.at

APPRAISING LAKATOS
Mathematics, Methodology, and the Man

Edited by

GEORGE KAMPIS
Eötvös University, Budapest

LADISLAV KVASZ
Comenius University, Bratislava

MICHAEL STÖLTZNER
Institute Vienna Circle, Vienna
University of Bielefeld

KLUWER ACADEMIC PUBLISHERS
DORDRECHT / BOSTON / LONDON

Library of Congress Cataloging-in-Publication Data is available.

ISBN 1-4020-0226-2

Published by Kluwer Academic Publishers,
PO Box 17, 3300 AA Dordrecht, The Netherlands.

Sold and distributed in North, Central and South America
by Kluwer Academic Publishers,
101 Philip Drive, Norwell, MA 02018, USA

In all other countries, sold and distributed
by Kluwer Academic Publishers, Distribution Center,
PO Box 322, 3300 AH Dordrecht, The Netherlands

Printed on acid-free paper

Gedruckt mit Förderung des *Österreichischen Bundesministeriums für Bildung, Wissenschaft und Kultur*
Printed with financial support of the *Austrian Ministry for Education, Science and Culture*

In Zusammenarbeit mit dem *Zentrum für überfakultäre Forschung der Universität Wien*
In cooperation with the *Center for Interdisciplinary Research of the University of Vienna*

Printed and bound in Great Britain by MPG Books, Bodmin, Cornwall

Lakatos in December 1954
(By kind permission of Éva Pap)

Table of Contents

III. LIFE AND POLITICS

IV. IMRE LAKATOS' HUNGARIAN DISSERTATION

Introduction

The present collection is the result of a cooperation between the Institute Vienna Circle and the Institute for History and Philosophy of Science of Eötvös University, Budapest, which was dedicated to the philosophy of science in the Austro-Hungarian context. Probably no other protagonist of 20th century philosophy of science fits better into such a framework than does Imre Lakatos (1922–1974). The book *Proofs and Refutations* which made him famous, declares itself as a continuation of György Pólya's works on mathematical heuristics and plausible reasoning and, accordingly, stands in the eminent tradition of Hungarian mathematics. When Lakatos subsequently embarked upon general philosophy of science and developed his methodology of scientific research programs (henceforth MSRP), he became involved in at times heated debates with three offshoots and opponents of the Vienna Circle: Karl Popper, Thomas S. Kuhn and Paul Feyerabend. Before giving a short overview of the present volume, we shall briefly sketch this historical setting in order to illustrate the motivation of the Institute Vienna Circle to start its new book series with a volume on Imre Lakatos.

After completing his Cambridge Ph.D. disseration in 1961, Lakatos came to the London School of Economics to work under Popper. He subsequently attempted to extend – though with nuances of his own – Critical Rationalism to mathematics. Seeking more and more an independent stance in the late 1960s, Lakatos' relationship with Popper developed into a thematic rift, culminating in personal tensions. Lakatos' recently published last lectures are quite telling in this respect. Nevertheless, both Lakatos and Popper stood side by side in struggling against the 'new epistemology' of the 1970s. While they emphasized the unavoidability of a criterion demarcating science and scientific rationality from non-science, the 'new epistemologists' – among them Thomas Kuhn and Paul Feyerabend became the most prominent – insisted on the indivisibility of the whole of scientific practice, in particular on the impact of societal factors, institutions, ideologies, etc., upon scientists and their research programs. They held that – even for philosophical purposes – the actual course of history could not be replaced by a rational reconstruction of history, and that there existed no methodology guaranteeing scientific success. Hence no absolute justification of scientific rationality could be reached.

G. Kampis, L. Kvasz and M. Stöltzner (eds.),
Appraising Lakatos: Mathematics, Methodology and the Man, ix–xiv.

During the last two decades, historical investigations into Logical Empiricism have brought to the fore various historical facts pertaining to the battles waged during the 1960s and 1970s. On the one hand, Popper's relations with the Vienna Circle were much more complex than his claims to the effect that he had simply 'killed' Logical Positivism suggest; in particular, the *Logic of Scientific Discovery* played its part in the movement's discussions during the 1930s. Victor Kraft, who was, together with Béla Juhos, the last local representative of the Vienna Circle after 1945, shared some of Popper's core ideas. In 1956, Kraft recommended Lakatos to Cambridge to obtain a second Ph.D. under Braithwaite. The correspondence between Lakatos and Kraft preserved in the Lakatos Archive at LSE indicates that Kraft was basically the only Vienna Circle member whose writings – apart from Carnap's of course – Lakatos had thoroughly studied. Lakatos especially appreciated Kraft's early philosophy of mathematics.

On the other hand, Kuhn's seminal book *The Structure of Scientific Revolutions* appeared in the *International Encyclopedia of Unified Science.* When Otto Neurath had launched this project in the mid 1930s, he did so in the attempt to replace *the* system of science by an ongoing cooperation of scientists in the Vienna Circle's spirit. Because of the war and Neurath's untimely death in 1945, a tiny core of 19 booklets, along with bibliography, were the only elements of his enormous plan to make it into print until 1970. A closer look at the writings of Logical Empiricists after 1945 reveals that many Kuhnian themes were not unfamiliar or outlandish to that movement. Yet neither Philipp Frank's nor Otto Neurath's respective activities became widely known until the fairly recent Neurath renaissance. Paul Feyerabend, a physics student in post-war Vienna, started his philosophical career in Kraft's circle. In a way, his "anti-system" and "anti-method" crusades can be seen as pointedly exaggerating Neurath's criticism of Popper's pseudo-rationalism.

***

A quarter of a century after Imre Lakatos' untimely death, and more than a decade after the hitherto immense popularity of Lakatos-style case studies has significantly decreased, it seems to be a suitable moment – and at a sufficient distance – to critically reevaluate the old debates: Is our historical picture of the classical controversy over scientific rationality still adequate? Are Lakatos' methodological proposals, above all the MSRP, still promising today, and can they be suitably refined so as to accommodate the lessons of the large number of detailed case studies ensuing from them? Or do they contain limitations and deficiencies in principle? Discussion, thus, of possible excess content – to phrase it in Lakatosian terms – will be one of the objectives of the first section of the present collection. Although the history of philosophy of science is itself presently becoming a research program, the goal of the papers assembled therein is not merely retrospection. To the contrary, the so-called 'science wars' ensuing from the Sokal hoax taught us that good arguments in defense of scientific rationality are still timely. Present combatants sometimes reveal a certain reluctance to draw lessons from classical debates. In this respect, this section might be a useful reminder of a former state of the art. Moreover, Lakatos' thinking still bears unexplored perspectives. He was a philosopher of many roots, some of which, such as the thinking of Hegel and Lukács, constitute a rather atypical background for a 20th century philosopher of science, especially for one who had started out in mathematics.

Recent publications on Lakatos in scientific journals demonstrate, above all, a revived interest in his philosophy of mathematics which represents one of those few approaches that are not centered around (or limited to) foundational issues, but center around mathematical practice as such. On this, the second section focuses. After all, Lakatos regarded himself primarily as a philosopher of mathematics. During the last years of his life, he was seeking to apply the MSRP to mathematics. Although in some papers from the 1970s one finds footnotes commenting upon *Proofs and Refutations*, Lakatos could not close the circle himself, and his papers do not provide a real clue about his plans. Today, it seems clear that major modifications in both the MSRP and his philosophy of mathematics are required to reach that goal. Mathematics is much more finely subdivided into sub-disciplines than are the empirical sciences. On the other hand, mathematical research is not so strongly concentrated around a small number of problems such as, say, hunting the Higgs particle or deciphering human DNA. More generally, mathematicians typically do not dispute a common set of facts. Instead, aesthetic criteria, such as conceptual simplicity or the beauty of a proof, essentially determine the quality of a mathematical argument; and they are typically at issue whenever mathematicians debate whether a theorem is not only true but really final, or whether we have fully understood a mathematical structure. Interestingly, the Hungarian tradition of mathematics to which Lakatos is indebted through György Pólya, Alfréd Rényi, and Arpád Szabó was much more problem-oriented than, for instance, the various French or German schools. Especially in Pólya's writings, heuristic beats justification by far. This stance not only opens many interesting perspectives for the connection between mathematics and the sciences, but is also most attractive in mathematics teaching.

Lakatos' philosophy emphasizes mathematical growth and conceptual evolution over the justification and consolidation of a theory's structure. But in his charges against Euclideanism, and through the all-pervasive fallibilism behind them, the Lakatosian dialectics of proofs and refutations visibly overshoots the mark. Although mathematical rigor is indeed often suspended in the early phase of a budding research program, and although understanding may play a role far more important than justification, the standards of rigor furnished by a commonly accepted metatheory unite the variety of mathematical sub-disciplines. Ultimately, all doubts will be cast on the axioms. Already the rift between the first and the second parts of *Proofs and Refutations* reveals that Lakatos was at odds with the axiomatic method, which utterly characterizes 20th century mathematics. For many modern concepts and theorems – even for those which have found applications in physics – there simply are no informal ancestors. Still, most mathematicians maintain that mathematical intuition must somehow precede proof. If this problem could be remedied, Lakatos' quasi-empiricism might be rather promising for assessing the spectacularly successful encounters between mathematics and theoretical physics.

In Lakatos' mature thinking, some apparently contradictory influences have merged, including Hegel and Popper, mathematics and critical fallibilism. Interestingly, some core themes, such as the idea of research programs, are foreshadowed already before his emigration. Both tendencies can well be seen from the paper "Modern physics, modern society" that was published in Hungarian in 1947. There, a rather standard Communist exercise in ideology suddenly develops into a discussion of L. Susan Stebbing's *Philosophy and the Physicists*. We provide here a full translation of this text mainly because it represents a major part of Lakatos' Hungarian Ph.D. dissertation defended in the same year. The dissertation as a whole is missing in the University archives at Debrecen, but the opinion of his supervisor Sándor Karácsony proves that "Modern physics, modern society" represented its first part. We thank Gábor Kutrovátz for translating both documents into English. The documentary section is rounded off by a fairly complete bibliography of Lakatos' Hungarian papers written prior to his emigration in 1956. Discussing the philosophical import of Lakatos' text, the different personal histories of the editors came to the fore. To a Western reader, both parts are somewhat separated, and historical distance renders them mainly just a document. Having been constantly imbued with – and often suffered under – the idea that there are deterministic laws governing class history in the same sense as Newtonian mechanics, the East-European reader gains quite another picture.

Until his imprisonment in 1950, Lakatos was a radical communist who considered his own deeds – "unforgivable" as they appear today – to be part of a historical process governed by dialectical laws, which could be rationally reconstructed. While such a parallel must not be overstressed, it does provide an interesting perspective from which to start studying Lakatos' biography. To be sure, only few members of his generation had the boring sort of biography that philosophers are alleged to have. Yet Lakatos' was exceptionally rich in curves. There have been many intellectual émigrés from the communist East, but only a few of them made such a significant shift from the far left to the far right. The rupture in his biography became almost tantamount to meeting two personalities. His earlier political involvement were stories he hardly ever told anybody in the West. As rumors today abound, a comprehensive volume on Lakatos has to tell them and contrast them to his political activity at LSE. But there is more at stake than just being comprehensive. The historical investigations into scientific philosophy in exile have set a high standard in contextualizing this tradition with the general history of intellectual emigration. As an Austro-Hungarian joint project, the final section of this volume intends to provide a certain basis for undertaking similar scrutiny with Lakatos in the future.

***

The present volume emerged from two workshops which were held in Vienna, from 12–14 September 1997, and Budapest, from 30–31 October 1997. The second was very close to what would have been Lakatos' 75th birthday, which provided an excellent occasion for assessing the genuinely Hungarian aspect of his personality. Many contemporaries of his are still alive; some of them joined forces with him, and some suffered from his deeds. We held a panel discussion moderated by Lee Congdon, in which took part Alex Bándy, Alex Bellamy, György Litván, Jancis Long, Sándorné Kántor, Gábor Palló, Éva Pap, László Ropolyi, Miklós Szabolcsi, and Gábor Vajda. Panelists also reported on their experiences and difficulties in securing biographical material on Lakatos. But at the core of the exchange were the recollections of Lakatos' political activities, and of science policy of the 1950s, in general. So this event, in a way, became a part of Hungary's coming to terms with its Communist past. The fact that our initial plan to hold this discussion in the Eötvös Collegium – an institution which was closed down under the influence of its former student Imre Lakatos – was first accepted and then turned down by its current Director, is a clear sign how present this past is still for many. For a certain time, the editors considered reproducing selected passages from the tape of his panel discussion in the present volume. We refrained from doing so because a substantial part of the facts presented there is contained in Jancis Long's contribution, and the 'event-like' character of the panel discussion cannot be put into print.

A series of photographs of Imre Lakatos from different epochs of his life completes the volume. The pictures show him with his second wife Éva Pap, in scientific and political company. We are very grateful to Éva Pap for granting us permission to publish these photos from her personal archive for the first time. She holds the copyright for all of them.

On July 26th, 1999, Professor John Watkins suddenly passed away. It had been a great honor for us that he accepted our invitation to the Budapest workshop and we well recall his talk and the most interesting personal remarks about the linkage of Imre Lakatos' two lives. There have appeared many thoughtful obituaries since. With some delay we present a collection of papers that also show the important role of John Watkins within post-war philosophy of science.

***

The cooperation that led to the present volume would have been impossible without the constant and generous funding of a bilateral organization specifically created to support Hungaro-Austrian contacts, the Aktion Österreich-Ungarn. We especially thank Elke Schmidt and Agnes Schnaider, of the Vienna and Budapest offices, as our respective institutional partners. In order to improve the quality of this volume, the editors have sent out the papers to other contributors for an internal refereeing. We are very indebted to them for their very constructive suggestions. Christopher Roth has improved the stylistic quality of those contributions written by non-native speakers. Marco Mertens and Jeanette Wette have prepared the index of names. We owe thanks to all those who helped us in organizing the workshops: the team of the Institute Vienna Circle: Robert Kaller, Helmut Ruck, Angelika Rzihacek, and Margit Kurka; and in Budapest: Jancis Long, Gábor Palló, Éva Pap, Attila Pók, György Litván, Gábor Zemplén, László Ropolyi, and Gábor Kutrovátz. Last, but not least, we are highly appreciated that this new series came to light, and we thank the series editor Friedrich Stadler of the Institute Vienna Circle, and Jolanda Voogd, Charles Erkelens, Phil Johnstone, and Linda Thomas of Kluwer, for making this volume possible.

Budapest, Bratislava and Vienna
The Editors

I.

In Defence of Scientific Rationality

Lakatos in England around 1960
(By kind permission of Éva Pap)

Lakatos in 1962
(By kind permission of Éva Pap)

John Watkins†

The Propositional Content of the Popper–Lakatos Rift

Imre Lakatos telephoned me on the morning of February 2, 1974. He was in a fury. He had received page proofs of his contribution to the Schilpp volume on the philosophy of Karl Popper. He had been promised that his notes would appear as foot-notes; he now found that they were collected together as end-notes. He urged me to join him in getting the volume held up while this was put right. The volume was already several years behind schedule and I declined. It was our last conversation. He died later that day, going down, one might say, with guns blazing. The volume appeared a few months later (with an editorial apology about the placing of his foot-notes).[1]

The title of his paper, 'Popper on Demarcation and Induction,' harked back to the two problems which the young Popper had taken as fundamental and which he claimed to have solved at a stroke. The previous orthodoxy, as Popper saw it, was that science is differentiated from metaphysics, pseudo-science and other kinds of non-science, by its use of an inductive method. But induction, as Hume's deadly analysis had showed, is without justification – the glory of science but the scandal of philosophy, as C.D. Broad (1952, p. 143) had put it in 1926. Popper sought to solve the demarcation problem by saying that science consists, not of verifiable hypotheses, but of falsifiable conjectures which it tests as rigorously as it can. However well they survive testing they are never verified or quasi-verified or probabilified or confirmed in any inductive sense. They can never be more than well-tested and unfalsified *conjectures*. A theory such as Freudian psycho-analysis, although it may seem to gain a great mass of confirmations, is only pseudo-science because (or so Popper claimed) it is untestable. This solution of the demarcation problem solves the problem of induction by eliminating it; science needs only deductive inferences, from premises to a prediction and from a falsified prediction to the falsity of the conjunction of premises which entail it. "There is no need even to mention 'induction'," Popper declared (1959, p. 315, italicized in the original); it plays no role in science.

Besides demarcation and induction, Lakatos' paper in the Schilpp volume took up a third issue; he brought in, and then discussed and amended with much appeal to the (largely unspoken) value-judgements of the scientific elite, a meta-criterion by which Popper's and others' demarcation criteria may be assessed and amended. My heart rather sinks when a discussion turns

G. Kampis, L. Kvasz and M. Stöltzner (eds.),
Appraising Lakatos: Mathematics, Methodology and the Man, 3–12.

to meta-criteria; I feel like protesting, "First provide me with the meta-meta-criterion by which I may judge your meta-criterion." In the present discussion I am going to stick to Lakatos' challenges to Popper over the two main issues. This will provide quite enough material.

When Popper read Lakatos' Schilpp volume paper, late in 1969, he responded angrily. The dispute escalated into a bitter quarrel whose reverberations continue to this day. Spectators have tended to take sides, some friends and pupils of Lakatos concluding that an old tyrant had been humbled, while others regarded his paper as a shameless work of deconstruction. Among the latter is Joseph Agassi. He writes: "Lakatos died suddenly in 1974 ... His posthumous contribution to *The Philosophy of Karl Popper* of 1974 says of Popper that he has made no contribution to philosophy worth mentioning" (1993, p. 9). Does it? Lakatos' paper opens with these sentences:

> Popper's ideas represent the most important development in the philosophy of the twentieth century; an achievement in the tradition – and on the level – of Hume, Kant, or Whewell. Personally, my debt to him is immeasurable: more than anyone else, he changed my life. I was nearly forty when I got into the magnetic field of his intellect. His philosophy helped me to make a final break with the Hegelian outlook I had held for nearly twenty years (1978, p. 139).

You may say that those are only opening compliments. Well, in the body of the paper he mentioned, quite truly, that some great scientists whose judgement had been warped by previous philosophies had been helped by Popper's philosophy (1978, p. 154). More relevant to an issue that will come up later is Lakatos' evaluation of Popper's theory of verisimilitude. He said that it was "an achievement marvellous both in its simplicity and in its problem-solving power. It became possible, for the first time, to define *progress* even for a sequence of false theories" (1978, p. 156). Agassi's memory let him down here.[2]

Popper's Reply contains a passage which is out of line with his critical rationalism. He said of a thesis of Lakatos: "Were the thesis true, then my philosophy of science would not only be completely mistaken, but would turn out to be completely uninteresting" (1974, p. 1005). The inference the reader was presumably intended to draw was that since his philosophy of science is not completely uninteresting Lakatos' thesis is not true. But why should finding his philosophy of science to be greatly mistaken render it uninteresting? According to his critical rationalism, the exposure by a pupil of a master's great mistake is the lifeblood of intellectual progress. He praised Thales in this connection. By contrast with dogmatic schools which seek to preserve a doctrine pure and unchanged, Thales created a new type of school based on a new relation between master and pupil (1963, pp. 149–150). Thales had explained the stability of our earth by saying that it is held up by water on which it floats (when it is rough, earthquakes occur). His pupil Anaximander explained it by saying that the earth "is held up by nothing, but remains stationary owing to the fact that it is equally distant from all other things." This implies that Thales' explanation is greatly mistaken; does that have any tendency to render it uninteresting? Not in Popper's eyes: he found it a 'beautiful theory.' Or consider Popper on Kant. Popper made a short but devastating analysis of confusions in Kant's idea of synthetic apriori truth (1963, pp. 47–8). Since that idea was central to Kant's whole philosophy, Popper was convicting him of a great mistake; but Popper would not have dreamt of saying that this renders Kant's philosophy uninteresting. In the Preface to *The Open Society* Popper said that great men may make great mistakes; there was no suggestion in that book that ideas of Plato and Marx that are mistaken are thereby uninteresting.

One can regard the Popper-Lakatos dispute as a World 2 contest and ask how the contestants fared: was there an outright winner, and if not, did either man win on points? But I want to view it as a World 3 contest, and see how the underlying propositions fared. More specifically, I want to find out what revisions, if any, to the propositional content of Popper's philosophy of science are called for by Lakatos' challenge. However, that is easier said than done. The propositions involved here are sometimes distorted or disguised or even hidden by the words of our two protagonists, and to get at the World 3 content we will need to work through a good deal of World 2 material.

The lack of a straightforward correspondence between written sentences and propositional content takes different forms with the two men. I begin with Lakatos. In his case the problem is often that he was writing in a coded way or using a kind of Doublespeak. In his Schilpp paper and elsewhere he talked a good deal about different falsificationists; for instance, 'the' naive or dogmatic falsificationist and 'the' methodological falsificationist. For a time he seemed to be pairing the former with a mythical 'Popper' invented by Ayer, Medawar and others, and the latter with the real Popper; thus he often brought in quotations from Popper in the course of presenting 'the' methodological falsificationist's position (1978, p. 24). Yet at one place he claimed to have exposed the main weakness of *Popper's* naive falsificationism (1978, p. 150), and at another place, where he referred to "our savage falsificationist" (1978, p. 26), the foot-note trail leads back to Popper. He also operated for a time with numbered Poppers, 'Popper_0' (invented by Ayer, Medawar et al.), 'Popper_1' (the naive methodological falsificationist), and 'Popper_2' (the sophisticated methodological falsificationist), and then agonized over which is the real Popper. It would have been much easier for the reader if he had discarded this proliferating apparatus in favour of unambiguous phrases like 'According to Ayer, Popper holds ...,' 'Popper holds ...,' 'In opposition to Popper I hold ...'.

Just what did Lakatos say about Popper and the problem of induction? After mentioning that Popper in his early philosophy had suggested a purely *negative* solution, he added that Popper's "later philosophy (based on the idea of truth-content and verisimilitude) involved a shift of the problem and also a *positive* solution of the shifted problem; but, to my mind, he has not yet realized the *full* implications of his own achievement" (1978, p. 140). It sounds as though Popper, with his theory of verisimilitude, had in his hands a positive solution though one he had not yet fully articulated. Lakatos also spoke of three prongs in Popper's anti-inductivist campaign, declaring himself in full agreement with Popper concerning the first two. Concerning the third he reported:

> I had long discussions with Popper in 1966–7 about these issues; I profited immensely from them. But I was left with the impression that on what I called the 'third prong of his anti-inductivist campaign' we may never see eye to eye. *The reason is not that our disagreement is too big; but that it is so very small* (p. 164, his italics).

What was this minuscule disagreement? It was nothing less than whether Popper, the 'scourge of induction' as Lakatos called him (1978, p. 161), should introduce a synthetic inductive postulate to link corroboration-appraisals to verisimilitude-appraisals. I suggest that Lakatos was using a kind of Doublespeak, flattering on the surface but less flattering underneath. I would decode the above sentences of Lakatos as follows:

> Popper's negative solution didn't work. He later discovered the idea of verisimilitude. Around 1966–7 I suggested to him that he could obtain a positive solution by linking corroboration and verisimilitude with the help of a synthetic postulate. He opposed this suggestion.

Lakatos' idea was that without such a link the corroboration a theory gains when it passes a searching test remains no more than an excellent move in the 'game of science.' Such a link is needed to turn it into an advance in our understanding of the world.

In Popper's case the problem of getting at the propositional content of his sentences is different. He was not in the habit of writing sarcastically or with his tongue in his cheek (though he did describe Hegel as a "master logician" employing "powerful dialectical methods"). Where an individual sentence of his should not be taken quite literally this is usually because of exaggeration. I quoted him earlier saying that if a certain thesis of Lakatos were true his philosophy of science would be completely mistaken; but it is not logically possible for any body of propositions to be *completely* mistaken. Or take another sentence of his: "A Marxist could not open a newspaper without finding on every page confirming evidence for his interpretation of history" (1963, p. 35): on *every* page, including the sports pages? More serious difficulties are sometimes raised by the ways he deployed sentences. It is an interesting exercise to ask whether Popper ever explicitly addressed Lakatos on the question of induction. His Replies in the Schilpp volume are in five sections; section II is entitled 'The Problem of Demarcation' and section III 'The Problem of Induction.' Lakatos was accorded fifteen pages at the end of section II. His name does not appear, apart from two incidental mentions, in section III, whose opening sub-section is entitled 'My solution of Hume's problem of Induction.' (Soon afterwards this was expanded into Chapter I of *Objective Knowledge* which opens with the words "I think that I have solved a major philosophical problem: the problem of induction.") And Popper did not refer to Lakatos in any subsequent publication. So the answer appears to be that Popper did not explicitly address Lakatos on the question of induction.

But that turns out to be incorrect. We found that the gist of the propositional content of the passages from Lakatos quoted earlier was: corroboration needs to be linked to verisimilitude. Now a corroboration appraisal as understood by Popper sums up how a theory has performed so far and says nothing about future performance; by contrast, verisimilitude appraisals have no temporal restriction. So a principle linking corroboration to verisimilitude in the required way would have an inductive character. And in the answer which Popper made to Lakatos under the heading 'The Problem of Demarcation' there occurs the following remark: "I *did* suggest in *Conjectures and Refutations*, Chapter 10, that the degree of corroboration may be taken as an indication of verisimilitude" (1974, p. 1011, his italics). That little sentence means that the propositional content of Popper's earlier publications already contained the inductive postulate called for by Lakatos. As I put it elsewhere (1984, p. 283), the whiff of induction which Lakatos invited Popper to introduce into his philosophy was already there.

Lakatos generated at least two pseudo-disputes with Popper as a result of what I call his Research-Programme imperialism. Let *p* be a proposition held by Popper and *q* be a proposition held by Lakatos, where *p* and *q* are, from a logical point of view, mutually compatible. A pseudo-dispute arises if Lakatos nevertheless asserts that *p* must be supplanted by *q*, although he makes no case for this and his reason for it is only that it is good propaganda for his Research-Programme methodology. Let *p* be the proposition that in science we appraise competing *theories* while *q* is the proposition that in science we appraise competing *research-programmes*. It might likewise be held that in architecture we appraise individual buildings and that in architecture we appraise streets. Neither kind of appraisal need exclude the other. But now imagine that within the architectural profession a militant Street Party starts up whose main plank is that street-appraisals must supplant building-appraisals since *streets* are the basic unit of appraisal in architecture. The main plank of Lakatos' Research Programme Party was, in his words, "The basic unit of appraisal must not be an isolated theory or conjunction of theories but rather a *research programme*."[3] I accept that scientific research programmes, about which he said interesting and important things, do constitute genuine units of appraisal. But what argument did he offer for his conclusion that theories, which he usually spoke of as 'isolated,' do not also constitute genuine units of appraisal? After saying, rightly, that a scientific theory is judged in relation to its predecessors, he immediately added: "Then, of course, what we appraise is a *series of theories*" (1978, p. 33). That is rather as if, after saying that a building has to be appraised in relation to neighbouring buildings, he had added: "Then, of course, what we appraise is *streets*." And in any case a series of theories is not yet a research programme: the series terminates with a last theory, whereas a research programme looks to the future. It is as if he had claimed that the basic unit of appraisal is not buildings, or streets, but plans for urban renewal. So far as I know Lakatos never attempted to justify the final step, from series-of-theories to research programmes. At one place, after saying that he had shifted "the problem of appraising theories to the problem of appraising historical series of theories," he casually added: "or, rather, of research programmes" (1978, p. 148). I conclude that Lakatos' campaign against theories as units of appraisal provides no case for a revision of the propositional content of Popper's philosophy of science.

Now to a related issue concerning acceptance and rejection in science. Popper had made it a main task of his philosophy of science "to answer the question: How and why do we accept one theory in preference to others?" (1959, p. 108; and see p. 22). He gave a two-stage answer. For a theory to be short-listed as possibly acceptable it must be internally consistent, and testable, and have been well tested without being refuted. Suppose that two or more possibly acceptable theories are in competition. For one of them to be selected in preference to the other short-listed candidates it must be better corroborated than them. In short, for Popper accepting a scientific theory means judging it to be the best (in a non-inductive sense) of the theories on the table so far. Acceptance is always provisional and will turn to rejection if the theory later turns out to be inconsistent, or is falsified, or is superseded by a better one. I will call this 'Accept' in the 'Judge best' sense. Within Lakatos' methodology of scientific research-programmes, by contrast, 'Accept' is employed in a 'Work on' sense; acceptance of a research programme means either the decision to work on it yourself or to regard working on it as rational, and rejecting a hitherto accepted research programme means either the decision to cease working on it or to regard continuing to work on it as irrational.

Let p now be the proposition that 'Accept' may rightly be used in the 'Judge best' sense, and let q now be the proposition that 'Accept' may rightly be used in the 'Work on' sense. These two propositions seem perfectly compatible. Indeed, the methodology of scientific research-programmes itself seems to require both. Let A be the latest theory produced within research programme Alpha and B be the latest theory produced within research programme Beta, where Alpha and Beta are rival research programmes and A and B are competing theories. Then according to Lakatos there may well be scientists who accept programme Alpha in the 'Work on' sense while accepting theory B in the 'Judge best' sense. (In Lakatos' terminology, they would be admitting that their research programme has fallen behind, but hoping that it will stage a comeback.)

In his propaganda against Popperian theory-appraisal Lakatos proceeded as if Popper too had been operating with the 'Work on' and 'Cease working on' construals of 'Accept' and 'Reject.' For instance, he wrote that for "our savage falsificationist" – and that, as we saw earlier, means Popper – a falsification amounts to the methodological elimination of the theory, *making further work on it irrational* (1978, pp. 26–7, my italics). This perverse reading of what Popper meant by rejection led to absurd results in connection with inconsistent systems. Lakatos said that, for Popper, working on an inconsistent system must invariably be regarded as irrational although

> some of the greatest scientific research programmes progressed on inconsistent foundations. Indeed, in such cases the best scientists' rule is frequently: 'Allez en avant et la foi vous viendra.' This anti-Popperian rule secured a sanctuary for the infinitesimal calculus hounded by Bishop Berkeley, and for naive set theory in the period of the first paradoxes (1978, p. 147).

On Lakatos' ill-chosen terminology it is sometimes right to *accept* contradictions! Perhaps his Hegelian past was staging a comeback here.

Issues to do with contradictions had been dealt with by Popper in a famous paper, 'What is Dialectic?' published in 1940. He there said: "Having thus correctly observed that contradictions ... are extremely fertile, and indeed the moving forces of any progress of thought, dialecticians conclude ... that there is no need to avoid these fertile contradictions." Popper's answer was that they are fertile "only so long as we are determined not to put up with contradictions, and to change any theory which involves contradictions" (1963, pp. 316). For Popper, working on the contradictions discovered in the infinitesimal calculus and in naive set theory was eminently rational.

Did anything said by Popper provide a semblance of justification for reading him as holding that it is irrational to work on a consistent but falsified theory? Well, he called his scientific method *the method of elimination* (1959, p. 131). But by 'eliminate' he seems to have meant, not 'annihilate,' but only to expel it from the class of short-listed candidates for best theory. He contrasted his method of elimination with Bacon's method of eliminative induction whose aim is to arrive at a true theory by a process of elimination of false rivals. On Popper's view we '*accept*' – his scare quotes and italics – the most testable of the surviving theories as the best theory in some non-inductive sense of 'best' (1959, p. 419). (Just what that sense could and should be I tried to spell out in my *Science and Scepticism*.) Lakatos seems to have understood Popperian 'elimination' in some more dramatic sense. So did Feyerabend; according to him a big difference between Lakatos' and Popper's methodologies is that the former allows unsuccessful hypotheses to hang around available to be worked on, but the latter demands that they be *removed* (1975, p. 186), which would presumably make them unavailable for working on. Feyerabend did not say how, according to Popper, the removal was to be carried out; was a falsified theory to be shredded, or vaporized, or sent to outer space? Popper had said that once a theory has been refuted "its empirical character is secure and shines without blemish" (1963, p. 240), which it could hardly do if it had been banished to some limbo. Popper's World 3 differs from Plato's world of Ideas in that the objects in it have not been there from eternity but have been put into it by human beings; but once in they cannot be removed by human beings; they remain for all eternity.

Very little revision to the propositional content of Popper's philosophy seems to be obligated by the criticisms by Lakatos we have so far investigated. Perhaps their main result was to provoke Popper into that veiled admission that he *had* introduced an inductive assumption. But let us now come back to the question of Popper's solution for the demarcation problem. Popper had famously denied scientific status to Freud's psycho-analytic theory on the ground that psycho-analysts had been unable to specify beforehand observable situations which, if actually observed, would refute "not merely a particular analytic diagnosis but psycho-analysis itself" (1963, pp. 37–8 and note). In his Schilpp paper Lakatos famously riposted: "Freudians have been nonplussed by Popper's basic challenge ... But what if we put Popper's challenge to the Newtonian scientist: 'What kind of observation would refute to the satisfaction of the Newtonian not merely a particular Newtonian explanation but Newtonian dynamics and gravitational theory itself?'" (1978, pp. 146–7). I discussed this dispute in my *Science and Scepticism* under the heading 'A storm in an ink-pot,' and I will use ideas which were argued for there rather dogmatically here. Popper seems to have treated a scientific theory as a large propositional chunk; it has a certain compactness or organicity which, however, resists analysis (1972, p. 197). He did at one place (a foot-note in *The Poverty of Historicism*) agree with Duhem that we can test only a huge and complex theoretical system (1957, p. 132n), but he never, so far as I know, entered into an analysis of the different components that make up such a testable theoretical system. He thought of scientific theories as approximating the ideal of axiomatized systems; but he never, so far as I know, sought to distinguish some axioms as more central than others.

Einstein used to speak about the *fundamental assumptions* of a theory. In my book I proposed a way of axiomatizing theories that enables us to segregate fundamental assumptions from auxiliary assumptions. I used 'core theory' to denote a theory's fundamental assumptions on their own, and 'fleshed out theory' to denote these conjoined with appropriate auxiliary assumptions. An advantage of using 'scientific theory' in the 'core theory' sense is that one can speak of *one* theory persisting through all sorts of tinkering with the auxiliary assumptions (or 'protective belt' in Lakatos' terminology). But for a falsificationist there is also a major disadvantage. As Einstein put it, the more science advances the more these fundamental assumptions distance themselves from what is directly observable (1949, p. 27). And at least since Newton's day, 'scientific theories' construed in the 'core theory' sense have not been falsifiable.

Now Popper in his Reply rightly took Lakatos to be making a claim about Newton's core theory, took him to be asserting that the conjunction just of Newton's three laws of motion and his law of gravitation is not falsifiable (1974, p. 987). So his answer should have been: "What you say is of course true; various auxiliary assumptions must be added to make it falsifiable." Instead of which he sought to meet Lakatos' challenge by imagining various observable situations which, if actually observed, would refute Newtonian theory *itself*, i.e. Newton's core theory. They nearly all involved planets behaving in remarkable ways. But Newton's core theory says nothing about what kinds of objects exist in our solar system; we must turn to his auxiliary assumptions if we want to rule the possibility that the planets are not inertial masses but giant rockets with powerful accelerators and capable of abrupt alterations of course.

What sort of revision of the propositional content is called for here? Should we conclude that, from a falsificationist point of view, Newton's core theory *is* on a par with Freud's? No; we get that answer from asking the wrong question, namely "Is Newton's core theory on its own any less unfalsifiable than Freud's core theory on its own?"; to that question the answer is indeed, "No, they are both unfalsifiable." The right question to ask here is, "Are Newton's fundamental assumptions, when married to appropriate auxiliary assumptions, more predictively fertile than are Freud's when married to appropriate auxiliary assumptions?" Whether Freud's fundamental assumptions, assuming we could identify them, have any such fertility I do not know, but Newton's certainly have enormous predictive fertility.

NOTES

†John Watkins died on July 26th, 1999.

1. See the editorial foot-note in Schilpp, 1974, p. 241.
2. Agassi's memory also seems to have let him down on the page before he made that statement about Lakatos where he makes a statement about myself. According to it I behaved in a nasty and mendacious way to him on a certain occasion in 1967, when he was passing through London. Having no recollection of so behaving, I looked up our correspondence, which was quite considerable around that time. There are two letters from him to me written shortly after the date in question; they are entirely friendly and make no reference to any such incident. I drew his attention to them in 1994 and asked him if is it possible that he would have been writing to me like that if I had recently behaved to him as he described. He answered *yes*, he has a very broad view of the possible!
3. 1978, p. 110; and see pp. 34, 46–7. I am silently dropping the single quotes he often put, for no good reason as far as I can see, round the words 'research programme' when he was not mentioning but using them.

REFERENCES

Agassi, Joseph (1993) *A Philosopher's Apprentice: In Karl Popper's Workshop*, Amsterdam: Rodopi.

Broad, C.D. (1952) *Ethics and the History of Philosophy*, London: Routledge.

Einstein, Albert (1949) 'Autobiographical Notes', in: Schilpp P.A. (ed.) *Albert Einstein: Philosopher-Scientist* (The Library of Living Philosophers); reprinted New York: Harper, pp. 2–95.

Feyerabend, Paul K. (1975) *Against Method*, London: NLB.

Lakatos, Imre (1974) 'Popper on Demarcation and Induction', in: Schilpp, pp. 241–273; reprinted in Lakatos 1978, pp. 139–167.

— (1978) *The Methodology of Scientific Research Programmes, Philosophical Papers*, Volume 1, eds. John Worrall and Gregory Currie, Cambridge: Cambridge University Press.
Popper, Karl R. (1957) *The Poverty of Historicism*, London: Routledge & Kegan Paul.
— (1959) *The Logic of Scientific Discovery*, London: Hutchinson (3rd edn. 1972).
— (1945) *The Open Society and its Enemies*, 2 volumes, London: Routledge and Kegan Paul (5th edn. 1966).
— (1963) *Conjectures and Refutations*, London: Routledge and Kegan Paul (4th edn. 1972).
— (1972) *Objective Knowledge: An Evolutionary Approach*, Oxford: Clarendon Press.
— (1974) 'Replies to My Critics', in: Schilpp, pp. 961–1197.
Schilpp, Paul Arthur (ed.) (1974) *The Philosophy of Karl Popper* (The Library of Living Philosophers), La Salle: Open Court, 2 volumes.
Watkins, John (1984) *Science and Scepticism*, Princeton: University Press and London: Hutchinson.

DONALD GILLIES

LAKATOS' CRITICISMS OF POPPER

1. INTRODUCTION

Imre Lakatos' thought is very fascinating, but it is also fascinating to study the development of his thought. This development is not at all linear and uniform. On the contrary it is full of great upheavals, and dramatic changes of direction. The works of philosophers who often change their opinion, usually give the impression of a lack of any coherent position. Strange to say, the works of Lakatos, despite his frequent changes of opinion, do not give this impression. The reason for this paradoxical situation is perhaps that at the centre of Imre Lakatos' philosophy there is a dialectical concept of continual change. His own changes of opinion are, consequently, compatible with his philosophical vision of the world.

In this paper, I want to discuss only one of Lakatos' great changes. In the years 1963–4 when he published *Proofs and Refutations*, he was a follower of Popper and a defender of Popperian philosophy. He wrote in the introduction:[1]

> The purpose of these essays is to approach some problems of the *methodology of mathematics*. I use the word "methodology" in a sense akin to Pólya's and Bernays' "heuristic" and Popper's "logic of discovery" or "situational logic."

It is obvious, moreover, that the full title of the work: *Proofs and Refutations. The Logic of Mathematical Discovery* refers to two of Popper's most famous books: *Conjectures and Refutations*, and *The Logic of Scientific Discovery*. Lakatos has the same attitude towards Popper in his 1968 article: *Changes in the Problem of Inductive Logic*. In this article with a title so characteristic of Lakatos, he defends Popper's theory of corroboration against Carnap's theory of confirmation. In 1973, only 5 years later, the situation was very different. In that year Lakatos gave his last lectures on method at the London School of Economics. These lectures were published for the first time in Italian translation in 1995 in *Sull'Orlo della Scienza*, edited by Matteo Motterlini. In them, Lakatos attacks Popper in a ruthless fashion. Lakatos still sees some merit in Popper's political philosophy, but says that there is nothing of value in Popper's philosophy of science. To quote Lakatos himself:[2]

G. Kampis, L. Kvasz and M. Stöltzner (eds.),
Appraising Lakatos: Mathematics, Methodology and the Man, 13–22.

Allegedly, Popper's three major contributions to philosophy were: (1) his falsifiability criterion – I think this is a step back from Duhem; (2) his solution to the problem of induction – where I think he is a step back from Hume ...; and (3) his literary masterpiece "*The Open Society* by one of its enemies" ... what is it called? *The Open Society and its Enemies*. ... *The Open Society* is frankly a literary masterpiece: not being a political philosopher I cannot comment on its contents, but I certainly think it is a marvelous book. So, in conclusion, two-thirds of Popper's philosophical fame is based mis-judgement.

To this Lakatos adds a little later:[3]

I think that the fact Popper's philosophy survived for so long is a sociological mystery. Popper's immortality is secured by this idiotic result.

How did such an enormous change in Lakatos' attitude to Popper take place? In fact I was working on my PhD under Lakatos' supervision when this change occurred. So I hope that it will be useful if I give, in the next section, some personal reminiscences of Lakatos at that time.

2. Some Reminiscences of Lakatos

My undergraduate degree was from Cambridge, where I studied mathematics for two years, and then philosophy for another two years. In my last academic year as an undergraduate (1965–6), I formed the plan of doing a PhD on the philosophy of mathematics. Consequently I read all the recent articles on this subject in the hope of discovering some interesting research line to follow. After this survey of the literature, I had no doubt that the most interesting recent article on the philosophy of mathematics was *Proofs and Refutations*, published two years previously in the British Journal for the Philosophy of Science. I decided to do for my PhD a historical/philosophical study of some branch of mathematics on the model of *Proofs and Refutations*. So I wrote to the author Imre Lakatos in the summer of 1966 to ask if he was willing to take me on as a PhD student. Lakatos replied suggesting that I should come to meet him at the London School of Economics.

My first meeting with Lakatos was certainly an occasion to remember. After a little philosophical discussion, I uttered the name of Wittgenstein. Lakatos replied: "Wittgenstein was the biggest philosophical fraud of the twentieth century." This statement really came as a surprise to me, since in those days the cult of Wittgenstein was still very strong at Cambridge, where I had been studying philosophy. Indeed my first year of philosophy at Cambridge consisted, for three quarters of the time, of reading Wittgenstein with care and attention. So I replied: "Dr Lakatos, what you say is truly surprising for me because I have just finished writing an essay in which I maintain that there are close links between your concept of mathematical proof and Wittgenstein's." The next time I met Lakatos, he said to me: "Regarding Wittgenstein, I looked through my copy of his *Remarks on the Foundations of Mathematics*, and I was surprised to find that I had written enthusiastic notes in the margins. But these notes were written in Hungarian which means that I must have been written them ten years ago, just after I had arrived in England." In fact Lakatos and I got on extremely well during the first few years of our acquaintance, although, unfortunately, we began to quarrel for various reasons later on. After I had known him for a few months, Lakatos said to me on one occasion: "Donald there is a problem about you. You are very difficult to understand because of your English accent." In fact in those days the philosophy department at the London School of Economics had a lively international atmosphere, and was full of Austrians, Americans, Hungarians, Italians, etc. So an English accent was really quite rare.

I began working on my PhD with Lakatos in October 1966. The first thing Lakatos told me was that I should begin by reading the entire works of Popper, because they were essential. Lakatos was truly enthusiastic about the philosophy of Popper in those days. Initially, however, I was reluctant to follow these instructions, because up to that time Popper had written little on the philosophy of mathematics. However, being a well-behaved student, I began to read Popper carefully, and, within a short while, there was no need for Lakatos to give me any further encouragement, because I was finding Popper's writings ever more interesting. At that time Lakatos was working on the article already mentioned: *Changes in the Problem of Inductive Logic*. As I said, this article is a defence of Popper's theory of corroboration against Carnap's theory of confirmation. For this reason, I also read Carnap's *Logical Foundations of Probability*, and so became more and more interested in the foundations of probability. As a result of all this I ended up by writing my PhD not on the history and philosophy of mathematics, but on the foundations of probability. A problem which particularly interested me in that area was one already discussed by Popper concerning the falsifiability of probability statements. In fact my original plan was only carried out in 1992, not in the form of a PhD, but in that of a collection of essays by various authors. I was the editor of this collection on *Revolutions in Mathematics*, and in the preface I made an acknowledgement to Imre Lakatos as the inspirer of this type of research.

Returning to 1966, the reason why Lakatos was engaged in a defence of Popper against Carnap at that time was the following. In the previous year (1965), Lakatos had organised a conference on the philosophy of science and mathematics in London. Many famous philosophers of the time attended, including Carnap, Kuhn and Quine. Lakatos was the editor of the conference proceedings, and decided to add to these proceedings two essays, both defending Popper, but against two different opponents, namely Carnap and Kuhn. All went well while Lakatos was writing against Carnap. But the defence of Popper against Kuhn turned into a critique of Popper, and the development of a new approach to scientific method – Lakatos' methodology of scientific research programmes. In the next section I will consider Lakatos' criticisms of Popper.

3. LAKATOS' CRITICISMS OF POPPER

In this context the key work of Lakatos is *Falsification and the Methodology of Scientific Research Programmes*, published in 1970. His criticisms of Popper are based fundamentally on the Duhem thesis. It is true that in the appendix of the article just cited, Lakatos speaks of the Duhem-Quine thesis, but in fact Quine's philosophy had little impact on Lakatos, while he used to study Duhem with the very greatest attention. It is not by chance that Lakatos, in one of the passages already quoted from his last lectures on method, speaks of[4] "his falsifiability criterion – I think is a step back from Duhem ...".

Duhem expounds his thesis as follows:[5]

> In sum, the physicist can never subject an isolated hypothesis to experimental test, but only a whole group of hypotheses; when the experiment is in disagreement with his predictions, what he learns is that at least one of the hypotheses constituting this group is unacceptable and ought to be modified; but the experiment does not designate which one should be changed.

If we accept the Duhem thesis, it would seem to be impossible to falsify an isolated hypothesis, and hence that the falsifiability criterion is unsatisfactory. The Duhem thesis poses the following further problem. If experience disagrees with a group of hypotheses, how can we know which of the hypotheses should be changed? Lakatos proposes his methodology of scientific research programmes as a solution to this problem of Duhem's. According to Lakatos, a scientist always works in the context of a research programme, which has a *hard core* or *negative heuristic*. Lakatos claims that[6]

> This "core" is "irrefutable" by the methodological decision of its protagonists: anomalies must lead to changes only in the "protective" belt of auxiliary, "observational" hypotheses and initial conditions.

So if experience disagrees with a group of hypotheses, the scientists do not change those hypotheses which constitute the hard core of their programme. This reduces the difficulty of choice, and, moreover, each scientific research programme has a *positive heuristic* which gives advice on the way in which the programme should be developed.

These are the fundamental ideas of the methodology of scientific research programmes. Here I will not give further details, because this methodology is now well-known. Instead I would like to present the reasons why I did not, and still do not, accept this Lakatosian account of scientific methodology.

4. SOME CRITICISMS OF THE METHODOLOGY OF SCIENTIFIC RESEARCH PROGRAMMES

When Lakatos began to develop his new methodology, I had, following his instructions, just finished reading the works of Popper. I must say that I had found Popper's philosophy very much to my taste. I do not have Lakatos' capacity for making sudden changes in my intellectual opinions, and I was, therefore, reluctant to give up the new faith which I had just acquired. Moreover, in my researches on the problem of the falsifiability of probability statements, I had discovered that the concept of falsification gives a reasonable foundation to the theory of statistical testing developed by Fisher and used by the majority of statisticians. Falsificationism therefore succeeds in providing a simple and satisfying explanation of probability which accords well with the most widely diffused statistical practice. This situation convinced me that there was a certain validity in the concept of falsification, despite the criticisms based on the Duhem thesis, and that Lakatos was therefore wrong to abandon falsificationism completely.

When Lakatos began to develop the methodology of scientific research programmes, the philosophy department at the London School of Economics was divided into two groups: those who accepted Lakatos' new ideas, and the 'old guard,' who tried to defend Popper's philosophy and criticize the new methodology. After Lakatos' death in 1974, I continued for many years to think about these problems. I finally reached the conclusion that Lakatos was correct in thinking that Popper's philosophy needs to be modified in the light of the Duhem thesis. However, I have come to accept the methodology of scientific research programmes only in part and with modifications, and so think that the problem created by the Duhem thesis needs to solved in a way different from that suggested by Lakatos. In the rest of this section, I will explain the reasons why I do not accept Lakatos' position, or, rather, why I accept it only in part. In the next and final section, I will give a sketch of my own position regarding falsificationism and the Duhem thesis.

At the most abstract level, the difference between my position and that of Lakatos can be explained as follows. I accept the distinction between the *discovery* of scientific hypotheses and their *justification*. Lakatos, by contrast, tried to reduce the problem of the *appraisal* of knowledge to that of the *growth* of knowledge. My criticism of Lakatos is that this reduction does not work, and that, we need some notion of the empirical confirmation of theories in addition to those of the progress and degeneration of research programmes. Of course this notion of empirical confirmation need not be the same as Carnap's – indeed I think it will be closer to Popper's concept of corroboration. This then is my general position. I will begin to expound it in more detail by giving a quotation from Lakatos where he explains his idea of the reduction of *appraisal* to *growth*.[7]

> But then two new problems arose. The *first* problem was *the appraisal of conjectural knowledge*. ... The *second* problem was *the growth of conjectural knowledge*. ...
>
> In this situation *two schools of thought emerged*. One school – *neoclassical empiricism* – started with the first problem and never arrived at the second. The other school – *critical empiricism* – started by solving the second problem and went on to show that this solution solves the most important aspects of the first too.

This passage is taken from the article in which Lakatos tries to defend Popper against Carnap. So 'neoclassical empiricism' means Carnap, and, theoretically, 'critical empiricism' means Popper. In practice 'critical empiricism' is more the position of Lakatos than that of Popper. Already at this stage Lakatos' position was begin to diverge from that of Popper.

My own position in contrast to Lakatos' is that we must solve both problems – that of the *appraisal* of knowledge, and that of the *growth* of knowledge. A solution to the second is not sufficient to solve the first. To be more precise, let us consider Lakatos' theory of growth, that is his theory of scientific research programmes. My thesis is that the appraisal of a scientific research programme can be very different from the appraisal of the scientific theories of which is is composed. To see this, let us define two scientific research programmes :

$$\mathbf{R}_1 = (T_1, T_2, \ldots, T_n)$$
$$\mathbf{R}_2 = (S_1, S_2, \ldots, S_m)$$

The following situation is possible:

(1) $\mathbf{R}_1$ makes very good progress, but the theory T_n is not very well confirmed. This case occurs (for n small) in the initial stages of many programmes, for example Bohr's programme.

(2) $\mathbf{R}_2$ degenerates, but S_m has a very high degree of confirmation. An example of this case is given by the hidden variables programme in quantum mechanics. The attempts to replace standard quantum mechanics by a new and better theory have hitherto failed. So $S_1 = S_m$ = standard quantum mechanics, and this shows a total stagnation of the research programme. But S_m has a very high degree of confirmation, partly because of those experiments, for example Aspect's experiment, which were carried out in the context of work on $\mathbf{R}_2$.

The conclusion which I draw from this argument is that philosophers of science should develop not only a theory of the growth of science, but also a theory of the appraisal of scientific hypotheses. Naturally there are connections between these two types of theory, but they are different. Lakatos' attempt to reduce everything to the theory of growth is, I think, the consequence of a notable aspect of his writings. Lakatos gives many examples of scientific and mathematical discoveries, but he never mentions the practical applications of science and mathematics. Reading only Lakatos, a Martian would have the impression that mathematics and science are intellectual amusements for humans similar to novels. He, she (or it!) would not be in the position to guess that mathematics and science are used in industry and commerce. Modern society, however, depends for its existence on the continual application of mathematics and science.

For the satisfactory application of science, we need a theory of the appraisal of scientific hypotheses. To give just one obvious example, it is not permitted to sell a new medicine until the hypothesis that is has no harmful side effects has been very well confirmed. Indeed most governments specify the tests which must be carried out with satisfactory results before a company is allowed to put a new medicine on the market.

Returning to Lakatos' theory of scientific research programmes, my view is that it forms part of the theory of the growth of science and mathematics. With a single modification, I accept this theory as a valid and important part of this branch of the philosophy of science and mathematics. The modification is the following. I prefer to substitute for the concept of the *hard core* of a research programme, that of the *aim* of a programme. Some critics of Lakatos have claimed that his theory of scientific research programmes is identical to Kuhn's theory, apart from a change of terminology. This opinion does not seem to me correct. I maintain on the contrary that Kuhn's concept of *paradigm* is different from Lakatos' concept of *scientific research programme*, and that both concepts are necessary for analysing the development of science and mathematics. I would further claim that Popper's concept of *conjectures and refutations* is also necessary. I will next give the example of Kepler as an illustration of the use of all these concepts.

When Kepler investigated the planetary orbits, he was working on a scientific research programme, whose aim was to find a simple mathematical expression for these orbits. Kepler's research was part of the Copernican revolution. This revolution replaced the Aristotelian–Ptolemaic paradigm which had dominated in the Middle Ages by the Newtonian paradigm. It is clear that Kepler's research programme was neither the Aristotelian–Ptolemaic paradigm nor the Newtonian paradigm. Thus the concepts of scientific research programme and of paradigm are distinct, and both are necessary. While following his research programme, Kepler created a series of mathematical models of the orbits of the planets. Some of these models were refuted by Tycho Brahe's data. So the detailed structure of Kepler's programme, as of many other scientific research programmes, was a series of conjectures and refutations in the manner of Popper.

After these brief observations on the theory of scientific growth, I turn to the theory of appraisal. Lakatos wanted to solve Duhem's problem in the context of the theory of growth. My plan is, instead, to try to solve it in the context of the theory of appraisal, and, in particular, of confirmation. I have tried to carry out this plan in Chapter 10 of my 1993 on *Philosophy of Science in the Twentieth Century*[8], and in the last section of this paper I will give a brief sketch of the principal ideas.

5. FALSIFIABILITY AND THE DUHEM THESIS – AN APPROACH DIFFERENT FROM LAKATOS'

The principal idea of my suggested approach to the problem of falsificationism and the Duhem thesis is contained in the 4 level model. These levels are the following.

Level 0. *Observation statement.* Truth-value determinable by observation. Example: Statement recording the position of Mars at a particular time.

Level 1. *Scientific statement.* Falsifiable and confirmable. Example: Kepler's first law.

Level 2. *Scientific statement.* Confirmable but not falsifiable. Example: Newton's first law.

Level 3. *Metaphysical statement.* Not confirmable. Example: Greek atomism.

Popper's model has 3 levels, our levels 0,1 and 3. In the 4 level model, we retain, as level 1, statements which are falsifiable and therefore scientific. This is because there do really seem to be statements of this type. An example is Kepler's first law, i.e the law that planets move in ellipses with the Sun at one focus. It is possible to imagine accurate observations showing that a particular planet moves in, let us say, a square rather than an ellipse. In such a situation Kepler's first law would be refuted by observations, and this shows that the law is falsifiable. An advantage of retaining a level of falsifiable statements is that we can include statistical hypotheses at this level. As I said earlier, falsificationism succeeds in providing a simple and satisfying explanation of probability which accords well with the most widely diffused statistical practice. This is one of the reasons for not abandoning completely the concept of falsifiability.

It remains true, however, that the Duhem thesis holds for some scientific statements. A statement of this sort, therefore, is not falsifiable in isolation, but only in conjunction with a group of other statements. An example of this kind of statement is Newton's first law. It is not possible to obtain from this law alone results which can be checked by observations. To obtain such results, it is necessary to add to Newton's first law his other three laws, and some further assumptions as well. Despite this fact, it seems to me obvious that Newton's first law is scientific rather than metaphysical. Consequently, there is a level 2 of statements which are scientific without being falsifiable. My suggestions is that these statements are *confirmable.*

This suggestion is rather vague, and to make it more precise it is necessary to develop the theory of confirmation. This development ought also to give a reply to Popper's objection that it is easy to obtain confirmations of any theory whatever, if we look for confirmations. In an attempt to solve these problems, I have proposed in chapter 10 of my 1993 book[9] a new principle for the theory of confirmation called *the principle of explanatory surplus*. This principle denies that if *e* follows logically from *h*, that automatically signifies that *e* supports *h*. It asserts that a given hypothesis is not supported by all the facts deducible from it, but only by a subset of these – a subset which constitutes an explanatory surplus. Suppose we have a number *n* of theoretical assumptions at level 2, and these are used to explain a number *m* of facts. The simplest and most direct estimate of the explanatory surplus is *m* – *n*. It follows that, if in order to explain a certain number of facts we use a greater number of theoretical assumptions, these facts do not give any confirmation to the theoretical assumptions. Only if we can explain, using our group of theoretical assumptions, a much greater number of facts, are these theoretical assumptions confirmed by the facts. Consequently it is not at all easy to find confirmations of a group of theoretical assumptions. We have to succeed in explaining many facts by these assumptions before they obtain any confirmation.

Returning to our example of Newtonian mechanics, Newton did succeed in explaining many facts with his 4 laws. He explained the approximate truth of Kepler's laws, and *also* some perturbations in the planetary orbits. He explained Galileo's law of falling bodies, the laws of impact, the tides, the paths of comets, irregularities in the Moon's motion, results relating to the figure of the Earth, and so on. This evidently amounts to a notable explanatory surplus, and so Newtonian mechanics obtains a high degree of confirmation by the facts. It is difficult, however, for a group of theoretical assumptions to have such a success, and so for the members of such a group to become scientific statements of level 2. So if we adopt the principle of explanatory surplus, the criterion of confirmability does not become empty.

NOTES

1. Imre Lakatos, *Proofs and Refutations. The Logic of Mathematical Discovery*. Cambridge: Cambridge University Press 1976, p. 3.
2. Matteo Motterlini (ed.), *For and Against Method*. Chicago and London: University of Chicago Press, 1999, pp. 89–90. This is the English version of Motterlini's original (1995) Italian edition.
3. *Ibid.*, p. 92.
4. *Ibid.*, p. 89.
5. Pierre Duhem, *The Aim and Structure of Physical Theory*, English translation by Philip P. Wiener. New York: Atheneum, 1962, p. 187. The book was originally published as a series of articles in French in the years 1904–5.
6. Imre Lakatos, "Falsification and the Methodology of Scientific Research Programmes", in: Imre Lakatos and Alan Musgrave (eds.), *Criticism and the Growth of Knowledge*. Cambridge: Cambridge University Press 1970, pp. 91–195. The quotation is from p. 133.
7. Imre Lakatos, "Changes in the Problem of Inductive Logic", in: Imre Lakatos (ed.), *The Problem of Inductive Logic*. Amsterdam: North-Holland 1968, pp. 315–417. The quotation is from p. 322.

8. Donald Gillies, *Philosophy of Science in the Twentieth Century. Four Central Themes*. Oxford: Blackwell 1993.
9. *Ibid.*, pp. 214–21.

Department of Philosophy
School of Humanities
King's College
London, UK

MATTEO MOTTERLINI

PROFESSOR LAKATOS BETWEEN THE HEGELIAN DEVIL AND THE POPPERIAN DEEP BLUE SEA

1. INTRODUCTION: A POP–HEGELIAN PHILOSOPHER

When Lakatos unexpectedly died in February 1974, Paul Feyerabend was invited to write an appreciation of his friend for the *British Journal for the Philosophy of Science.* He portrayed Lakatos as "a fascinating person, an outstanding thinker and the best philosopher of science of this strange and uncomfortable century"; as "a *rationalist*, for he thought that man had the duty of using reason in his private affairs as well as in any enquiry concerning the relation between himself, nature, and his fellow men"; and as "an *optimist*, for he thought that reason was capable of solving most of the problems arising in the course of such an inquiry" (Feyerabend, 1975b, p. 1).

Just a few years before Feyerabend had claimed, in a rather different style, that his fellow was a "*big bastard* – a Pop–Hegelian philosopher born from a Popperian father and an Hegelian mother" ("Lakatos–Feyerabend Correspondence," fothcoming in Motterlini, ed., 1999).[1] In fact, some of Lakatos' most fruitful contributions such as his dialectical conception of mathematical heuristic, the idea of rational reconstruction and, more generally, the emphasis on the role of criticism in the progress of knowledge come directly from a peculiar philosophical conflation of Hegelian and Popperian ideas. I shall argue, however, that for the same reason, an tension in Lakatos' thought cannot be ultimately resolved. Like a seaman in the famous pirates' ballad caught "between the devil and the deep blue sea," that is between his captain who held near-dictatorial powers on the one side, and the dangerous boundless forces of nature on the other,[2] Lakatos too was caught between the devil of Hegelian historicism and the deep blue sea of Popperian fallibilism.

I shall provide an analysis of the roots and objectives of Lakatos' philosophical programme especially in the light of the material in the *Archive of Professor Imre Lakatos* at the British Library at the London School of Economics for Political and Economic Science (henceforth Archive).[3] Instead of giving a complete description of the available material, I shall emphasise those items which illustrate most clearly Lakatos' method of "Proofs and Refutations," his revision of Popper's falsificationist approach, the shift that occurred in the conception of methodology from his early writings to his later papers, his criticism of the neoauthoritarian philosophies such as Toulmin's, and finally his struggle to defend "Reason" against Feyerabend's neo-sceptical challenge.

G. Kampis, L. Kvasz and M. Stöltzner (eds.),
Appraising Lakatos: Mathematics, Methodology and the Man, 23–52.

2. A CRITICAL MARXIST POLEMICIST

The contributions written by Lakatos for Hungarian literary and academic journals in the early Fifties show how Lakatos' enquiry into science, mathematics, history and method has always been firmly linked to *pedagogical* and *socio-political* issues. The Hegelian-Marxist influence is clear in the paper entitled "'Le Citoyen' and the working class" (Archive 1.1). Lakatos contrasts the abstract figure of the *Citoyen* with the reality of the working class. By analogy, he contrasts the abstract principles of the philosophy of science with the substantive scientific practice of working scientists. In 1956, just before leaving Hungary, he was co-author of the Declaration of the National Committee of the Hungarian Academy of Science calling for "the freedom of science from political and moral pressure," and in particular for "the freedom of the Hungarian scientific life from its Stalinist shackles" (Archive 1.10). In a passionate speech Lakatos delivered at the Petőfi Circle pedagogy meeting the same year, he argued for encouragement of a critical attitude, absence of censorship and science as a guide to the party instead of the other way round (Archive, 1.9).[4]

These writings possess the sharpness, originality, forcefulness, clarity and, at the same time, ambiguity, which was to distinguish all of Lakatos' later works. Here is for example how Lakatos deals with the problems concerning the education of a new generation of scholars claiming a role for talent, curiosity, original thinking, autonomy, right to doubt and dissent, demand for proofs and respect for facts.

> The first question is that of *talent*, its sociological role and evaluation. [...] A counter-selection has been going on for years at a national scale on this basis. Talented, courageous men of initiative were pushed more and more into the background of so-called "simple, colourless, decent, disciplined" men. When a post had to be filled or a prize given, it was always the latter type who moved up a rung, while the former moved down one. At the same time, when it came to sacking or even arresting someone, the same selective principles were at work, only this time operating in the opposite direction. [...] Education, if it is to produce scholars of whatever field, must have, as one of its central elements, the training for *original thinking*, must help develop a reliance on individual judgement, sense of justice and truth, and conscience. In the past years, however, there has been an ideological campaign against original thinking and for preventing us from believing our own sensory organs. It is enough to refer here to the unfortunately misunderstood or misinterpreted slogan: "The Party is our mind." Another vital quality of future learned men was also put in the dock, "petty bourgeois" branded on its forehead: *curiosity*. Curiosity and interest were restricted most brutally within narrow, brain-stifling limits. In libraries the pile of strictly confidential stock was getting higher and higher. [...] It is the basis of scientific education to train students and research students to respect facts, to demand exact thinking and proofs. Stalinism, on the other hand, branded these very demands as "bourgeois objectivism." Under the banner of party-minded science, a large (even, we could say, world-scale) attempt has been made to create fact-free and proof-free scholarship or science. (For example, Lysenko's and Lepichinskaia's biology.) The extermination of facts was often carried out under the pretext of a "Marxist" fight

> against empiricism – an invisible and frequently non-existent "salient feature" was given first importance over the miserable and mostly unpalatable "phenomena." The victims of this fervid fight against formalism included logic; and many branches of applied mathematics (biometrics, econometrics) were anathematised. Dialectics was corrupted into scholastic sophism. The history of science indicates that we ought to teach the future scholar to be modest, to be humble in his scientific claims, to be averse to all kinds of fanaticism. He ought to learn that what he does not understand, or disapproves of, still has a right to exist, and that no scientific theory, no theorem can conclude anything finally, in the history of science. [...] New, hitherto unfamiliar chapters ought to be included in pedagogical textbooks, such as "Methods for stimulating curiosity and developing it into interest," "How to teach to think scientifically," "How to teach people respect for facts" and – God forbid! – "How to teach people to doubt." [...] At the last Party Congress in China, Teng Xiao Ping talked about guaranteeing the *right to dissent* and remarked that if, perchance, truth happened to be on a minority side, this right would facilitate the recognition of that truth. This principle has enormous significance in science, where new conceptions are formulated at no instance by "the demand of the masses," but always by the single, solitary voice of a fragile scholar. It often takes many decades for his opinion to become that of a majority. That is to say, it would be good if our pedagogical textbooks devoted a chapter to "*How to teach respect for the right to dissent*." (Bearing in mind that he who tramples upon a dissenting individual opinion is usually not interested in the opinion of the majority either.) (Tudományra Nevelésről – "On Rearing Scholars," English translation by Ninon Leader, Archive, 1. 9).[5]

In the same period, during a discussion with friends, he is reported to have lost his temper when the question turned to defending Marxism: "You are talking about scientific method, why do you keep calling it Marxism?." It is clear that, instead of preaching the dogmas of communist orthodoxy, Lakatos claims an active role for dialectics as an instrument of criticism, rather than as mere rhetoric for empty scholasticism. In his defence of "dialectical rationality" as opposed to "irrationalistic mystification" Lakatos was probably influenced by György Lukács.[6] The appeal to dialectics has to be seen in connection with the prevalence of vulgar Marxism in organised working class movements and pedestrian mechanistic materialism in an age of totalitarian systems in which mankind was repeatedly menaced by self-destruction.[7] Hence, asserting the validity of dialectical rationality was, according to Lakatos, an attempt, on behalf of detractors who had not grasped the point of Hegel's logic, to condemn all forms of irrationality and decadentism. Following this line of argument, history is not brought into the picture to "explain" the necessary realisation of the present society and, therefore, to vindicate the *status quo*, but rather to recognise that knowledge is fallible (for "no scientific theory, no theorem can be eternally estabilished in the history of science").

Leaving his country for Cambridge, after the Uprising in late 1956, Lakatos would not entirely give up the outlook in which he was brought up, rather he would take with him the "forbidden brew" of Hegelian–Marxist dialectic to employ in a creative way.

3. THE ENTANGLED ROOTS OF LAKATOS' PHILOSOPHICAL PROJECT

In the Acknowledgements of his Ph.D. thesis, Lakatos claims that his work was born from the aversion to a conception of mathematics as static and authoritarian. In fact, he aims at showing that "mathematics is *dialectics* and that it cannot exist without *criticism*." Lakatos also remarks that

> The three major – and *apparently quite incompatible* "ideological" sources of the thesis are Pólya's *mathematical heuristic*, Hegel's *dialectic*, and Popper's *critical philosophy* (Ph.D. thesis, emphasis added, Archive, 3. 4)

The importance given to the "movements of concepts," i.e. the "unfolding" of mathematical developments seen as a product largely independent of the producer's psychology, is a clear reference to Hegel; whereas when Lakatos refers to Popper he is taking a position against any account of mathematics as certain and definitive knowledge.[8] Combined to this is also the reference to Pólya: mathematics is a problem solving activity.

With reference to Hegel, Lakatos never specified the kind of works and contributions he regarded as fundamental for his education. This is why such a source of inspiration is merely "ideological." It is even possible that Lakatos never read Hegel's works and that, like many others, he knew of Hegel what he read in Marx. Lakatos had certainly studied Marxism at the time he took part in Szabó's seminar on Plato at Debrecen University (Szabó for example recalls that: "Lakatos was more interested in Marxism than in philosophy"), and he later attended Lukács' lectures on Aesthetic centred on Kant's *Critique of Judgement* and Hegel's *Phenomenology of Spirit*, at Budapest University. It is worth noticing that "Lakatos' mathematical Hegelism" does not endorse Hegel's dogmatically undialectical philosophy of mathematics. On the contrary, Lakatos criticises precisely that kind of "deductive style" and static rationality which is typical in Hegel's idea of mathematics as proposed in the *Phenomenology of Spirit*. In this work, Hegel regards mathematics as the "inert and lifeless" realm of "rigid, dead propositions," i.e. the very opposite of the dynamic self-movement of concepts which constitutes the subject matter of philosophy.[9]

With reference to Pólya, it is worth quoting a passage from the "Preface" of his *How to solve it* (a book Lakatos translated from English into Hungarian):

> Studying the methods of *solving problems*, we perceive another face of mathematics. Yes, mathematics has two faces; it is the rigorous science of Euclid, but it is also something else. Mathematics presented in the Euclidean way appears as a systematic, deductive science, but *mathematics in the making* appears an *experimental*, inductive science (Pólya, 1945, p. vii).

The idea that observation may also play a role in pure mathematics goes back at least to the great mathematicians of the seventeenth and eighteenth centuries, who had shown that inductive procedures are often present where least we would expect them; in geometry, for instance, or in the theory of numbers etc.,[10] although ultimately the reliability of results is guaranteed by a rigorous (Euclidean) proof.[11] Lakatos progressively separates *reliability* from *certainty* in mathematics. Suppose we express the proof of a theorem in a given axiomatic-formal system; if we accept that the latter is consistent, we could thereby exclude the possibility of formalising any counter-example in terms of the given system. But mathematics in the *making*, mathematics in its *growing process*, rarely expresses itself in axiomatic-formal theories; instead, mathematicians too make progress through conjectures, *experiments* and refutations. In line with Arpad Szabó's classic works,[12] Lakatos considers *informal proof* as just another name for *thought experiment.* Broadly speaking, just as *in physics* we have to deal with an entire experimental set-up in order to guess why a theoretical system has failed, and to find the possible ways out, so *in mathematics* we have to analyse "proof-thought experiments" in order to find the hidden assumption from which a paradoxical result or contradiction follows. Similarly, just as it is not always easy to deal with an anomaly of a scientific theory, so it is not always easy to deal with a counter-example in mathematics.

In both cases we have to direct the refutations towards some identified auxiliary lemmas in order to save the "hard-core" of our research. *Feedback* from counter-examples is particularly crucial in mathematics because, in calling for a further analysis of the primitive conjecture and of the proof, it suggests *where* the amendments have to be made and which (no longer hidden) lemma has to be replaced. The whole process is not just a matter of conjectures and refutations, but rather of conjectures, *proofs* and refutations. This implies a fundamental *unity* between the *context of discovery* and *the context of justification*. Proofs are the engines of discovery.[13]

As is well known, according to Popper (and to Reichenbach), there is, strictly speaking, no "logic" of discovery. The psychological process of having a new idea or arriving at a new conjecture cannot be rationally analysed. Rationality is a matter of testing. It operates only in the context of "justification." But Lakatos does not follow Popper here. According to him, a third alternative between "mechanical rationalism" and the "irrationalism of blind guessing" is possible: a rational and non psychologistic heuristic providing a guideline, a set of instructions, from the criticism of an old conjecture to the "discovery" of a new, improved one.

From Pólya, Lakatos took the idea that mathematical discovery follows some patterns that can be rigorously analysed. But it is mainly because Lakatos did not give up his Hegelian background that he was able to look at the process of discovery in a different way than both Popper and Pólya. In fact, the growth of mathematical knowledge is *autonomous* and *objective* and so must be its heuristic. As the Hegelian influence suggests, *growth* is not just a feature of mathematics and science, but their very *essence*. What Lakatos himself refers to as a "Hegelian conception of heuristic" follows:

> Mathematical activity is human activity. Certain aspects of this activity – as of any human activity – can be studied by psychology, others by history. Heuristic is not primarily interested in these aspects. But mathematical activity produces mathematics. Mathematics, this product of human activity, "alienates itself" from human activity which has been producing it. It becomes a living, growing organism, that *acquires certain autonomy* from the activity which has produced it; it develops its own autonomous laws of growth, its own dialectic (1976, pp. 145–146).

Heuristic is therefore concerned with the autonomous dialectic of mathematics. Furthermore, for Lakatos, Euclidianism and formalism endorse a static conception of rationality. ("Euclidism is antispeculative and puritanical"). Heuristic, on the contrary, relates to the *dynamic movement of concepts.* Incidentally, in 1966 "Proofs and Refutations" was translated into Russian in 70,000 copies. Lakatos himself was very proud of the success it met. Some hand-written *Notes* in the Archive are useful in explaining this success: the Hegelian dialectical triad lurks behind the historical reconstruction of Euler's theorem:

> THESIS: Primitive conjecture
> ANTITHESIS: Counterexample
> SYNTHESIS: Theorem and proof generated concept (+ lemma incorporation)
> (Archive, 3)

Russian readers of Lakatos' *Dokazatelstva i Oproversheniа* encountered no problem in detecting a certain familiarity linking mathematical heuristic and dialectic.[14]

4. THE HEGELIAN DEVIL

As the title of this paper suggests, Hegelian philosophy also has some diabolical effects. Lakatos thought that the Hegelian language might "be generally capable of describing the various developments of mathematics," but he was aware of both "its dangers as well as its attractions." (1976, p. 145)

Lakatos had survived the dramatic experience of the Recsk punishment labour camp where he was imprisoned between 1950 and 1953, a place where Stalinist Hungary sadly tried to emulate the Soviet Union. Even though his unorthodox Communist faith remained apparently more or less intact after this experience, he later had changed his mind after having come across the classics of Western liberalism[15] and, above all, Popper's *Open Society.*[16] He was therefore aware that the impressive all-explanatory power of the Hegelian and Marxist dialectic might easily overshoot. Lakatos' Ph.D. thesis reads:

> My concept of the mathematician as an imperfect personification of mathematics is closely analogous to Marx's concept of the Capitalist as the personification of Capital. *Unfortunately* Marx did not qualify his conception by stressing the imperfect character of this personification, and that *there is nothing inexorable about the realisation of this process.* On the contrary, human activity can always *suppress* or

distort the autonomy of the alienated process and can give rise to new ones. The neglect of this interaction was a central weakness of Marxist dialectic (Ibid. p.148 and footnote 1, emphasis added).[17]

Lakatos here criticises Marx and therefore Hegel in the light of Popper's fallibilism. Recall that Hegel regarded the aim of philosophy as delivering the "rule of Reason" in the "unfolding of Spirit" through world history. Hegel's logic can thus easily be twisted (by the Cunning of Reason) to justify anything that happens in history. The unfortunate implication is that whatever has been successful is thereby also somehow "right" and superior to what has been unsuccessful. The dangers of the dialectical approach lie obviously in the *authoritarian* attitude and in the *inexorability* of the process as formulated by Hegel first and then by Marx. On the contrary, Popper's critical philosophy suggests that this process is never ending, and that our syntheses today are our theses of tomorrow.[18]

In sum, Lakatos made use of Popper's fallibilism to wring authoritarianism out from Hegel's dialectical process and made use of the Hegelian idea of a dynamically unfolding rationality underlying the growth of knowledge to reject the alleged irrationality of the context of discovery. At the same time, he discarded Pólya's subjectivism in favour of Hegel's "alienated process." To grasp the last point, recall that for Pólya (1945) heuristic mainly refers to "*mental operations typically useful for the solution of problems*" (p. 2). His well-structured rules of discovery (e.g. "Look at the unknown! And try to think of a familiar problem having the same unknown," "Separate the various parts of the condition," "Is the condition sufficient to determine the unknown?," "Could you restate the problem?," "Did you use the whole condition?" etc.) are intended as tactics for finding solutions to problems and, to a lesser extent, for finding proofs of theorems. The heuristic rules of *Proofs and Refutations*, on the other hand, belong to the realm of the objective growth of knowledge (in the sense of world 3) rather than to mathematicians' minds (in the sense of world 2).[19] Here is how writing to his former student, Pólya explained the main difference between his works in the pedagogy of mathematics and Lakatos' enquiry into rationality:

Dear Imre, [..:] I can see clearly how *Proofs and Refutations* relates to my work. The basic difference is this: I myself would hardly be able to say anything on 'epistemology' which would deserve the attention of the public. Had I been able to say anything about it, even then I would have refrained from it. It is difficult enough to have the public accept heuristic, and I would not have wanted to make this even more difficult by combining it with other controversial things. This main point of "Proofs and Refutations" is, at least according to me, to call attention to the possible connection between heuristic and epistemology" (15th December 1965, Archive, 12.9 item 236, translated from Hungarian).

5. FROM DIALECTICS TO METHODOLOGY

Lakatos' works in mathematics were mainly carried out in Cambridge in the late Fifties and early Sixties, whereas his philosophy of science was developed at LSE initially under Karl Popper. We have seen that Lakatos' philosophy of mathematics is a product of many influences, one of which was certainly Popper's falsificationism. So, on the one hand Popper's theory of scientific method influenced Lakatos' philosophy of mathematics. On the other, Lakatos' ideas in philosophy of science, which improve on Popper's approach, are clearly based on his previous research into the growth of mathematical knowledge. Lakatos' programme therefore develops along a single route, but in the two ways: from the philosophy of science to the philosophy of mathematics, and then back from the philosophy of mathematics to the philosophy of science.[20] As we shall see, however, the goal of these enterprises is quite different in one respect. Lakatos' philosophy of mathematics aims to challenge dogmatism (i.e. Euclidianism and formalism) rather than to defend fallibilism from the attack of sceptical irrationalism. Within an Hegelian framework, the *rationality* of the development of mathematics (and science) being the very premise does not need to be argued for. Lakatos' concern here was simply to show that knowledge in mathematics is not static but dynamic, that it cannot exist without criticism (i.e. dialectic). On the other hand, Lakatos' philosophy of science aims to grasp the unfolding rationality within the history of science and to defend it from any attack. These being anarchic (Feyerabend), elitist (Polanyi, Merton, Kuhn) or even Hegelian in disguise (Toulmin). If forced to choose one side in the battle between dogmatists and sceptics, this time Lakatos would have probably sided with the former camp.[21]

Section 5 of the Archive allows us to reconstruct Lakatos' own pattern of discovery. I refer in particular to a couple of files catalogued by Lakatos himself under the significant titles: "Some Philosophical Implications of the Method of Proofs and Refutations" and "Research Programmes as a Continuation of the Method of Proofs and Refutations." Here is how Lakatos was extending the method of *Proofs and Refutations* to the empirical sciences:

In science the same heuristic pattern [one frequently comes across in mathematics] has an important role:

1. A problem is proposed.
2. A solution is put forward in the form of a naive conjecture
3. The naive conjecture is being *explained and refuted.*
4. *The explanation is analysed into lemmas, and lemmas are incorporated in the naive conjecture. The result is an irrefutable theorem.*
5. *The global counterexamples are tentatively traced to lemmas*
6. The lemmas – in particular those pointed out in 5 – are refuted (local counter-example) and replaced by more general ones, and the theorem is correspondingly generalised. Refutations lead to rival theories.

7. Total local refutations lead to rival theories.
8. After saturation point: rejection.
   (Hand-written notes, "The Logic of Explanations and Refutations," Archive 5.8)

These notes should be seen in the context of Lakatos' struggle with the "Duhem problem" and the related difficulties this implies for a falsificationist account of science.[22] As is well known, Duhem had claimed that physics, far from being a "machine which lets itself be taken apart" is rather "an organism in which one part cannot be made to function except when the parts that are most remote from it are called into play, some more so than others, but all to some degree."[23] In this sense, Lakatos noticed that

> In physics, if you have a global counter example you do not know what you have refuted (Poincaré, Duhem, Quine). Deductive model – sorting out possible lemmas – devise pinpointing tests. Heuristic versus deductive model (Hand-written notes Archive 5).[24]

Lakatos goes on from his consideration of the role of *criticism* in mathematics to the methodology of scientific research programmes (MSRP) as solution of the "Duhem problem." It is peculiar that Duhem had always set aside mathematics from the context of his challenge – and yet the passage in *La théorie physique* in which this occurs "is the only one in Duhem's great book that has aged since was written."[25] It remains valid for mathematics too that one should examine a series of propositions rather than a single one. Since 1961, Lakatos had claimed that when handling a counter-example to the initial conjecture ("global counter-example") or to any of the lemmas ("local counter-examples") one has to *choose* between various lines of behaviour each characterised by promises and risks peculiar to itself. The main case study of *Proofs and Refutations* illustrates the different possible strategies available in dealing with polyhedra that are exceptions to Euler's conjecture. A possible choice involves "eliminating monstrosities" by refining the concept of polyhedron; another consists in "surrendering to the counter-examples" thereby declaring the conjecture to be false; yet another in making certain "hidden lemmas" from background knowledge explicit and inserting them into the theorem enunciation, etc. The first kind of tactic reduces the content of Euler's conjecture and turns it into a "miserable convention."[26] The second move is that of "naive" falsificationism in mathematics. The point is, of course, that one must recognise the element inherent of decision both in *restricting* the meaning of basic terms (such as polyhedron, vertex, edge, face etc.) and in *extending* it. There is growth of mathematical knowledge in those same "adjustments" which allow what in the empirical sciences is an increase in (corroborated) content. I.e. those "stratagems" which allow one to invent/discover new problems, solve some of them and give a more elegant formulation of a simple conjunction of restrictive clauses. Only this third way corresponds to the "sophisticated falsificationism" in the empirical sciences.

Here is how Lakatos summarises up the whole matter by separating $Popper_1$, the naive falsificationist, and $Popper_2$, the sophisticated falsificationist:

[The *critical deductive model of explanation*] is untenable. For this model is *mono-theoretical*: it confronts *one single* theory with a combination of initial conditions and a prediction provided by an authoritative experimental scientist. But the initial conditions and the prediction are based not just on the uninterpreted experiment but on certain background theories, which are involved in the experimental techniques. This is fatal for $Popper_1$: his "refutation" – and even his concept of "testing" – are mono-theoretical concepts. In a pluralistic model there is no such "naive" refutation and $Popper_1$ can give no guidance as to *which* theory is tested and knocked out by the prediction. Moreover, if the premises of the extended deductive model (extended to include some clearly articulated background theories) contain, in addition, an "etc." or "ceteris paribus," the situation becomes still worse, and mono-theoretical refutation becomes utterly irrelevant. But this catastrophe for $Popper_1$'s position can be turned into a victory for the Popperian programme. For, in the spirit of $Popper_2$, we can admit that the premises form an indefinite – or even infinite – conjunction; we can suggest that we try to articulate (or even *invent*) new premises; for we shall solve the problem by suggesting that the "guilty lemma" was the one whose replacement leads to the most progressive problem-shift. But in this model there is no "refutation": that has gone with the mono-theoretical model. We may of course still retain the time-honoured word "refutation" in the expression "*refutation of T given B*" for a proposition which negates *T* in the light of *B*.[27] In a mono-theoretical model we regard the higher-level theory as an *explanatory theory to be judged by "facts"* delivered from outside: in case of a clash we reject the "facts" as "monsters." There is no other possibility. Of course, in a pluralistic model we have more than these two alternatives. Neurath was unhappy about this and this is why he suggested the use of "encyclopaedias" instead of "deductive models." But he did not appreciate the importance of criticism based on deductive articulation and he elaborated no rational alternative to mono-theoretical methodology.

Lakatos goes on to put forward the main features of his own *heuristic model of explanation* as an attempt to meet Duhem's challenge:

In this argument, of course, I take the "Duhem-Quine thesis" for granted in the sense that any refutation undermines a large bulk of our knowledge and not a uniquely specified part of it. But Duhem and Quine give no sufficient indication of how to make a reasoned guess *which* part of our knowledge is responsible for the inconsistency; *indeed, they insinuate that no such reasoned guess can be proposed.* This is the variant of the Duhem-Quine thesis that Popperians reject. Also, the Duhem-Quine thesis does not lay sufficient emphasis on the "implicit" or "hidden" character of the bulk of the background knowledge. For instance, let us take Newton's theory of gravitation *G* with suitable initial conditions *I*; and let us consider an anomalous phenomenon described by *A* such that *G*, *I* and *A* are inconsistent (since, on our assumption, the conjunction of *G* and *I* implies $\neg A$). Let us now introduce a (content-increasing) auxiliary theory *M*, say about magnetic fields which slightly perturb the spin of a planet: it may then seem that *G*, *I* and *M* in conjunction do imply *A*. Then it would seem that while *G*, *I* and *A* were inconsistent, the addition of a new proposition turned the inconsistent theory into a consistent one. But according to elementary logic, if a theory is inconsistent, so are all its extensions. The solution of this paradox is that *G* says not only that there is a Newtonian field of gravitation, but also that the whole field is *nothing but* that of Newtonian gravitation. When we add *M*, we delete *G* and replace it by a weaker *G'*. "Addition" of a hidden premise is not simply incremental; it is coupled with a modification in the extent premises. But if we formulate the hidden lemma as "there is no magnetic field," we may be asked: where has the hidden lemma been "hidden"? One may say: in the Platonic world of ideas which we gradually (but *not* cumulatively) invent/discover. The concrete, positive version of a hidden lemma is always invented/discovered under critical pressure. Imagination and criticism unfolds – slowly and with frequent hitches – ever more of the deductive structure. One may say that *the target of the arrow of refutation is shaped while the arrow is already in the air.* Criticism does not *assume* a fully articulated deductive structure: it creates it. The true deductive model of explanation is an ever-changing one; one in which propositions keep being added and deleted. *One may not explain what one has set out to explain; one may not refute what one has set out to refute* ("On the so-called 'deductive' model of explanation," Archive 5. 4).

Since theories are not fully-fledged deductive systems, counter-examples are not entirely negative as far as they help to reveal the hidden assumptions and unfold thereby a *new* deductive structure. In the very spirit of *Proofs and Refutations*, this long passage is nonetheless only a declaration of intentions. In fact, in promoting the MSRP in analogy to the method of *Proofs and Refutations,* Lakatos found himself giving too much credit to the theoretical autonomy of research programmes thereby undermining the role of counter-examples and also the intrinsic unity of the context of discovery and the context of justification. He thus lost sight of the main question – *How do you improve your conjecture* –, and did not fully exploit the possibilities he had outlined in his programme.[28]

From the method of *Proofs and Refutations* to the MSRP an important shift in the concept of *heuristic* occurs.[29] This took place during Lakatos' career at LSE, when Lakatos decided he wanted to *escape* the Hegelian devil by moving towards the Popperian blue sea. Lakatos, the philosopher of mathematics, claims that he uses the

word "methodology" in a sense akin to Pólya's and Bernays' "heuristic" and Popper's "logic of discovery" (1976, p. 3).

According to Pólya, *heuristic* is a set of strategies for solving mathematical problems to learn, to teach and to reconstruct mathematics. Discovery and invention are mainly considered in their *psychological* aspects. According to Popper, the *logic of discovery* (or, better, the "Logik der Forschung") in the sense of a theory of scientific method, is both descriptive and normative. Popper's demarcationist project not only evaluates scientific products but also offers *standards of intellectual honesty* that scientists have to meet in order to fulfil the aim of science. According to the author of *Proofs and Refutations*, the role of heuristic-methodology is strictly related to its object of inquiry. As we have seen, mathematics is a product of human activity that 'alienates itself,' it is autonomous and objective. So the purpose of heuristic-methodology is to grasp the *logic* of the development of mathematics, the *dialectical* pattern of growth, the rationality of mathematics in the making. From this point of view, the heuristic-methodology looks *backward* to identify the rules that made such a growth possible in the past, and at the same time it looks *forward* to advise on how to obtain progress in the future. Heuristic-methodology, although fallible, is both evaluative and normative. Generally speaking, from the Hegelian-Marxist view the aim of philosophy is not a contemplation of eternal truths, but rather an effort to interpret the present in the light of the past with a view to shaping the present for a better (utopian) future.[30] Thus, for the Hegelian philosopher of mathematics, methodology, heuristic and logic of discovery are synonymous. But this is not the case for the author of the MSRP. For the Professor at the LSE, the hope that methodology "would provide scientists with a mechanical book of rules for solving problems has been given up: modern methodologies or logics of discovery consist merely of a set of [...] rules for the *appraisal of ready articulated theories*. [...] The term "normative" no longer means rules for arriving at solutions, but merely directions for the appraisal of solutions already there"; it follows that

> methodology is separated from heuristic rather as value statements are from "ought" statements (1971a, p. 103 footnote 1).

Methodology no longer concerns the set of rules and strategies to be adopted in the context of discovery. Only heuristic does. But it does so in a different way than before. Heuristic principles (as separated from methodological ones) are not "objective" and "autonomous." They instruct scientists on how to go on *within* a particular research programme. They are therefore subjected to change alongside with changes in science. (For example, we have an Aristotelian Ptolemaic heuristic, a Copernican heuristic, a Newtonian heuristic and an Einsteinian heuristic and so on.)

The last nail in the coffin of the early Lakatos is that methodology (divorced from normative heuristic) "presumes to give advice neither about how to arrive at good theories nor even about which of two rivals programmes the scientist should work on":

> Whatever scientists have done, I can judge: I can say whether they have made progress or not. But I cannot advise them – and I do not wish to advise them – about exactly what to worry and in which direction they should seek progress (Lakatos, 1971b, p. 178).

6. The Understanding of "Human Understanding"

At the time of his death, Lakatos was working on a review of Stephen Toulmin's daring book *Human Understanding* (Oxford, 1972). He had already written and thrown away three increasingly detailed versions of his comment. A fourth longer version, set out in draft form in the summer of

1973 was never completed. Lakatos' intention was to place Toulmin's approach in the more general framework of comparison and conflict between the great traditions in the history of epistemology: *scepticism, demarcationism and élitism.*[31] Lakatos assimilates the enquiry on "human understanding" with the teaching of the "second Wittgenstein" from which Toulmin clearly descends, were it not for an important specification. From the point of view of Lakatos, the Wittgenstein of the *Philosophical Investigations* is an intellectual defender of the *status quo.* The mission of new Wittgensteinian philosophers is to discourage every incursion from outside and attempt to overthrow from inside a "linguistic game" or a "form of life." Toulmin, however, also supports the usefulness of change. Thus, in Lakatos' view, he avoids this reduction of philosophy to a mere "thought police" but at the cost of appealing to the "Hegelian Cunning of Reason," which justifies *change* in the name of *progress.* (If progress is guaranteed by the cunning of reason, the description of change is the description of progress. But what – Lakatos asks – if there is disagreement within the scientific community over some proposed change? Answer: only history will decide. See Lakatos, 1973/1976, p. 237).

Toulmin therefore recurrently finds himself "between the Wittgensteinian devil and the Hegelian deep blue sea," which is precisely the title of an earlier draft of the same paper on Toulmin (Archive 8.5). Lakatos had surely thought about this, since, in another version, "Professor Toulmin" was caught "between the Hegelian devil and the Wittgensteinian deep blue sea." The uncertainty here is understandable. The outcome of both of Toulmin's stands were diabolical to Lakatos, since he considered a last ditch defence of the "closed society":

> Following the tradition of Wittgenstein, Polanyi and Kuhn [...] the picture described [by Toulmin] is that of a society without radical alternatives, where one can only "improve" but not replace "the current repertory of concepts," a society whose membership depends on oaths of loyalty to specific doctrines ("commitment to collective ideas") and where only "professional forums" can judge the implications of these doctrines for specific cases. In this closed society critical reappraisal and modification are allowed only if done by "qualified judges." The layman is powerless, the elite self-perpetuating (1973/1976, p. 241).

However, it would be possible for Lakatos' rhetoric to *backfire* at this point, since *he* would himself remain undecided between the devil (of Hegelian authoritarianism) and the deep blue sea (of Popperian fallibilism).[32] In particular, the early Lakatos takes for granted that "After a saturation point: we reject the theory" (clause 8 of the passage cited above at p. 31). Lakatos (1970) states, on the other hand, that there is "no such a thing as a "natural" saturation point" for a research programme.[33] The MSRP does not in fact fix any *time limit* for the final assessment of the empirical progressiveness or degeneration of a programme. At the dawn of a new and ambitious scientific idea a certain methodological tolerance is called for, and this applies to research programmes whose heuristic has "run out of steam." It is "not irrational" for supporters to defend their theory with ingenious ad hoc stratagems and to hold out for a long time even without any empirical success. "Crucial" experiments are hence seen to be crucial only decades later, "after long hindsight," and

> rationality works much more slowly than most people tend to think and, even then fallibly. Minerva's owl flies at dusk (1970, pp. 72, 87).

Apart from the stress on fallibility, Hegel lives on.[34] "Absolute knowledge" in form of complete "self-consciousness" and "self-possession of spirit" is only available at the end-point of the thinking process:

> Philosophy always comes to the scene too late to give instruction as to what the world ought to be. As the *thought* of the world, it appears only when actuality is already there, cut and dried, after its process of formation has been completed. [...] It is only with the fall of the dusk that the owl of Minerva spreads its wings (Hegel, *Philosophy of Right,* "Preface," p. 23).

It may be the case that (the rationality of) science can be understood backward, but it has to be done forwards.[35]

Having set aside Popper's "instant rationality" in order to bestow on methodology the role of the owl of Minerva, Lakatos has been accused by Feyerabend of oscillating between a *conservative* use of the MSRP, which would eventually lead to benefits for the *status quo*, and a *revolutionary* use which amounts to nothing but "anything goes."[36] Lakatos betrays his *authoritarian* nature by strengthening his methodological standards not on an argumentative level, but by shaping a historical and social situation that renders it difficult, *in practice*, to cultivate a degenerating programme. For example, he claims that people who support a degenerating research programme should not enjoy as much freedom as might appear: "they can do this mostly in private. Editors of scientific journals should refuse to publish their papers, which will contain either solemn reassertion of their positions, or absorption of counter-evidence by *ad hoc* linguistic adjustments. Research foundations, too, should refuse money" (1971a, p. 117). On the other hand, taken by themselves, these standards are incapable of forbidding the most outrageous behaviour. For example, Lakatos claims that "there is freedom [...] in creation and over which programme to work on [...] . *Appraisal* does not imply *advice*"(1973, p. 110).[37] Thus, according to Feyerabend:

> Combining the common sense standards of scientists with the methodology of scientific research programmes, Lakatos utilises the intuitive plausibility of the former to support the latter: a splendid Trojan horse that can be used to smuggle real, honest (a word so dear to Lakatos) anarchism into the minds of our most dedicated rationalists (Lakatos–Feyerabend correspondence, letter dated 20th March 1973, Archive, 12.9).[38]

Whether or not Lakatos has really gone a long way towards epistemological anarchism is a problem we can better solve in the more general framework of Lakatos' late struggle against the "modern intellectuals' betrayal of reason."

7. HISTORICISM: i.e. DEALING WITH EXPLOSIVES

In his review of Toulmin, Lakatos goes on distinguishing his position from both the élitist "thought police" and the Cunning of Reason:

> I agree with Toulmin that no demarcation criterion is absolute. I am a fallibilist with regard to demarcation criteria, just as I am a fallibilist with regard to scientific theories. They are both subject to criticism and I have specified criteria not only by which one programme can be judged better than another, but also criteria by which one demarcation criterion can be judged better than another. But I do not draw the Wittgenstein inference from fallibility of propositions to their dismissal. I do not panic: I do not switch from articulated propositions to inarticulable skills of doing and judging science. For to do so is to reintroduce through the back door a pragmatist version of justificationism with the help of the Hegelian Cunning of Reason. I want clear theses in both science and philosophy of science where logic can assist criticism and help to appraise the growth of knowledge. [...] It is largely because of my conviction that without deductive logic there can be no genuine criticism, no appraisal of progress, that I stick to old fashioned Popperian-type *Criticism and the Growth of knowledge* (1973/1976, pp. 242–243).

Three relevant claims need to be distinguished here. *Firstly*, any demarcation criterion is fallible. Yet Lakatos maintains that we can avoid collapse either to historical relativism or élitarist authritarianism, only by appealing to a "core" of ("old fashion Popperian-type") standards of appraisal. In particular, to avoid historical relativism, "rational appraisal must *precede* and not *follow* full-scale empirical history" (i.e. "internal" normative history is primary and external descriptive history is secondary.) Similarly, to avoid the regressive problem-shift from the problem of demarcation to the problem of knowledge expressed "in the form of skills and activities" (the latter leading to the Orwellian world where history is for the winners), a "universal criterion of progress" is called for.

Secondly, such a criterion of progress is provided by MSRP in the form of very general principles of theory-appraisal. These will include "the basic tenets of deductive logic and intuitive rules for weighing evidence (especially the principle that special weight is to be given to a theory's predictive success)."[39]

Thirdly, as we appraise scientific theories, we should be able to appraise the standards of appraisals. The tool for this task is history. The meta-criterion is provided by the methodology of historical research programmes (MHRP). In particular, we accept a methodological proposal if it can be shown that it was effective in paradigmatic cases of growth of knowledge. In fact, even though there has been no general agreement concerning a methodological criterion,

Lakatos maintains there has been considerable agreement about whether a particular step in the game was scientific or crankish, or whether a particular gambit was played correctly or not. Thus Lakatos' meta-criterion appraises methodologies on their ability to provide historical reconstructions minimising the influence of "external" factors and maximising "internal" explanations.

Yet Feyerabend objected that the very idea of a "common scientific wisdom," i.e. a wide range of accepted "basic" value judgements on singular scientific achievements, is but a chimera.[40] The strength of scepticism flowed from the realisation that, together with single results, the *criteria* to assess them change as well. Thus, revolutions do actually challenge all the ideas born in connection with those procedures, including "basic" value judgements. It might therefore be the case that the better a methodology seems to capture the rationality of science, the greater its mystification. Moreover, according to Feyerabend, the fact the new astronomy of Copernicus, Kepler and Galileo took root, or the fact that witch hunts came to a stop, happened because independent thinkers resolved to introduce obsolete theories and defend them even in spite of all the traditional methodological rules. If this is the case, then putting forward a "theory of rationality" is a tyrannical intellectual act which assumes (without argument) that progress has occurred, that it has been obtained thanks to such a normative rule and, finally, it constitutes the best possible advancement, i.e. that of science as it has *de facto* developed "in the last three hundred years."

Feyerabend's criticism needs to be handled carefully. Firstly, I shall clarify the aim of rationally reconstructing history, and, secondly, in the next sections, I shall raise the related question of the stability of some core of methodological principles. To start with, there is no history of science without methodology, and no methodology without history of science. Lakatos' parody of Kant's maxim is well known: philosophy of science without history of science is empty, history of science without philosophy of science is blind. This attitude to history reveals once more the philosophical mix of Hegel and Popper. The '*historicist*' Lakatos starts from the unquestionable premises that knowledge (at least in the 'most advanced' sciences) does grow, suggesting that we need to *extract* rationality from its historical development. The '*critical*' Lakatos claims that we have to *test* our theory of rationality against history. Lakatos' appeal to rational reconstructions of single historical cases should thus be taken as an attempt to grasp the (objective) reasons and strategies that have brought new ideas.[41] Despite Feyerabend's criticism, there is nothing "mystifying" in appraising past beliefs according to given methodological standards. On the contrary, such judgements lead to historical data that are not easily obtainable in other ways, and allow one to outline and explain the whole process. Any appraisal of this kind is twice desirable: the historian may discover (old) "new" facts, the philosopher tests his own standards.[42] In this way, the *circularity* implicit in checking an epistemological proposal against the history of science is not vicious; rather it may turn out to be a *virtuous move*.[43] On the other hand, Lakatos concedes to Feyerabend that one has to be moderately sceptical with regard to an "a priori statute law." This is why he advocates a "pluralistic system of authority" thanks to which the authority of scientists' judgement on particular cases criticises the general authority of the immutable rules and *vice-versa*. In this way only, the proliferation of different points of view, the comparison between different "rational" reconstructions, the awareness of local strategies and the reasons behind researchers' moves specify how we can learn from history and, especially, how we can escape from the influence of the "bad" philosophies (i.e. theories of rationality).

Once more, however, it should be noticed that Lakatos' balance between Hegel and Popper is not stable. In particular, the "whiff" of historicism may turn into a full-blown storm. Lakatos was probably aware of the breaking power of his own stand already in 1961. At that time he wrote:

> I am afraid that some ardent Popperite may already be rejecting all that I am about to say [but] I am quite convinced that *even the poverty of historicism* is better than the complete absence of it – always providing of course that it is handled with the care necessary in dealing with any explosives (1959–1961, p. 61).

Yet, Feyerabend would probably have added, ... *providing it is placed under the "right" targets*. But here comes the main disagreement between Lakatos and Feyerabend, for they held different opinions regarding what the "right" targets should be. This brings us to the question, what changes in "The Changing Logic of Scientific Discovery"?

8. BACK TO THE FUTURE: FEYERABEND'S NEO-PYRRHONIAN CHALLENGE AND BEYOND

Lakatos left unfinished his reply to *Against Method* and the related proposed book "The Changing Logic of Scientific Discovery." Sections 5 and 8 of the Archive contain Lakatos' *Notes* (from 1963 up to his last days) on his life-long project of writing a history of the critical (and not so critical) interaction between *methodological standards* and *scientific achievements*. Lakatos had partially attempted this task in his paper "Newton's Effect on Scientific Standards,"[44] which precisely reconstructs the battle between Cartesians and Newtonians about the very standards of (scientific) proofs and (scientific) criticisms. In an unpublished paper of the same period, Lakatos puts forward an excellent analogy to depict the problem-shift brought about in epistemology by the rise and downfall of the Newtonian research programme:

> Let us imagine that mankind has set up moral standards and maintained them for centuries without anybody being able to live up to them. Some claimed, hypocritically, that moral acts – and, indeed, moral men, did exist; others kept unmasking these reports and either claimed that the high moral standards will necessarily remain utopian forever, or set up programme after programme for making man moral and

> noble. Then, suddenly, some actually started behaving in a new way which soon struck most people as truly moral. First they hailed them as having finally realised the old dreams. But then slowly it dawned on them that although the new men's acts were undoubtedly virtuous, they did not in fact comply with the old standards. They tried gradually to weaken the standards, but no matter how one weakens the standards for they remained unsatisfied. [...] People had two ways out. The sceptics celebrated the end of all morality. But a new school of dogmatists accepted the *new moral facts* and devised *new moral standards* in the light of which their morality – a new morality – could be seen. This was then *a revolution* – first in morality, then in moral theory. Not necessarily the last one: inconsistencies remained, new types of moral acts (expressed in "moral basic propositions") further upset the new standards. The problem of morality was shifted. Moreover, the revolutionaries set up a theory of moral progress which sets up standards for standard change; a theory in the light of which their revolution was not just a change in subjective fashion, but progress towards moral truth ("On the Intellectuals' betrayal of Reason," Archive, 8.2).

Lakatos' intention was to recapitulate the problem-background and to assess the problem-shift within the dispute between Cartesians and Newtonians. This would have helped to highlight the reason for the success that Newtonian physics achieved *regardless* of its violation of the standards commonly accepted by the scientific community. In particular, Lakatos points out the discrepancy between Newton's *actual* procedure (i.e. his *method*) and his *Regulae philosophandi* (i.e. his *theory of method*) put forward in explicit form to defend his discovery and to defeat his rivals. Thanks to a peculiar historical paradox, the following generations were to accept not only the richness of Newton's results, but also his theory of what made up his scientific achievement, despite the latter's poverty:

> *In this sense one may say that Newton's theory of method created modern philosophy of science.* But this turn separated science and philosophy of science from 1686 to 1905 [when Einstein's theory had superseded Newton's] or, rather, till 1934 [when Popper had first looked at the problem in this way]. *Science*, the schizophrenic genius, marched from victory to victory. *Philosophy*, unaware of the split between the wonderful Newtonian method as practised and the mad Newtonian method as professed, tried to clarify the professed method, and thus turned into *a mad study of madness*. ("The Rise of Defensive Positivism," Archive 5.5, this passage is a different version of the concluding section of Lakatos, 1963–1964/1978).

Underlying this paradox is the fact that Newton took his methodological standards from past centuries dominated by the all-pervading idea that religious knowledge was certain and indubitable; by analogy, science was expected to respect similar standards. In fact, the split brought on by the Reformation (though against the original intentions of the founding fathers) raised the problem of *fallibilism*. How is the Church expected to be the authority of its own infallibility, given that the point in question is precisely whether the Church is the 'true' authority on religious matters? This question was to unleash a sceptical crisis not only in theology but also, shortly thereafter, in the sciences and in all other areas of human knowledge.[45] Thus, sustaining Newton's views was the inconclusive and politically destructive competition of rival theological stands as well as the belief that science could achieve the certainty that theology at the time seemed painfully denied.[46] A century later, the stunning success of modern natural sciences, and primarily of Newton's research programme was to breathe new life into the battle between the dogmatics and the sceptics. On the one hand "some radical dogmatists, undeterred by the long series of defeats of reason in human affairs, have been trying to generalise Newtonian method to social, ethical and political problems." On the other, "some radical sceptics, undeterred by the long series of successes of Newtonian science, have been trying to show that all these successes were sham successes and even the best theories of the exact sciences were nothing more than irrational (if possibly 'great') beliefs" ("The Intellectuals' betrayal of Reason," *ibid.*). Lakatos refuses to side either with the radical dogmatists or with the radical sceptics ("the dogmatists tried to prove too much, the sceptics tried to explain too much"). Yet he regards as the "modern betrayal of reason," consisting in "the intellectual attack on the objective epistemological value of the exact sciences," as "criminal." If, on the one hand, Feyerabend has shown many different ways in which any principle of rationality can turn into a prison, and has consequently claimed along with the sceptics that the betrayal of 'reason' by man is preferable to the betrayal of man by reason; on the other hand, Lakatos has challenged his rival's anarchism insisting on the bias underlying the sceptico-dogmatist's point of view, by showing how radical relativism ultimately leads to forms of intellectual surrender which maintain that *might is right*.

In order to propose his (fallibilist) stand as an alternative to dogmatism and scepticism, therefore, Lakatos himself – a worthy theologian of rationality – has to face the challenge of the Greek Pyrrhonians, whose arguments after lying forgotten for centuries had suddenly come to the forefront especially in regard to the primary epistemological problem brought up by the Reformation. Luther's denial of the authority of the Church and his assertion of a new "rule of faith" (i.e. 'true' is what conscience is compelled to believe when reading Scripture) for determining religious truths, constitutes a rather neat example of the "problem of the criterion" as it appeared in Sextus Empiricus:

> In order to decide the dispute which has arisen about the criterion, we must possess an accepted criterion by which we shall be able to judge the dispute; and in order to possess an accepted criterion, the dispute about the criterion must be first decided. And when the argument thus reduced itself to the form of circular reasoning the discovery of the criterion becomes impracticable, since we do not allow the Dogmatic philosophers to adopt a criterion by assumption, while if they offer to judge the criterion by a criterion we force them to a regress ad infinitum (Sextus Empiricus, *Outlines of Pyrrhonism*, in Popkin, 1979, p. 3).

Feyerabend's criticism of Lakatos is clearly a revival of this classical Pyrrhonian challenge. The problem of justifying the standards of theory appraisal does not arise as long as there is an unchallenged criterion. But once scientific revolutions are brought into the picture epistemological scepticism may be back again.

In order to meet the Pyrrho-Feyerabendian challenge, let us distinguish three levels of commitment of the scientific community: (1) the factual level of accepted general theories, (2) the methodological level of accepted standards of appraisals, (3) and the axiological level of the aims of science (cf. Worrall, 1988). According to Feyerabend, history reveals diversity in method as well as diversity in science. Hence, change is not restricted to substantive scientific theories but involves methodological rules, and possibly even the aim of science. Once such a "big picture" or "holistic" view of scientific change is held, *relativism* is unavoidable. (How can changes be rationally accounted for if even the basic principles of rationality are subject to change?) The Lakatos-Feyerabend quarrel "For and Against Method" can be regarded as a dispute about the scope of the concept of "methodology." Feyerabend categorises as methodological any general metaphysical and heuristic principle. In his battle against "Reason" he overstretches the elastic term "methodology" to such an extent that the dispute on the criterion can never be "first decided," since no criterion is ever fixed and any attempt to assert it is prey of the circularity argument well outlined by Sextus Empiricus. If we do interpret methodology in the very wide sense of "Weltanschauung," "forms of life" or general heuristic principles, no doubt "methodology" is subject to change as science changes. Lakatos would have no objection to Feyerabend on this respect. But this is not at stake in Lakatos' defence of scientific rationality. Lakatos allegedly works on a narrower concept of "methodology." He maintains that heuristic norms change alongside with research programmes, but he stands for some "core" of invariant standards of theory appraisals (see Worrall 1988, 1989). The idea underlined here is that even though we have "learned how to do science better alongside doing better science," we maintain that the (implicit) standards of theory appraisal are not themselves historical (*ibid*). Bearing in mind the distinction between professed (and changing) methodologies and implicit standards of appraisal, the task of Lakatos' *Changing Logic of Scientific Discovery* would become perfectly clear: to grasp and to articulate sharply and clearly the unchanging standards (or, to put it against the Hegelian background, to grasp the "unfolding of reason" and present it "cut and dry," after its process of formation has been completed).

Since Lakatos had a higher opinion of historicism than Popper, he was aware that this task cannot be realised *a priori*, but rather by looking at science in its progress. At the same time, since he had a lower opinion than Hegel of the Cunning of Reason, he was aware that these standards should not be themselves subject to change. Lakatos could thus *consistently* defend his attempt to rationalise changes in professed methodologies only because progress in science is characterised by standards which are independent of them. Of course, the burden of the proof is with Lakatos. That is, he has to show that an "invariant core" of standards of appraisal, which is common to all different methodologies, and which "maximally satisfied the unchanging, abstract formal principles of good science,"[47] actually *exists.*

Lakatos perhaps made up his mind only near his end. Here is for example how he writes to Feyerabend on the 10th January 1974:

> Dear Paul, I was amused by your suggestion that the scientific revolution was a revolution in the standards. This is of course the story I encapsulated in the announced title: *The Changing Logic of Scientific Discovery.* A chapter of this book would have been my Newton paper in which I discuss this change in standards in detail. The trouble was that Worrall and Zahar persuaded me that the standards which I ascribed to the seventeenth century were already there at the end of the sixteenth. Bashi Sabra almost beat me up since according to him these standards were there in the age of Ptolemy. Now your letter caused a conversion effect, and now I think that my Newton paper is perfectly correct and can be made consistent with the Copernicus paper, and now I am going to publish it (Archive 12.8).

Lakatos might eventually agree with Feyerabend that the history of scientists' methodological beliefs is irrational. After all, scientists may know as much about doing science as fish do about hydrodynamics.[48] This is why philosophy of science that confines itself to the history of professed methodologies may easily result into a "mad study of madness." But, *contrary to Feyerabend,* he maintains that we should be able to rank Copernicus' research programme better than Ptolemy's and Einstein's better than Newton's with respect to some methodological standards implicitly endorsed by the scientific community. Only in this way rationality can be defended and the collapse into either sceptical relativism (disguised as anarchism) or the cunning of reason (disguised as authoritarian elitism) can be avoided.

One question still remains whether Lakatos' methodological appraisals intended to have practical import. His stand on this point can hardly be misunderstood. Here is for example how he once replied to his alleged 'fellow anarchist.'

> I do not mind your anything goes, but when it comes to moral theory I even make mincemeat of Pyrrho. You certainly would not hurt a fly – as you put it. The question is what the anarchist does when he is in a position to hurt either one fly or another but is bound to hurt one. Will he commit suicide? You may remember that before I started off on research programmes, I discovered that I had to substitute *acceptance* and *rejection* of theories by *preferring* one theory to the other. And this of course also applies to ethics and politics. For instance, I would not like to hurt the North Vietnamese, but if not hurting the North Vietnamese hurts the South Vietnamese I am faced with a problem which *I* am willing to face but *you* are not (14th August 1972, Archive 12.8).

So much for Lakatos' attitude concerning the socio-political relevance of any enquiry into the method of science.[49] Let us finally turn to the question of the backward looking character of Lakatos' standard of appraisal and to related concern about the relationship of methodology and its aim. Once again, I shall deal with this last point against the background of Lakatos being between Hegel and Popper.

9. THE POPPERIAN DEEP BLUE SEA

Ian Hacking wrote an influential paper (1979) in which he stressed a tension within Lakatos' philosophy. According to Hacking, Lakatos oscillates between realistic metaphysics and the idea of "methodology as an objective surrogate for truth." Educated in an Hegelian–Marxist tradition, Lakatos accepted the Hegelian demolition of the correspondence theory of truth and tried to develop a methodological account in the context of which the assumption of a realistic metaphysics seems to have no place. Hacking's "neo-Hegelian Lakatos" dispenses with the notion of truth as correspondence to the facts and *replaces truth by method.* The starting point of this enterprise is the simple fact that knowledge grows. What is then required is an analysis of growth. This being given, the aim of science is not to increase verisimilitude but to meet the requirements that the scientific method suggests.

It is my view that Hacking's story is just a part of the whole story. The late Lakatos could not have regarded methodology as a substitute for truth because he explicitly demanded a connection between the "game of science" (method) and its "rational" end (truth):[50]

> [We need] to posit some extramethodological inductive principle to relate – even if tenuously – the scientific gambit of pragmatic acceptances and rejections to verisimilitude. Only such 'inductive principle' can turn science from a mere game into an epistemologically rational exercise; from a set of light-hearted sceptical gambits pursued for intellectual fun into a – more serious – fallibilist venture of approximating the Truth about the Universe (1971a, p. 101; cf. also *ibid.*, pp. 97, 108–9 and 1974a, pp. 154–159).[51]

We have seen that the Hegelian-Marxist influence on Lakatos is articulated and important. However, contrary to Hacking's interpretation, Lakatos had eventually made the convergence of science and truth the matter of a separate metaphysical conjecture. Popperian *fallibilism* rather than the post Kantian demolition of the copy theory of truth is therefore the correct key to understand Lakatos' view in this context.

Of course, for a fully coherent fallibilist "there is no ultimate proof that [...] we have been heading the Truth" . We can only (non-rationally) believe, or rather hope, that we have been. Unless hope is a 'solution,' there is no solution to Hume's problem." (Lakatos, 1974b, p. 213). Having disentangled himself from the Hegelian devil, Lakatos is doomed to find the Popperian deep blue sea waiting at the end of the road he travels.

10. Final Remarks

On the one hand, Lakatos, allegedly is the Hegelian in disguise who advocates method as an objective surrogate for truth; *on the other hand*, he openly appeals to some sort of metaphysics to bridge the gap between method and truth. *On the one hand*, Lakatos claims a normative role for the heuristic (as synonymous of methodology) which he associates with the objective dialectical pattern of the growth of knowledge*; on the other hand*, he separates (normative but research-programme laden) heuristic from (retrospective) methodological appraisals. *On the one hand*, Lakatos appeals to Hegel and criticises Popper for leaving history outside the picture; *on the other hand* he denounces Hegel's misuses of history (in its cunning of reason version) and appeals to Popper's fallibilism. *On the one hand*, Lakatos claims there is no such thing as an "immutable statute law" to distinguish good and bad science. Taken literally, this would imply that our theory of rationality is not eternal and a priori, but rather submitted to continuous change under criticism (as is suggested by the title "The Changing Logic of Scientific Discovery"). *On the other hand*, he distinguishes between an implicit theory of scientific method and explicit methodologies, whereas the former has to be fixed if we want to account for the changes in the latter (as is also suggested by "The Changing Logic of Scientific Discovery"!). *On the one hand*, in the century-long debate between dogmatists and sceptics, Lakatos would have sided with the former camp against any intellectual surrender to positions equating *might* with *right*. *On the other hand*, he remains a *sui generis* dogmatist who relies to some extent on sceptical arguments. (For example, referring to a glorious theological image, he claims that "the sentences in God's Blueprint of the Universe [are of] infinite length" and that, consequently, "there can be no human statement expressing Natural Law.")

I have shown that the gap between these different stands can be only partially bridged because of the very nature of Lakatos' philosophical programme between conflicting parents. This is also why, from time to time, Lakatos concedes something to historicism, anarchism and elitism. It is, of course, possible that Lakatos' proposed book would have helped in narrowing the gulf between the present "reconstruction" and the "real thing." Unfortunately Lakatos never lived to complete his plan of a normative theory of scientific rationality that takes history at its face value. But it might also be the case that the task is in fact impossible and that the elusive God who drew the "infinite Blueprint of the Universe" has teased our contemporary theologian of rationality.

NOTES

* I should like to express my thanks to Francesco Guala, Donald Gillies, Giulio Giorello, Brendan Larvor, John Worrall for detailed comments on draft material. I am also grateful to all the participants in the Vienna and Budapest Symposia for their valuable criticisms and suggestions, particularly to Ladislav Kvasz and Michael Stöltzner. This paper has been written under the financial support of the University of Ferrara. The Centre for Philosophy of Natural and Social Science at the London School of Economics and the Imre Lakatos Memorial Fund are also gratefully acknowledged.

1. Cf. in particular the letters dated 11.12.69, 1.1.70 and 10.3.70.
2. On the seaman's dilemma, see Rediker, *Between the Devil and the Deep Blue Sea*, Cambridge, 1987, p. 5.
3. Lakatos' Archive was listed by Michael Hallet in 1979 and updated by Sue Donnelly in 1995. It is arranged in the following thirteen sections: 1. Papers published in Hungary; 2. Early Notes on Mathematics and the Philosophy of Mathematics; 3. Essays in the Logic of Mathematical Discovery; 4. Other Papers in the Philosophy of Mathematics; 5. Early Papers in the Philosophy of Science; 6. Middle-Period Papers in the Philosophy of Science; 7. Later Papers in the Philosophy of Science; 8. Papers on General Philosophy; 9. Lectures; 10. Notes on Feyerabend, Kuhn and Popper and Miscellaneous Notes; 11. Miscellaneous; 12 Selected Correspondence (includes exchanges of letters with George Polya, Victor Kraft, Rudolf Carnap, Richard Popkin, Paul Bernays, Alan Musgrave, Adolf Grünbaum, Thomas Kuhn and Karl Popper; 13. General correspondence.
4. Recall that Lakatos had worked at the Ministry of Education (Culture and Religion) between 1945 and 1948 dealing with education reform. (See Jancis Long, this volume).
5. About fifteen years later – in "A Letter to the Director of the LSE" (1978b, ch. 12) – Lakatos comments on the principle that students should determine academic policy defending academic autonomy on a similar line. Lakatos' political and pedagogical views are challanged by Feyerabend throughout their correspondence. See in particular the crucial years of students revolt, 1968 and 1969, when Lakatos and Feyerabend where at the centre of the event, respectively at the London School of Economics and at Berkeley.
6. On Lukács' concept of dialectic, cf. Mészáros (1972). As we shall see, Lakatos' later criticism of his contemporary "merchants of irrationality" such as Feyerabend and Toulmin will mirror the same kind of arguments. This point has also been raised by Larvor (1998), pp. 74–75 and Val Duseck (forthcoming).
7. Lakatos' adventurous personal history clearly shows the danger to his life especially during the three years period of confinement in the Stalinist prison camp in Resk. (See Long, this volume).
8. In fact, Lakatos here extended to mathematics the falsificationist approach Popper had put forward in connection to empirical sciences. Popper remained an "infallibilist" as far mathematics and logic are concerned. In general Popper seems to be rarely interested in mathematics, apart from his considerations on the historical development of infinitesimal calculus in his (1956/1983).
9. For a comparison of Hegel's view of mathematics and Lakatos', see Larvor (1999) and Kadvany (2001), ch. 1.
10. See, for example, Truesdell (1984), p. 496.
11. See, for example, Giorello (1992).
12. Arpad Szabó had taught Lakatos at Debrecen University in 1943. Lakatos (1976) refers to Szabó (1958) when he considers "thought experiments" as the pattern of mathematical proofs in the pre-Euclidean Greek mathematics. From Szabó (1960) Lakatos takes the idea that, in Euclid's time, "postulates" and "axioms" meant propositions in the critical (dialectic) dialogue put forward to be tested for consequences without being admitted as true by the discussants. Szabó showed his appreciation of his disciple by dedicating the English version of his *The Beginning of Greek Mathematics* (1978) "To the memory of my friend Imre Lakatos."
13. This can be regarded as another piece of the Hegelian inheritance not only for its emphasis on the objectivity of the process as opposed to irrational subjectivism, but also for the fact

that Hegel's *Phenomenology* theorises that in the unfolding of rationality the genesis is never separated from the result, and the result from the genesis.

14. Cf. also Appendix 2 of Lakatos (1976). Lakatos presents Seidel's theorem as a synthesis of the Leibniz continuity principle (thesis) and Fourier's counterexamples (antithesis). "In Hegelian jargon, the counterexamples do not stand in 'bare opposition' to the Leibniz principle, but rather offer a 'determinate (i.e. specific) negation' of it. Synthesis in this three-step does not simply unite the best of the thesis and antithesis. Rather, the synthesis solves the problem posed by the antithesis and the thesis" (Larvor, 1999).
15. After being released Lakatos gained asylum in the Hungarian Academy of Science where he had run the library thus having access to books censored to general public.
16. Cf. Lakatos (1974a): "Popper's ideas represent the most important development in the philosophy of the twentieth century; [...] Personally, my debt to him is immeasurable: more than anyone else, he changed my life. [...] His philosophy helped me to make a final break with the Hegelian outlook which I had held for nearly twenty years" (p. 139).
17. John Worrall and Elie Zahar, the editors *of Proof and Refutations,* have added footnote to this passage in which they comment as follows: "We feel sure that Lakatos would have modified this passage in some respects, for the grip of his Hegelian background grew weaker and weaker as his work progressed." (footnote 2*) I agree only partially. Since I am quite convinced that that the Hegelian background is the central *starting point* of Lakatos' research, I do not think that Lakatos would have changed this specific passage; a passage, moreover, which is a criticism of Hegel, and a criticism based on very good reasons. Nonetheless, and we shall come to this later, I agree that Lakatos' Hegelianism "grew weaker and weaker" and that the key understanding Lakatos' late attitude concerning the aim of science is Popperian fallibilism.
18. Probably at the same time Lakatos was putting the last corrections to his doctoral thesis, he discussed some philosophical implications pertaining to the method of proofs and refutations. This is how he described his *dialectical view point*: "a mathematical theory grows by new decomposition of the constituent conjecture: the theory has no level hypotheses only for the given moment. There are no axioms but the axioms of today turn into theorems tomorrow." ("The Philosophical Implications of the Method of Proofs and Refutations," Archive, 4.1)
19. Cf. Lakatos (1976), pp. 50, 58, 76. Unlike Lakatos, Pólya (1945 and 1956) does not raise doubts on the certainty of mathematics and does not raise any foundational issue. For a more extensive and critical comparison of Pólya and Lakatos' heuristic, see Feferman (1978).
20. For the analogies between Lakatos' logic of mathematical discovery and his MSRP, see Zheng (1990).
21. See Koetsier's well articulated distinction between $Lakatos_1$ and $Lakatos_2$ in *this volume*.
22. The Archive (5.4) includes three different contributions to Popper's philosophy: "The Popperian historiography," "The so-called deductive model of explanation," and "The metaphysical presuppositions of Popperian methodology." Lakatos was probably aiming to write a single article, which however was never put together. This material eventually leads to Lakatos (1970) and (1974a).
23. Duhem (1906), pp. 187–188.
24. There follow a few notes in which Lakatos shows his intention of articulating the method of *Proofs and Refutations* in order to give an explanation of instances taken from the history of science such as, for example, Bohr's discussion of the structure of the atom (see Lakatos, 1970). Among the papers present in the section there are also other draft notes on the *case study* of the alleged "deduction" of Newton's theory from Kepler's laws. Lakatos meant to present it in dialogue form, in analogy with his doctoral dissertation. The title: "From Facts to Empirical Law." The same problem has been tackled from a Lakatosian point of view by Worrall (mimeo) and Zahar (1983).
25. Truesdell (1984), p. 490.
26. One can read into it Poincaré's (1902) conservative stand on relevant cases taken from the history of physics.
27. In a footnote Lakatos adds: "Let me say again that Popper himself came very near to this position in the 1950s; but he did not draw some of the logical implications of the fallibility of background knowledge with sufficient emphasis."

28. This point has also been made by Fine (1978) who has stressed the difference between playing the game "save the proof" (by stretching the concept) and the game of "advancing our science by developing new theories." See also Forrai (1993), pp. 170–174.
29. Forrai (1993) has already raised this point and also suggested the importance of the Hegelian inheritance in this context.
30. Cf. Larvor (1998), p. 1.
31. Lakatos was not fully satisfied regarding the way he accomplished this task. (This is why the editors of *Philosophical Papers II* have chosen to divide the material between Chapter 6 "The problem of scientific theories: three approaches," and Chapter 8 "Understanding Toulmin," i.e. those pages dedicated to analysing Toulmin's idea of human understanding.) A reconstruction of the problems of epistemology around different traditions such as inductivism, probabilism, verificationism, conventionalism, elitism, neoscepticism is also provided in the "Lectures on Scientific Method" Lakatos gave in the same period at LSE (Archive 9, in Motterlini, ed, 1999).
32. In fact, in his late years, Lakatos became more and more disappointed about Popper's achievements to such an extent that, in this case too, reverting the terms (i.e. "Professor Lakatos between the Popperian devil and the Hegelian deep blue sea) could have worked just as well. To Lakatos, Popper's lack of tolerance with respect to the criticism directed towards his own falsificationism had to appear as diabolical as the old Hegelian authoritarianism. (See Lakatos-Feyerabend Correspondence, in particular, the letter dated 26.16.1972, where Popper is mentioned as "Al-Poppuni," "the great tyrant of Reason," Archive, 12.8) Here is for example how Lakatos taught his students about Popper's philosophy while lecturing at LSE in 1973: "Popper's three major contributions to philosophy were: 1) his falsifiability criterion – I think this is a step back from Duhem –; 2) his solution to the problem of induction – where I think he is a step back from Hume, and 3) his literary masterpiece 'The Open Society by one of its enemies'."
33. Cf. Lakatos (1970, p. 72 footnote 1): " [...] in my (1963–4) I was more of a Hegelian, and I thought there was [a natural 'saturation point']; now I use the expression with ironical emphasis. There is no predictable or ascertainable limitation on human imagination in inventing new, content-increasing theories."
34. It is worth noticing that retrospective rationality and ontological fallibilism come out jointly in Engels' *Anti Dürhing*, which Lakatos quotes in his (1960): "knowledge which has an unconditional claim to truth is realised in a number of relative errors; neither the absolute truth of knowledge nor the sovereignty of thought can be fully realised except through an endless eternity of human existence."
35. This is actually nothing but Kierkegaard's criticism to Hegel; simply substitute "life" for "science."
36. Feyerabend's criticisms of Lakatos are freely taken from his (1975a), Ch. 17, and (1976).
37. On the alleged existence of "two Lakatos," the supporter of a more or less close relation between appraisal and advice; and the one who, denying that MSRP does not give any advice at all, has gone a long way towards epistemological anarchism, see Motterlini (1995).
38. Cf. also *from* Feyerabend *to* Lakatos, June 1971: "It is evident that you are intellectually superior to me on account of being a better irrationalist. For your method of arguing, though speedier than mine and certainly more elegantly applied, is identical with mine in *spirit*: you do not argue, you *annihilate* with a slap of your tongue. This, incidentally, will lead to a situation where the more you get ahead of me, the more you are behind: you may overpower people, one after another, you may even overpower me, but in doing so you prove the weakness of *argument* (e.g. mine) and the force of *personality* (yours). So while *reason* may be on my side, the cunning of reason is on yours and thereby also on mine. So in the end I am really the better rationalist than you for I admit the force that makes people agree, while you use it with great effect without acknowledging its irrational character" (Archive 12.8).
39. Worrall (1989), p. 377. On the value of "novelty" for theory-confirmation, see Worrall (1985) and (1989), Zahar (1989). Lakatos' criteria of progress have been applied in many different case studies both in the natural sciences and social sciences. It goes beyond the scope of this paper to assess them. The question of course remains whether these principles are stable or change alongside changes in substantive theories. I shall deal with this problem in the next section.

40. In presenting Feyerabend's criticism I shall quote liberally as well as paraphrasing from Feyerabend (1975a), (1976) and the Lakatos-Feyerabend correspondence.
41. This does not imply that we need necessarily to pry into scientists' psychology in order to assess the 'reasons' or peculiar aversions which have governed their choices, but simply that we should analyse and evaluate the case we are faced with in the light of some explicit heuristic, see Zahar (1983).
42. Writing to Lakatos on the 18th April 1973, Feyerabend acknowledges that "research programmes are good for history [...] because they are connected with a *historical* method of evaluation while paradigms etc. etc. are not connected with any method of evaluation at all. Research programmes make you look for things in history which are much more interesting than what historians or philosophers look for. They are excellent for history, much better than anything that went on before (except, of course, Hegel)." But, he continues, "they do not solve the problem of rationality, that is, even the greatest success in history does not refute anarchism" (Archive 12.8).
43. Lakatos thus takes into account that "inclination towards history" which Federigo Enriques (1936) considered essential in a philosophical understanding of success in science and scientific practice.
44. Early draft of this paper were written in 1963–1964, but Lakatos returned to it several times in the following years.
45. Popkin (1979), ch. 1. Popkin's works on the historical passage from the epistemological (constructive) scepticism of the *crise pyrrhonienne* to the antireligious (dogmatic) scepticism, which through the Enlightenment brought the idea of science we today have, had an important influence on Lakatos. See, in particular, Lakatos (1963–1964). It is thus not surprising that Lakatos had chosen Popkin for the opening speech at the 1965 International Colloquium in the Philosophy of Science held at Bedford College in London. See Popkin (1968). Popkin, on the other hand, was interested in Lakatos' fallibilist approach to mathematics (refer to Popkin's letter to Lakatos of June 18th, 1962, Archive 12.2, item 31), and when he later extended his *History of Scepticism* up to Spinoza (1979), he dedicated it "to the memory of Imre Lakatos."45.
46. Consider, for instance, the shift in the criterion of truth from Erasmus to Descartes, as described in Popkin (1979).
47. See Worrall (1988), p. 272.
48. On the "implausibility" of Lakatos' resort to scientists' "false (methodological) consciousness" to account for the rationality of theory-change, see Newton-Smith (1981, p. 71), Laudan (1989, p. 321); for a reply Worrall (1985) and (1989).
49. Lakatos' vigorous interest with regard to religious, ethical, social and political issues as closely connected with the general problem of demarcation and scientific appraisal is a peculiar feature of his Lectures (Archive 9).
51. I have argued this point extensively in my (1995). Cf. also Larvor (1998), pp. 62–65.
52. Of course, according to Lakatos (1968a), "there is nothing wrong with fallible and speculative metaphysics, but only with interpreting some such metaphysical statements as infallible inductive principles" (pp. 186–187).

References

Duhem, P. (1906) *The Aim and Structure of Physical Theory*, English Translation of 2nd edn., 1914, Princeton: Princeton University Press 1954.

Engels, F. (1894) *Anti Dühring*, 3rd edn., London: Lawrence and Wihart 1955.

Enriques, F. (1936) *Il significato della storia del pensiero scientifico*, Bologna: Zanichelli.

Feferman, S. (1978) "The Logic of Mathematical Discovery Vs. Logical Structure of Mathematics", *PSA*, Volume 2, pp. 309–327.

Feyerabend, P.K. (1975a) *Against Method. Outline of an Anarchistic Theory of Knowledge*, London: New Left Books.

Feyerabend, P.K. (1975b) "Imre Lakatos", *The British Journal for the Philosophy of Science*, 26, pp. 1–8.

Feyerabend, P.K. (1976) "On the Critique of Scientific Reason", in: Howson (ed.) (1976).
Fine, A. (1978) "Conceptual Change in Mathematics and Science: Lakatos' Stretching Refined", *PSA* 1978, vol. 2, pp. 328–341.
Forrai, G. (1993) "From the Method of Proofs and Refutations to the Methodology of Scientific Research Programmes", *International Studies in the Philosophy of Science*, 2, pp. 161–175.
Gillies, D. (ed.) (1992) *Revolutions in Mathematics*, Oxford: Oxford University Press.
Giorello, G. (1992) "The 'Fine Structure' of Mathematical Revolutions: Metaphysics, Legitimacy and Rigour. The Case of the Calculus from Newton to Berkeley and Maclaurin", in: Gillies (ed.) (1992), pp. 134–168.
Hacking, I. (1979) "Imre Lakatos' Philosophy of Science", *British Journal for the Philosophy of Science*, 30, pp. 381–410.
Hegel, G.W.F. (1807) *Phenomenology of Spirit*, Oxford: Claredon Press 1977.
Hegel, G.W.F. (1820) *Elements of the Philosophy of Right*, Cambridge: Cambridge University Press 1991.
Howson, C. (ed.) (1976) *Method and Appraisal in the Physical Sciences*, Cambridge: Cambridge University Press.
Hyppolite, J. (1946) *Genèse et Structure de la Phénoménologie de Hegel*, Paris.
Kadvany, J. (2001) *Imre Lakatos and the Guises of Reason*, Durham and London: Duke University Press.
Koetsier, T. (this volume) "Lakatos' Mitigated Scepticism in the Philosophy of Mathematics".
Lakatos, I. (1959–1961) "What Does a Mathematical Proof Prove?", published as chapter 4 of Lakatos (1978b).
Lakatos, I. (1960) "Necessity, Kneale and Popper", published as chapter 7 of Lakatos (1978b).
Lakatos, I. (1963–1964/1978) "Newton's Effect on Scientific Standards", published as chapter 5 of Lakatos (1978a).
Lakatos, I. (1963–1964) "Proofs and Refutations", *The British Journal for the Philosophy of Science*, 14, pp. 1–25, 129–139, 221–243, 296, 342. Repubished in revised form as part of Lakatos (1976a).
Lakatos, I. (1968) "Changes in the Problem of Inductive Logic", in: Lakatos (ed.) (1968), republished as chapter 8 of Lakatos (1978b).
Lakatos, I. (1970) "Falsificationism and the Methodology of Scientific Research Programmes", in: Lakatos, Musgrave (eds.) (1970), republished as chapter 1 of Lakatos (1978a).
Lakatos, I. (1971a) "History of Science and its Rational Reconstructions", in R. Buck and R. Cohen (eds.) (1971) *P.S.A. 1970, Boston Studies for the Philosophy of Science*, 8, republished as chapter 2 of Lakatos (1978a).
Lakatos, I. (1971b) "Replies to Critics", in: R. Buck and R. Cohen (eds.) *PSA. 1970, Boston Studies for the Philosophy of Science*, 8, Reidel, Dordrecht, pp. 174–182.
Lakatos, I. (1973) "The Problem of Appraising Scientific Theories: Three Approaches", published as chapter 6 of Lakatos (1978b).
Lakatos, I. (1973/1976) "Understanding Toulmin", *Minerva*, 14, pp. 126–134, republished as chapter 11 of Lakatos (1978b).
Lakatos, I. (1974a) "Popper on Demarcation and Induction", in: P.A. Schilpp (ed.) (1974) *The Philosophy of Karl Popper*, La Salle, Illinois: Open Court, republished as chapter 3 of Lakatos (1978a).
Lakatos, I. (1974b) "Anomalies versus 'Crucial Experiments' (a Rejoinder to Professor Grünbaum)", published as chapter 10 of Lakatos (1978b).
Lakatos, I. (1976) *Proofs and Refutations. The Logic of Mathematical Discovery*, edited by J. Worrall and E. Zahar, Cambridge: Cambridge University Press.
Lakatos, I. (1978a) *The Methodology of Scientific Research Programmes: Philosophical Papers*, vol. I, edited by J. Worrall and G. Currie, Cambridge: Cambridge University Press.
Lakatos, I. (1978b) *Mathematics, Science and Epistemology: Philosophical Papers*, vol. II, edited by J. Worrall and G. Currie, Cambridge: Cambridge University Press.
Lakatos, I. (1999) "Lectures on Scientific Method", in: Motterlini (ed.) (1999).
Lakatos, I. Feyerabend, P.K. (1999) "Correspondence 1968–1974", in: Motterlini (ed.) (1999).

Lakatos, I., Musgrave, A. (eds.) (1970) *Criticism and the Growth of Knowledge*, Cambridge: Cambridge University Press.
Lakatos, I., Musgrave, A. (eds.) (1968) *The Problem in the Philosophy of Science*, Amsterdam: North Holland.
Larvor, B. (1998) *Lakatos: An Introduction*, London: Routledge.
Larvor, B. (1999) "Lakatos' Mathematical Hegelianism", *The Owl of Minerva*, Vol. 31, 1 (Fall, 1999), pp. 23–44.
Laudan, L. (1989) "If It Ain't Broke, Don't Fix It", *British Journal for the Philosophy of Science*, 40, pp. 369–375.
Lukács, G. (1962) *The Destruction of Reason*, London: The Merlin Press 1980.
Mészáros, I. (1972) *Lukács Concept of Dialectic*, London: The Merlin Press.
Motterlini (ed.) (1999) *For and Against Method: Including Lakatos' Lecture on Scientific Method and the Lakatos–Feyerabend Correspondence*, Chicago: Chicago University Press.
Motterlini, M. (1995) "Has Lakatos really gone a long way towards epistemological anarchism?", *Epistemologia*, XVIII, 1995, pp. 215–232.
Newton-Smith, W. (1981) *The Rationality of Science*, London: Routledge.
Poincaré, H. (1902) *La science et l'hypothèse*, Paris: Flammarion. English authorized translation by G.B. Halsted, *Science and Hypothesis*, in Poincaré (1913) *The Foundation of Science*, Washinghton: University Press of America 1982.
Pólya, G. (1945) *How to Solve it*, Princeton: Princeton University Press 1973.
Pólya, G. (1954) *Mathematics and Plausible Reasoning*, 2 vol., Princeton: Princeton University Press.
Popkin, R.H. (1968) "Scepticism, Theology and Scientific Revolution in the Seventeenth Century" in: Lakatos, Musgrave (eds.) (1968).
Popkin, R.H. (1979) *The History of Scepticism from Erasmus to Spinoza*, California: Berkeley University Press.
Popper, K.R. (1956/1983) *Realism and the Aim of Science*, From the *Postscript to the Logic of Scientific Discovery*, edited by W. Bartley III, London: Hutchinson.
Rediker, M. (1987) *Between the Devil and the Deep Blue Sea*, Canto, Cambridge: Cambridge University Press.
Szabó, A. (1958) "'Deiknymi' als mathematischer Terminus für 'Beweisen'", *Maia*, 10, pp. 1–26.
Szabó, A. (1960) "Anfänge des Euklidischen Axiomensystems", *Archive for the History of Exact Sciences*, I, pp. 27–106.
Szabó, A. (1978) *The Beginning of Greek Mathematics*, Boston-London: Reidel.
Toulmin, S. (1972) *Human Understanding*, Oxford: Oxford University Press.
Truesdell, C. (1984) *An Idiot's Fugitive Essays on Science. Methods, Criticism, Training, Circumstances*, New York-Berlin-Heidelberg-Tokyo: Springer.
Wittegentein, L. (1958) *Philosophical Investigations*, Oxford: Oxford University Press.
Worrall, J. (1984) "The Background and the Forefront: A Response to Levi and Shapere", *PSA* 1984, vol. 2, pp. 672–682.
Worrall, J. (1985) "Scientific Discovery and Theory-Confirmation", in: J. Pitt (ed.), *Change and Progress in Modern Science*, Dordrecht, Reidel.
Worrall, J. (1988) "The Value of a Fixed Methodology", *British Journal for the Philosophy of Science*, 39, pp. 263–275.
Worrall, J. (1989) "Fix It and Be Damned: A Reply to Laudan", *British Journal for the Philosophy of Science*, 40, pp. 376–388.
Worrall, J. (1990) "Scientific Revolutions and Scientific Rationality: The Case of the 'Elderly Holdout'", in: W. Savage (ed.), *The Justification, Discovery and Evolution of Scientific Theories*, Minneapolis, University of Minnesota Press.
Worrall, J. (mimeo) "How to Deduce Theories from Phenomena: Newton on Scientific Method", mimeo.
Zahar, E. (1983) "Logic of Invention or Psychology of Research?", *British Journal for the Philosophy of Science*, 34, pp. 234–261.
Zahar, E. (1989) *Einstein's Revolution. A Study in Heuristics*, La Salle, Illinois: Open Court.

Zheng, Y. (1990) "From the Logic of Mathematical Discovery to the Metodology of Scientific Research Programmes", *British Journal for the Philosophy of Science*, 41, pp. 377–399.

Dipartimento di Economia e Laboratorio di Scienze Cognitive
Università degli Studi di Trento
via Inama 5
I-38100 Trento
Italy

MARTIN CARRIER

EXPLAINING SCIENTIFIC PROGRESS: LAKATOS' METHODOLOGICAL ACCOUNT OF KUHNIAN PATTERNS OF THEORY CHANGE

Theory choice decisions were at the focus of the debate on theory change that dominated the philosophy of science in the 1960's and 1970's. Comparative evaluations of competing theoretical approaches were supposed to form the nucleus of scientific progress and, at the same time, the nucleus of scientific rationality. A theory choice decision singles out the methodologically most qualified alternative from among a set of rival theories. A sequence of such decisions is supposed to generate a series of theories with increasing methodological virtue; and a series of this kind constitutes scientific progress. Moreover, scientific rationality is tied to the nature and justification of the criteria brought to bear on theory choice decisions. Rational theory choices are made relying on objective and epistemically significant criteria. The intertwinement of rationality, theory change and theory choice is among the outstanding commitments underlying the entire debate on "theory dynamics"; in particular, it is constitutive of the methodological approaches of Larry Laudan and, above all, Imre Lakatos. The backdrop, against which this debate unfolded, is provided by the methodological challenge involved in Thomas Kuhn's philosophy of science. My objective is to elucidate more clearly Lakatos' attempt to neutralize methodological threats he assumed to be inherent in some of Kuhn's claims on theory change.

I begin by giving a brief sketch of Kuhn's model of scientific change and continue by highlighting the limitations it entails for any account of theory choice decisions appealing exclusively to objective and epistemically significant methodological criteria. Subsequently, I outline Lakatos' methodology and elaborate the criteria of theory choice involved in it. Finally, I develop the implications of Lakatos' model for scientific change and address, in particular, its bearing on those features of scientific change, which Kuhn thought defied the grip of systematic methodology. It is well-known that Lakatos attempted to provide a sort of rational reconstruction of some of Kuhn's allegedly descriptive generalizations about scientific change. My aim is to spell out the nature and import of this endeavor.

G. Kampis, L. Kvasz and M. Stöltzner (eds.),
Appraising Lakatos: Mathematics, Methodology and the Man, 53–71.

1. SOME BASICS OF KUHN'S "PARADIGM THEORY"

Kuhn introduced a distinction between two levels of scientific theorizing that was retained in all later conceptions; the distinction, namely, between an overarching theoretical framework, on the one hand, and its more specific elaboration, on the other. This framework he called "paradigm"; it was supposed to embrace theoretical principles, methodological or metaphysical commitments, and a collection of exemplary solutions to problems (whence derives the appellation "paradigm"). For example, the paradigm of 19th-century wave optics proceeded from the assumption that light is to be conceived as a state of elastic oscillation of a pervasive medium. Specific versions of the paradigm consisted in more elaborate explanations of optical phenomena such as refraction, diffraction or interference. A scientific discipline that is dominated by one particular paradigm has entered the stage of "normal science." The shared commitment to an overarching framework relieves the scientists from the need to defend their basic orientation and thus allows them to focus on more productive, technical work. In normal science, a paradigm rules monopolistically and unquestioned. Its principles are not liable to empirical testing. If an application of a paradigm fails unexpectedly, i.e. if an anomaly emerges, the blame is not attributed to the paradigmatic principles themselves. Rather, additional unrecognized influences or lack of ingenuity on the part of the scientists are held responsible. That is, either the situation is considered more complex than anticipated, or the scientists' creativity and technical skill are found to be wanting. The paradigm is maintained, in spite of empirical counterinstances. In sum, Kuhnian normal science is characterized by *paradigm monopoly* and *paradigm immunity* (Kuhn, 1970a, pp. 77–80; Kuhn, 1970b, p. 6).

It is obvious that the central traits of Kuhn's normal science stand in marked contrast to Popper's methodological advice to heed counterinstances. Scientists informed by Popper's falsificationism must not ignore empirical problems. Rather, they are called upon to either modify the theory in a methodologically acceptable fashion or to drop it entirely. A theory is improved in an acceptable way if the anomaly is resolved and at the same time the theory's empirical content (i.e. the number of possible observations conflicting with the theory) is expanded (Popper, 1935, §§ 20, 31). Nothing of the kind is required from Kuhnian normal scientists. They are licensed to shelve unsolved problems and go ahead undauntedly. The difference between Popper and Kuhn cannot be traced back to the ubiquitous contrast between lofty normative principles and sloppy practice. Kuhn, namely, gives epistemological reasons for the nonchalant attitude toward anomalies he assumes to be characteristic of normal science. The chief reason is based on the historical observation that no theory ever gets rid of anomalies. This Kuhnian tenet of the "omnipresence of anomalies" rules out assessing each of them as a serious threat to the theory. Taking – in the Popperian spirit – each anomaly to be a potential refutation amounts to closing down the business of scientific theorizing altogether (Kuhn, 1970a, pp. 79–82). By contrast, the immunity conferred to paradigms in normal science provides a basis for the tenacious pursuit of theories, which is in turn a necessary precondition for overcoming recalcitrant difficulties.

However, when anomalies pile up and the paradigm proves incapable of guiding a successful tradition of normal science, commitment to the fundamental principles is increasingly weakened and finally lost. In the course of such a "crisis," alternative theoretical options are considered and pursued. The emergence of a crisis follows from the principles of Kuhn's account. As a result of the sophisticated and highly specialized work done in normal science, anomalies are bound to turn up (barring the extremely improbable eventuality that a theory gets everything right). As unsolved problems pile up, each tradition of normal science sooner or later slides into crisis. Such crises frequently result in a "paradigm shift" that is characteristic of a "scientific revolution." It is one of Kuhn's central historical claims that a theory is never given up, unless it can be replaced by an alternative approach. Abandoning a paradigm is tantamount to adopting a new one. In contrast to the smooth development of normal science, Kuhnian cataclysms amount to a wholesale substitution of the former conceptual framework. This means, in particular, that revolutions are non-accumulative, in that they involve taking back problem solutions that were formerly accepted as correct. What counted as trustworthy scientific knowledge before, is at least drastically reinterpreted and frequently rejected as misleading or false.

The non-accumulative character of scientific revolutions becomes manifest in four features, namely, in changes: of the relevant concepts, of the problem-situations, of the criteria for evaluating theoretical achievements and, finally, in the occurrence of so-called "Kuhn-losses." The assumed conceptual change constitutes the notorious doctrine of meaning variance, which grows out of the assumption that meaning is determined by the pertinent theoretical context. Drastic alterations of this context lead to significant changes in the meaning of the concepts involved, which in turn may vitiate the translatability of concepts from different theories. This result constitutes the "incommensurability thesis," which denies that the substantive claims of disparate theories can be translated into one another (Kuhn, 1983). I won't go into this matter here. The reason is, first, that addressing this contentious issue deserves a full-scale treatise in itself, and that, second, it is of no relevance for the methodological problems I wish to discuss. The latter problems arise irrespective of any potential further aggravation generated by the breakdown of translation. Thus, I proceed from the assumption – as Lakatos does – that the substantive content and the empirical consequences of rival theoretical approaches can be compared.[1]

The second major shift occurring during a revolution concerns the change of problems. This is unsurprising at first sight. After all, it conforms well to the traditional picture of scientific progress that old problems are solved and new problems crop up. Kuhn does not deny that problem changes of this kind appear in the course of a revolution; on the contrary, the solution of long-standing anomalies within the new framework constitutes one of the chief reasons for the shift of allegiance. However, as Kuhn stresses, an additional pattern of problem change turns up which amounts to "dissolving" a problem rather than solving it. That is, the legitimacy of the problem is rejected by the alternative approach (Kuhn, 1970a, p. 103). Around 1770, for instance, one of the major challenges in chemistry was to explain the role of phlogiston in the release of hydrogen from the solution of metals in acids. The claim of the rival oxygen theory was that phlogiston doesn't exist at all, and that it consequently plays no role whatsoever in these processes. Accordingly, the question wasn't answered; it was rejected as misguided, instead.

The third important alteration refers to the criteria invoked for judging problem solutions. Such criteria are frequently tied up with – and specific to – a given paradigm, and thus change upon paradigm substitution. One of Kuhn's examples is again taken from the Chemical Revolution. Within the phlogistic framework, it was considered the chief task of chemical theories to account for the properties of chemical substances (such as hardness, combustibility, volatility and the like), along with their changes during chemical reactions. Consequently, chemical explanations are to be judged according to their capacity to afford such an account. In the course of the switch to the oxygen theory, these problems were shifted into the background, whereas the challenge of accommodating reaction weights was moved to center stage. Chemical theories are to be assessed according to their ability to meet this challenge. As a result of the paradigm shift, the standards for judging the adequacy of theoretical achievements are altered as well (Kuhn, 1970a, p. 107; Kuhn, 1977, pp. 335–336).

Fourth, scientific revolutions frequently go along with what is now called Kuhn-losses. A new paradigm may be accepted in spite of the fact that some of the former explanatory achievements are thereby lost. More specifically, some of the phenomena accounted for previously are re-transformed into open problems. Already explained data once again become anomalous. To be sure, Kuhn-losses are only tolerated as long as they do not exceed a low-level threshold. But the salient point is that they do occur, and that their mere existence vitiates any claim to the effect that the new paradigm reproduces all the explanatory achievements of the old one. One of Kuhn's favorite examples of a Kuhn loss again refers to the Chemical Revolution. In the phlogistic framework, a metal was regarded as a compound of a specific component (the "calx") and phlogiston. Since phlogiston was assumed to be present in all metals, the theory could explain why they resembled one another to a much greater extent than the corresponding calces (the oxides, in modern terminology). The oxygen theory, by contrast, considered metals to be elementary, and thus lacked any resources to account for their similarity. The adoption of phlogiston theory thus reopened an empirical problem that had been considered settled before (Kuhn, 1977, p. 323; Kuhn, 1970a, pp. 132, 157, 170; Kuhn, 1970b, p. 20).

On the whole, then, and due to these four features, revolutions are characterized by a fundamental theory change, which admits of no reconstruction to the effect that the earlier theory is approximately retained by the later. The contrast between theories separated by a revolution are far-reaching and unbridgeable (Kuhn, 1970a, pp. 5–6, 97–110).

2. KUHN'S ACCOUNT OF THEORY CHOICE DECISIONS

Within the framework of the paradigm theory, theory choice decisions are only made in the course of revolutions. Naturally enough, Kuhn's account of such decisions focuses on such cataclysmic periods. The central claim is that the best choice is not fixed by appeal to the available facts and to standards of evaluation traditionally deemed "rational." Rather, there is room left for subjective factors, and their supplementary influence is not detrimental to scientific progress, but constitutes a methodological virtue.

The first contention is that the evaluation of theories cannot be made by relying solely on the data. This claim follows directly from the basics of Kuhn's paradigm theory. On the one hand, the old paradigm is afflicted with a particularly large number of anomalies; otherwise a crisis wouldn't have occurred in the first place. On the other hand, the new paradigm has, due to its nascent state, not yet reached a level of elaboration and articulation comparable with the former monopolist. A freshly invented approach inevitably suffers from gaps and lacunae, most of which are unlikely to appear in the older competitor. The conclusion is that both rival approaches are anomaly-ridden, so that the evidential situation fails to give unique preference to one of them. Empirical adequacy cannot be the sole criterion for theory choice.

Second, this shaky empirical ground necessitates the invocation of additional, non-empirical standards. However, the catch is that these standards depend on – and vary with – the paradigm candidates at issue. As mentioned above, the contenders typically provide different standards for judging the appropriateness of problem solutions. As a result of these disparate criteria, each competitor appreciates its own assets and its rival's liabilities, drawing on its own specific measures of adequacy. Naturally enough, the adherents of the contrasting paradigms will fail to convince one another (Kuhn, 1970a, pp. 109–110).

Third, not all criteria of appraisal are tied to one of the rival candidates. There are shared methodological values, including explanatory power, precision, consistency or simplicity. The problem is, though, that these standards are imprecise, and can be made precise in disparate ways. If the simplicity of a given theory is to be assessed, different results are likely to turn up. Moreover, the application of more than one of these standards to a specific case may easily engender conflicts among them. One of the candidates may have a wider scope, while the other may furnish more precise explanations within its more restricted domain. Due to the necessity of making precise and weighting standards of evaluation, even transparadigmatic agreement on such standards in no way guarantees the unanimous appraisal of particular theories.

This contention can be condensed into what I call Kuhn's "uncertainty thesis": If the methodological judgment of theories is based exclusively on a consideration of the facts and the application of a canon of overarching, epistemically justified methodological rules, no clear comparative evaluation will emerge. To be sure, clear judgments could be made relying on theory-specific criteria; owing, however, to their linkage with one of the competitors, they fail to carry much force. This restricted bearing of objective, overarching standards of evaluation makes room for the possible intrusion of subjective factors. Theory appraisal is a delicate affair, in which a multitude of objective and subjective elements are intertwined in ways that may well differ among different scientists. And none of them could sensibly be blamed for violating the canon of rationality (Kuhn, 1970a, pp. 151–153).

In Kuhn's opinion, the room left for subjective factors in no way threatens the rationality of theory judgment. To be sure, free play of this sort is in conflict with a model of scientific rationality that regards theory evaluation as a quasi-algorithmic procedure of applying strict methodological rules. But such a view flies in the face of the real uncertainties involved in any such appraisal. There can be no reliable methodological measure of future achievements of a theory. No methodology can possibly be able to tell how well a theory will be able to cope with future challenges – after all, some of the relevant problems may not even be known at the time in question. Methodological standards suitable to unambiguously giving preference to one of the competing approaches, would lead to the latter's unanimous adoption and pursuit. Given the real uncertainties of the situation, any methodological selection of this kind involves a high risk of being wrong. But as a result of the general acceptance of one of the candidates, no alternative is available anymore. Consequently, a possibly mistaken theory choice couldn't be corrected. The approach singled out earlier has become the only game in town. By contrast, the rational way of dealing with the risks involved in theory choice is to pursue the entire spectrum of options. In a crisis, theoretical pluralism is the order of the day. This can be achieved by splitting up the relevant scientific community. Some scientists follow one theoretical route, and some trace the consequences of another. And the most natural way to split up a community is to appeal to subjective factors (such as aesthetic considerations) which, as a rule, vary strongly among its members. Methodological uncertainty thus leads to a division of theoretical labor which constitutes the most appropriate way of coping with the risks involved in theory choice decisions. In sum, Kuhn's claim is that methodological uncertainty of the kind specified in no way vitiates scientific rationality, but rather brings out more clearly its core and kernel (Kuhn, 1977, pp. 325–332).

However, this account raises a problem for Kuhn's account of scientific development. Kuhn gives an epistemological argument in favor of paradigm monopoly and the subsequent emergence of a crisis. Therefore, one should also expect an epistemological argument for the eventual completion (or abortion, as the case may be) of a scientific revolution. It should follow from Kuhn's principles that a crisis grows into a new tradition of normal science. But in light of his epistemological defense of methodological uncertainty, quite the reverse is true. Given the virtue of theoretical pluralism implied by Kuhn's argument, one anticipates that a crisis should endure indefinitely and that monolithic dominance of a paradigm is never reestablished. Normal science is unstable but crises are not. The completion of a revolution and the corresponding onset of a new tradition of normal science thus remain puzzling within Kuhn's framework. There are two options for dealing with this historical anomaly. One could either deny that theoretical pluralism is confined to crisis situations and disappears after these specific periods, or one could claim that the methodological uncertainties are less marked than Kuhn would have it. In the latter case, all parties would eventually place their bets on a common candidate. Both options, however, would seriously undermine the fundamentals of the paradigm theory. Actually, they form part of Lakatos' methodology of scientific research programs (see below Sec. 5).

Let me summarize the most important historical claims entertained by Kuhn. Periods of normal science or theoretical monism are followed by revolutions or periods of theoretical pluralism. In normal science, the theoretical principles are retained entirely; no significant theoretical innovation is found. Theory choice decisions are only made during revolutionary periods, and these decisions are not fixed by the available data along with transparadigmatic methodological standards. This means that there is no way of objectively singling out the most qualified paradigm candidate. Furthermore, a revolution always involves the substitution, and never the mere abandonment, of a paradigm. And the new paradigm cannot be reconstructed to contain the claims of its predecessor "in the limit"; revolutions proceed in a thoroughly non-accumulative fashion.

3. LAKATOS' METHODOLOGY OF SCIENTIFIC RESEARCH PROGRAMS

Lakatos is chiefly concerned with the methodological explanation of theory choice decisions. In particular, he accepts some of the historical generalizations advanced by Kuhn and takes them as a sort of evidential basis of methodology. I wish to clarify the nature and import of Lakatos' endeavor. This involves a pinch of "rational reconstruction" of his approach – that is, some sort of improvement (as I hope) where the real wording is given in the footnotes.[2] I begin by sketching Lakatos' criteria for evaluating theories or research programs, and subsequently bring them to bear on the explanation of the growth of knowledge.

Lakatos presupposes the viability of the so-called Duhem-Quine thesis. Roughly speaking, it says that experience alone is insufficient for conclusive confirmation or refutation of any given piece of a theoretical network. First, conclusive empirical confirmation is beyond the reach of hypothetico-deductive testing because mistaken principles may entail truthful empirical consequences. It follows that true observational results do not license inference to the truth of the principles from which they were derived. Second, conclusive empirical refutation is vitiated by the fact that it is not a single theoretical assumption, but always a complex network of theoretical hypotheses that gives rise to observational consequences in the first place. As a result, an anomaly can never be attributed to a specific axiom. To a significant extent, it is up to the scientists to distribute praise and blame for empirical success or failure within the system of theoretical principles at issue (Lakatos, 1970, pp. 97–98; Lakatos, 1971, p. 111).

If theoretical assumptions cannot be judged relying on the facts alone, the only alternative that conforms to the received notion of rationality is to also draw on objective, epistemically relevant standards. According to this notion, rationality is essentially characterized by guidance through general and justified rules. However, Kuhn's uncertainty thesis for methodological evaluation blocks this auxiliary procedure. It is this result that constitutes the major threat to scientific rationality as it emerges from Kuhn's approach. Lakatos' intention is to demonstrate that this threat can be countered. His approach can be reconstructed to the effect that Kuhn's uncertainty thesis does not capture the essentials of theory choice, but rather arises from Kuhn's failure to consider appropriate standards of theory evaluation.

Lakatos' basic unit of scientific progress is the "research program." Science is considered to be an ongoing competition among research programs. The anatomy of a research program is characterized by two elements, namely, its "hard core" and its "positive heuristic." The hard core contains the fundamental postulates of the program; they are retained during the program's lifetime. The positive heuristic spells out guidelines for the program's future theoretical development. It singles out significant problems, and sketches pathways to their solution. Thus, it directs scientists through the maze of confusing difficulties by giving them an idea where to drive at. The positive heuristic draws a picture of the program's "better self." One of Lakatos' historical assertions is that the development of a qualified program is determined by its heuristic. It is a hallmark of superiority if a program proceeds autonomously, and does not merely respond to conceptual and empirical difficulties (Lakatos, 1970, pp. 47–52, 68–69).

The hard-core postulates are irrefutable within the pertinent program. Abandoning them is tantamount to giving up the program. However, their irrefutability has nothing to do with the substance of the claims involved; it arises, rather, from the firm commitment of the program adherents not to admit the refutation of their claims. Instead, anomalies are accommodated by adapting the "protective belt" of auxiliary hypotheses. According to the empirical needs, this protective belt is expanded, modified, restructured or replaced entirely. It follows from the Duhem-Quine thesis that an empirical problem never bears directly on a specific theoretical hypothesis. Thus, there is always room left for the scientists to shield their pet principles from the refuting impact of anomalies, and to redirect the latter's force to assumptions from the protective belt. The belt thus constitutes a sort of buffer zone, in which the required adaptations are made (Lakatos, 1970, pp. 48–50).

Modifications within the protective belt may concern, first, observation theories, second, initial and boundary conditions, and third, additional assumptions within the respective theory itself. Take the program of Newtonian celestial mechanics as an example. Its hard core is made up of Newton's laws of motion and the law of gravitation. These laws are intended to be kept intact, come what may. Suppose the astronomical data regarding planetary positions don't match the theoretical expectations. Instead of regarding this finding as a falsification of the theory, one might pick one of three responses to follow. A modification of the relevant observation theories consisted, e.g. in tampering with the laws of optics upon which the inference from a spot observed in a telescope to the planet's position relies. An adaptation of assumptions concerning the relevant initial and boundary conditions could be made by postulating a hitherto unknown planet and by attributing the anomalous behavior of the observed planet to the unrecognized influence of this additional celestial body. An auxiliary adjustment of the theory itself could involve the introduction of a non-uniform mass distribution of the sun, which would lead to a predicted deviation from the inverse-square drop of intensity. In this case, the original law is not abandoned; rather, a supplementary gravitational effect is additionally invoked.

The hard core and the positive heuristic characterize a program in its entirety; they constitute its identity. A specific version of a program is made up of these invariant elements, together with the changing protective belt. The development of a research program can be represented as a chain of subsequent versions, each of which arises from its predecessor by some theoretical modification, usually performed on the protective belt. It is obvious that some sort of grading of such modifications is needed; if each were supposed to be as good as any other, the coherence of a research program would soon be lost. However, one could conclude from the above description that Lakatos doesn't place much faith in the constraining force of facts. In order to distinguish between acceptable and inappropriate theoretical changes, constraints of a different nature are called for. Lakatos appeals to methodological standards, instead. Such standards primarily express conditions for a justified transition from one program version to its successor. They stipulate which theoretical changes within a program are methodologically sound: a program version that complies with these standards counts as superior to its predecessor. I confine myself to stating and elaborating Lakatos' criteria in their original form. Shortly before his death in 1974, Lakatos started to attenuate his standards, and in the years to follow, a large number of other suggestions cropped up within the Lakatosian school. I cannot address all these distinctions and ramifications, and so I will stick to the original.[3]

Lakatos requires that a theoretical alteration within a research program is justified if the following four conditions are satisfied:

(1) The modification conforms to the positive heuristic of the program.

(2) The follow-up version is able to account for all those phenomena that are successfully explained by its predecessor.[4] This is not to say that all corroborated derivations are to be reproduced by the successor. Rather, the relevant phenomena may be accounted for in different ways.

(3) The successor "predicts some novel, hitherto unexpected fact" (Lakatos, 1970, p. 33). Such "novel facts" are empirical generalizations that were not to be expected on the basis of "previous knowledge" (Lakatos, 1970, p. 34). That is, Lakatos demands the prediction of observational regularities that were unknown to science before.[5]

(4) The predicted effects are found empirically. If the first three conditions are satisfied, the program is "theoretically progressive"; if the fourth demand is also met, the program is "empirically progressive." Lakatos' standard of progress requires that all justified transitions within a program be at least theoretically progressive and that empirically progressive shifts be accomplished "occasionally." If a program version keeps up with the explanatory achievements of its predecessor, and brings forth new successful predictions, it possesses "corroborated excess content," and it is this kind of surplus performance that is the most significant justification of its acceptance. The empirical confirmation of theoretical anticipations is the stuff of which methodological superiority and scientific progress are made (Lakatos, 1970, pp. 31–34; Lakatos, 1971, p. 112).

Rival research programs can be evaluated analogously. The condition of progress within a program is transferred – with minor variations – to the comparative judgment of entire programs. The first relevant question is whether each of the two competing programs is progressive in and of itself. If one is progressive and the other isn't (i.e. if the latter program is "degenerating"), the first is superior to the latter. Second, the conditions (2) to (4) are brought to bear on the rival program. These conditions require coping with all known phenomena, as well as the anticipation of new ones. In judging entire programs, the relevant standard of comparison is not the earlier version of the same program, but the most recent version of the alternative program. It is demanded that it predict empirical regularities that were not known before and are thus, in particular, novel with respect to the competing program. Evaluating two progressive programs thus involves comparing the respective explanatory achievements in light of the methodological criteria given.[6]

It is worth repeating, that the demand to reproduce the explanatory achievements of another theory (preceding version or alternative program) does not amount to the requirement to reproduce the explanations themselves. Retention of the principles and the modes of derivation is not necessary; the relevant phenomena may well be accommodated in a disparate fashion. It is thus permitted, that a modification involve a fundamental change in the theoretical treatment. For this reason, Lakatos' condition of progress is compatible with non-accumulative change in the Kuhnian sense of a complete replacement, rather than an approximate retention, of a theory by another one.[7]

4. THE RATIONALITY OF THEORY CHOICE DECISIONS

It is glaringly obvious that some of Lakatos' concepts are modeled on Kuhnian notions. A research program corresponds roughly to a paradigm, and a program change approximately resembles a scientific revolution. The retention of the hard core and the positive heuristic in pursuing a program reproduces the continuity of normal science. Finally, the methodological license to give up accepted explanations and to accommodate the relevant phenomena in a distinct fashion, captures the non-accumulative character of revolutions.

However, Lakatos' methodology in no way simply mimics Kuhn's notions within a different conceptual framework. It aims, rather, at solving methodological problems raised by the paradigm theory. The first relevant claim concerns the rejection of Kuhn's uncertainty thesis. This alleged uncertainty arises from Kuhn's approach to basing methodological judgment on a collection of alternative criteria. Lakatos replaces Kuhn's heterogeneous collection by a small number of clear requirements; namely, to reproduce explanatory content and to anticipate novel regularities. Moreover, conflicts between these demands cannot arise, because satisfaction of all of them is required. A superior theory needs to accomplish reproduction *and* anticipation. The Kuhnian process of considering and weighting alternative standards is dropped. Lakatos' procedure of methodological comparison thus removes the uncertainties characteristic of Kuhn's approach.

This procedure can be spelled out as follows. The relevant difficulty, as pointed out by Kuhn, is that each theory at the same time agrees and conflicts with a large number of facts, so that the empirical situation is confusing. Namely, any theory which is at all seriously discussed in a scientific community is buttressed by some successful explanations. On the other hand, Kuhn's thesis of the omnipresence of anomalies entails that any theory is also liable to a number of unsolved problems. Judging or comparing the empirical adequacy of rival theories is thus a highly intricate matter, in which an unanimous outcome cannot be expected.

Lakatos' way out of this difficulty amounts to granting only a small number of phenomena any relevance for the methodological assessment of a theory. Most of the facts with which a theory agrees, and most of its anomalies do not have any bearing on the theory's qualification. Rather, the evaluation of a theory is, essentially, based on the few cases of corroborated excess content, i.e. on the confirmed novel predictions. Lakatos' methodological criteria specify demands for those facts which a theory can rightly count in its favor. These criteria operate by singling out "supporting facts"; that is, facts explained in a particularly qualified manner. The appraisal of a theory is to be based exclusively on these facts. A theory is only confirmed by its supporting facts, not by all of its correct empirical consequences. If consideration is restricted to a selected group of outstanding explanations, it is much easier to reach agreement on how well the theory is doing empirically. The "qualified empirical adequacy" of a theory is based on facts singled out as important by the methodology, and this guidance facilitates tremendously the comparison of empirical performance (Lakatos, 1970, p. 36; Lakatos, 1971, p. 112).

Kuhn's theory-specific criteria, which depend on and vary with the respective paradigm, become part of Lakatos' positive heuristic. The positive heuristic determines which kinds of theoretical means and procedures are acceptable within a program, or which theoretical aims are to be followed. The positive heuristic plays the same role in a program as does Kuhn's set of theory-specific standards in a paradigm. In contrast to Kuhn's, however, these standards are denied any influence on the comparative evaluation of programs. After all, compliance with the positive heuristic is only required within a program, and the judgment of entire programs relies exclusively on overarching, theory-neutral – or "transparadigmatic" – criteria.[8]

Whether or not Lakatos' conception is actually able to overcome the uncertainties indicated by Kuhn is a matter of empirical scrutiny. It can only be shown by studies in the history of science whether Lakatos' criteria of theory evaluation are actually suitable for determining theory choice uniquely, for one thing, and for singling out theories deemed superior in retrospect, for another. However, there is a second strand in Lakatos' approach that addresses more general tenets of the paradigm theory. It is the overarching characteristics of theory change, as advanced by Kuhn, that are at Lakatos' focus. Lakatos' conception allows for a methodological derivation of some of Kuhn's descriptional generalizations about theory change. The first relevant finding concerns the immunity of paradigms to anomalies. Its analog for research programs can be derived as follows. Supporting facts are constituted by correctly predicted empirical regularities. Such facts have not been accounted for by the competing program; they would otherwise not be "novel." Thus, Lakatos' condition of progress implies that only those facts are suitable for buttressing a program which cannot be explained by its rival.[9] Conversely, the only facts militating against a program are those that favor this program's competitor. It follows that only those anomalies, which can be solved by the rival in a qualified fashion (i.e. by predicting unknown phenomena at the same time), count as failures. There is no refutation without confirmation (Lakatos, 1970, p. 37). This entails that the mere inability to accommodate this or that observational regularity does not bring a program into difficulty (Lakatos, 1970, p. 92). And it is precisely this conclusion that coincides with the Kuhnian immunity claim, which is thereby derived from Lakatos' notion of supporting facts. Research programs are – rightly – immune to "mere anomalies." If a research program is to be criticized effectively, it is not sufficient to expound its liabilities. Rather, what hurts a program is the qualified solution to its problems within a different theoretical perspective.

Another historical regularity stressed by Kuhn is that paradigms are never given up unless an appropriate alternative is available. Scientific revolutions always involve theory-substitutions. The Lakatosian analog is: No program abandonment without program replacement. A program is only disqualified methodologically if a superior competitor is at hand. This condition can be derived from a corollary to the immunity argument. This argument states that the liabilities of one theory are the assets of the other. There are no significant failures without an alternative solution. And obviously enough: if a theory is not in trouble, it should not be given up. It follows that a program can never be rated as deficient unless there is a contender attacking it with some success. Disqualification of a program is brought about by a corroborated rival (Lakatos, 1970, p. 35; Lakatos, 1971, pp. 112–113).[10]

Kuhn's claims with regard to paradigm immunity and paradigm substitution are leveled as descriptional objections at methodological requirements entertained by Popper. They are advanced as historical counterexamples to Popper's demands. It is true: a historical description cannot refute a methodological norm. But Popper and Kuhn share the commitment that historical theory change, at least concerning its fundamental and long-term aspects, is rationally justified and constitutes growth of knowledge. In light of this commitment, historical arguments cannot be dismissed offhand. Within the framework of Lakatos' methodology, by contrast, the two features of immunity and substitution constitute theorems rather than objections. They follow from Lakatos' conception of how theories are to be evaluated. The important result is that if theory change occurs the way Kuhn says, it does not violate methodological rules but, rather, conforms to such rules. Lakatos' conception is thus able to provide a methodological explanation of these Kuhnian characteristics of scientific change.

Lakatos' conception also manages to solve a long-standing problem of theory evaluation. Since the days of Pierre Duhem, it has been stressed repeatedly that the application of tough methodological criteria to nascent theories would defeat them prematurely, thereby tending to thwart scientific progress. A freshly invented theory needs room to grow before it is subjected to the full force of demanding standards. The traditional solution to this problem is to require temporary suspension of methodological judgment. For instance, Duhem argues that only full-fledged theories should be blamed for empirical difficulties in their domain (Duhem, 1906, 10.§7). That is, the condition of empirical adequacy is weakened during the early stages of a theory's development.

As Kuhn stresses, no such explicit suspension is requisite within his approach. Since even the application of shared methodological standards is subject to an individually varying process of making them precise and weighting them, it is to be expected that promising, newly conceived theories come out as qualified on at least some such application. So it is not necessary to resort to any specific rule to weaken the full force of methodology in a theory's early stages (Kuhn, 1977, pp. 331–332).

Lakatos' approach likewise entails such an initial restriction of methodological evaluation – automatically, and without any specific rule to this effect. But the mechanism by which this restriction is achieved is quite different from Kuhn's. In Lakatos, the postponement of judgment immediately follows from the notion of supporting facts. First, mere anomalies are not troublesome for a program. Consequently, fresh programs are not liable to the suffocating grip of having to cope with each possibly relevant phenomenon. In this vein, Lakatos derives from his methodology that budding programs are to be treated leniently (Lakatos, 1970, p. 92; see also ibid. p. 70). Second, empirical support arises from confirmed predictions. But in order to assess whether a prediction is successful, a period of time is necessary. As a rule, experimental work is required, in order to examine the correctness of a theory's anticipation, and experiments usually take their time. As a consequence of Lakatos' notion of what kind of evidence backs a theory, the performance of a theory cannot be assessed instantly. A program cannot be judged by drawing only on the evidence available at the time of its inception. This delay in judgment Lakatos calls "the end of instant rationality" (Lakatos, 1970, pp. 68, 87). Its salient aspect is that it entails an initial suspension of judgment, thus sparing one the need to postulating it separately. An isolated methodological rule thus becomes a methodological theorem.

On the other hand, this built-in delay considerably limits the possible scope of Lakatos' methodology. It follows, namely, that the methodology is of virtually no use to the working scientist. Empirical progressiveness can only be identified in retrospect, which means that Lakatos' criteria are not overly helpful in the actual process of theory choice: They typically fail to establish an unambiguous ranking of the rival approaches.[11] Theories can only be judged when the relevant evidence is finally in, and this means that they can only be judged with the benefit of hindsight.

However, this restriction does not vitiate the viability of Lakatos' methodology for two reasons. First, as pointed out by Kuhn, the uncertainty of methodological evaluation is not detrimental to scientific progress but, rather, promotes it. The division of theoretical labor is the most fruitful strategy for coping with the objective risks involved in theory choice decisions, and this division is achieved most easily if distinct criteria of evaluation are brought to bear (see above Sec. 2). But this strategy can only be pursued if none of the rivals comes out looking clearly superior. And precisely this room for consideration is supplied by the delayed applicability of Lakatos' criteria. This means that Lakatos rejects Kuhn's uncertainty thesis in letter but retains it in spirit. Second, one of Lakatos' chief aims is to defend the rationality, or the rule-guided nature, of theory choice decisions. For the success of this endeavor, it is immaterial which is the right perspective for reconstructing such decisions as directed by objective, epistemically relevant criteria. What counts is that such a perspective exits. It need not be identical to the point of view of the working scientist; it may also be the retrospective position of the historian of science (Hacking, 1979, pp. 387–388). On the whole, then, Lakatos' claim is that theory evaluations can be reconstructed as rule-guided and hence rational, without at the same time running into Kuhn's difficulty of premature decisions. Once in a while, it is possible to have one's cake and eat it, too.

5. COMPARING METHODOLOGICAL RESEARCH PROGRAMS: KUHN VERSUS LAKATOS

Given the historical bearing of methodologies, their performance is subject to quasi-empirical scrutiny.[12] Lakatos' conception entails consequences in terms of rational theory choice decisions. Decisions of the scientific community about particular theoretical alterations or achievements (that is, Lakatos' "basic value judgments") constitute the evidential basis of methodological theories (Lakatos, 1971, pp. 123–124, 132–134). This opens up an avenue for reconstructing the approaches of Kuhn and Lakatos as methodological research programs, and applying Lakatos' criteria to their evaluation.

In order for a program to outperform its rival, it is necessary that it reproduce the rival's confirmed explanatory content. Lakatos' conception is able to accommodate the following features of scientific change as brought forward by the paradigm theory: first, the non-accumulative character of scientific change; second, paradigm immunity; third, paradigm substitution; and fourth, the room left for theory choice decisions, i.e. the initial suspension of judgment (see above Sec. 4). The further major stipulation for program replacement is the requirement of producing "novel predictions," or corroborated excess content. This is achieved with respect to two features of scientific change. First, Lakatos' methodology entails that science is characterized by a thorough theoretical pluralism, rather than by the Kuhnian monopolistic rule of paradigms punctuated by occasional cataclysmic eruptions. Judging from the historical record, Lakatos' picture captures scientific change more accurately than Kuhn's. In particular, Lakatos' commitment to theoretical pluralism is apt to solve the historical anomaly of the paradigm theory that in view of Kuhn's epistemological arguments in favor of pluralism, the return to paradigm monopoly is puzzling (see above Sec. 2). Judging by Lakatos' lights, no such return to theoretical monism should occur.

Second, Lakatos' demand that the empirical achievements of an abandoned program be reproduced by its successor (albeit perhaps in a theoretically disparate fashion) also matches the historical record better than Kuhn's view of unbridgeable gaps between successive paradigms. For instance, one of the objections to Kuhn's view is that the very first laws enunciated in modern science, namely, Snell's law of refraction and Galileo's law of free fall, still form part of latter-day textbooks. To all appearances, theoretical accomplishments are retained to a far greater extent than Kuhn would have it. So far, then, Lakatos' methodological research program comes out superior with respect to Kuhn's. On the other hand, there are clear historical violations of Lakatos' demand to reproduce completely the explanatory achievements of replaced programs. These violations become manifest in the existence of Kuhn losses (see above Sec. 2). Program substitutions in spite of Kuhn losses are incompatible with Lakatos' standards. But they do, in fact, happen.

Moreover, in contrast to Lakatos' picture of a linear chain of subsequent program versions, several conflicting versions of the same program do frequently co-occur in historical reality. Accordingly, programs resemble clusters rather than linear chains (Laudan, 1977, pp. 78, 97–100; Gholson and Barker, 1985, 762–763). However, the latter empirical problem likewise turns up in the paradigm theory; Kuhn's normal science is also characterized by theoretical unanimity. Barring crises, no multiplicity of theoretical approaches occurs.

This means that it is only the existence of Kuhn losses which prevents Lakatos' methodology from superseding the paradigm theory as an account of scientific change by its own methodological lights. Actually, the problem of Kuhn losses was faced within Lakatosianism. The attempt was made to adapt Lakatos' standards of theory evaluation so as to allow for a clear methodological comparison even in their presence (Worrall, 1978, 63–64). However, these adaptations also have an influence on the success of Lakatos' methodology in the explanation of features of theory change. These repercussions make it impossible to trace their consequences here.

NOTES

1. The comparability of the empirical consequences of different paradigms is also admitted by Kuhn; see e.g. Kuhn, 1970a, pp. 97, 150; Kuhn, 1977, p. 339.
2. This is analogous to Lakatos' recommendation to reconstruct the history of science; Lakatos, 1971, p. 120.
3. For a brief overview of post-Lakatos elaborations on the crucial notion of "novel facts," see Carrier, 1988. See also Musgrave, 1974.
4. Rendering the preservation condition this way involves what I take to be a charitable reading of Lakatos. Lakatos misleadingly – or inappropriately – demanded that the successor version reproduce the entire "unrefuted content" of its predecessor (Lakatos, 1970, pp. 32, 33) – that is, including the claims that were never tested empirically. But this is a strange demand. Why should a theory be forced to account for claims that may quite well be mistaken? Moreover, the demand to retain a theory's unrefuted content is at odds with Lakatos' own paraphrase of this very condition; namely, to explain its "previous success" (Lakatos, 1970, pp. 32, 68). Untested claims can hardly be grouped among the cases of previous success. The formulation adopted here is intended to capture the intuition of preserving a theory's successful applications. It amounts to replacing "unrefuted content" by "corroborated content." See also Rott, 1994, pp. 33–36.
5. I adopt a temporal reading of "novel fact": A fact is novel if its theoretical derivation precedes its empirical detection. A lot of passages in Lakatos support this reading; in addition to the ones given, see e.g. Lakatos, 1970, p. 36 (which involves contrasting predicted facts that solved anomalies), Lakatos, 1971, pp. 114–116 (which stresses the significance of the temporal relations between theoretical derivation and empirical detection). On the other hand, Lakatos also gives a comparative interpretation of novelty. Facts count as novel when they are "improbable in the light of, or even forbidden by," a rival theory (Lakatos, 1970, p. 32). In that view, a theory's explanation of a rival's known anomaly would qualify as novel fact, although it is "old" in the sense of being familiar to science. To make matters worse, Lakatos also advances an interpretative reading of novelty. As he says, an alternative explanation of a known fact should in the early stages of theory development also be accepted as novel (Lakatos, 1970, pp. 69–70). That is, the mere reinterpretation of a known and explained phenomenon makes a program progressive. The comparative and the interpretative reading involve an attenuation of the temporal one. I adopt the strongest, i.e. temporal, version, since it is the only one that actually provides all the methodological explanations I attribute to Lakatos in Sec. 4.
6. It is true, Lakatos' remarks on program replacement are brief and obscure (Lakatos, 1970, p. 69; Lakatos and Zahar, 1976, p. 179), and he never specifies the details of comparative program evaluation. It is clear, however, that his evaluation procedure involves a transfer of the criteria for progress within a program to the comparative appraisal of entire programs. The reconstruction given here represents what I take to be the most direct or most immediate such transfer. With the exception of the positive heuristic, which is specific to a program and thus cannot directly be applied to program comparison, it brings to bear all the hallmarks of a progressive program, so as to identify a superior program.
7. As indicated above with respect to paradigms (see note 1), the feasibility of this procedure requires that the empirical achievements of distinct research programs can indeed be compared. This is tantamount to rejecting a strong version of incommensurability, according to which the occurrence of meaning variance excludes such a comparison. It is now generally accepted that the latter claim involves an exaggeration, and is unfounded (see for instance Laudan, 1977, pp. 143–144; see also Carrier 2001a,b). Conversely, Lakatos' condition for inter-program comparison can indeed be put into practice.
8. To state the matter somewhat more carefully, all what Lakatos really does is deny theory-specific criteria such as Kuhn's any right to rule on the acceptability of explanations (Lakatos, 1970, p. 68). However, precisely because such criteria are intimately tied to the conceptual essentials of the respective paradigm (or research program, for that matter), they are apt to guide its future development. Kuhn's criteria determine what an exemplary explanation is supposed to accomplish in a theoretical tradition, and this is also part of the

task Lakatos assigns to his positive heuristic. Thus, the latter provides the proper home for these criteria.

9. It is true: this does not follow rigorously. It might happen that the prediction of one program is accommodated without any conceptual adaptation by another program. In cases of this kind, a fact supporting one program is not anomalous for its rival. Although historical examples of this pattern can be given, it is still a somewhat remote eventuality. In the vast majority of cases, one theory's support is another theory's pitfall.
10. This important feature of Lakatos' methodology has completely escaped Kuhn's notice. Kuhn objects to Lakatos in that for him, just as for Popper, "the existence of a potential replacement theory appears pure coincidence" (Kuhn, 1980, p. 191). This criticism is certainly unjustified.
11. In addition, Lakatos does not regard an eventual rank-order as a recommendation to the scientists to pursue the best-rated theory to the exclusion of the rest (Lakatos, 1970, pp. 71–72; Lakatos, 1971, p. 117). I leave this further restriction out of consideration.
12. See Laudan et al. (1986) for a collection of the relevant historical claims.

REFERENCES

M. Carrier (1988) "On Novel Facts. A Discussion of Criteria for Non-ad-hoc-ness in the Methodology of Scientific Research Programmes", *Zeitschrift für allgemeine Wissenschaftstheorie* 19, 205–231.

M. Carrier (2001a) "Changing Laws and Shifting Concepts: On the Nature and Impact of Incommensurability", in: P. Hoyningen-Huene and H. Shankey (eds.), *Incommensurability and Related Matters (Boston Studies in the Philosophy of Science)*, Dordrecht: Kluwer, 65–90.

M. Carrier (2001b) "Shifting Symbolic Structures and Changing Theories: On the Non-Translatability and Empirical Comparability of Incommensurable Theories", in: M. Ferrari and I.O. Stamatescu (eds.), *Symbol and Physical Knowledge. On the Conceptual Structure of Physics*, Berlin: Springer, 125–148.

P. Duhem (1906) *The Aim and Structure of Physical Theory*, New York: Atheneum, 1974.

B. Gholson and P. Barker (1985) "Kuhn, Lakatos, and Laudan: Applications in the History of Physics and Psychology", *American Psychologist* 40, 755–769.

I. Hacking (1979) "Imre Lakatos's Philosophy of Science", *The British Journal for the Philosophy of Science* 30, 381–402.

T.S. Kuhn (1970a) *The Structure of Scientific Revolutions*, Chicago: University of Chicago Press.[2]

T.S. Kuhn (1970b) "Logic of Discovery of Psychology of Research?", in: I. Lakatos and A. Musgrave (eds.), *Criticism and the Growth of Knowledge*, Cambridge: Cambridge University Press, 1–23.

T.S. Kuhn (1977) *The Essential Tension. Selected Studies in Scientific Tradition and Change*, Chicago: University of Chicago Press.

T.S. Kuhn (1980) "The Halt and the Blind: Philosophy and History of Science", *The British Journal for the Philosophy of Science* 31, 181–192.

T.S. Kuhn (1983) "Commensurability, Comparability, Communicability", in: P.D. Asquith and T. Nickles (eds.), *PSA 1982. Proceedings of the Biennial Meeting of the Philosophy of Science Association*, East Lansing Mich.: PSA, 669–688.

I. Lakatos (1970) "Falsification and the Methodology of Scientific Research Programmes", in: Lakatos (1978), 8–101.

I. Lakatos (1971) "History of Science and its Rational Reconstruction", in: Lakatos (1978), 102–138.

I. Lakatos (1978) *The Methodology of Scientific Research Programmes (Philosophical Papers I)*, eds. J. Worrall and G. Currie, Cambridge: Cambridge University Press.

I. Lakatos and E. Zahar (1976) "Why Did Copernicus's Research Programme Supersede Ptolemy's", in: Lakatos (1978), 168–192.

L. Laudan (1977) *Progress and its Problems. Toward a Theory of Scientific Growth*, Berkeley: University of California Press.
L. Laudan et al. (1986) "Scientific Change: Philosophical Models and Historical Research", *Synthese* 69, 141–223.
A. Musgrave (1974) "Logical versus Historical Theories of Confirmation", *The British Journal for the Philosophy of Science* 25, 1–23.
K.R. Popper (1935) *The Logic of Scientific Discovery*, London: Hutchinson, 1959.
H. Rott (1994) "Zur Wissenschaftsphilosophie von Imre Lakatos", *Philosophia naturalis* 31, 25–62.
J. Worrall (1978) "The Ways in which the Methodology of Scientific Research Programmes Improves on Popper's Methodology," in: G. Radnitzky and G. Andersson (eds.), *Progress and Rationality in Science*, Dordrecht: Reidel, 45–70.

Abteilung Philosophie
Universität Bielefeld
D-33501 Bielefeld
Germany

Gábor Forrai

Lakatos, Reason and History[1]

1. The Rationalist Program

Lakatos' philosophy of science is puzzling and can be interpreted in widely varying ways. According to the most widespread interpretation, his aim was to reconcile a basically Popperian outlook with the historical facts discovered by Kuhn and Feyerabend.[2] The purpose of this paper is to use his work to critically assess the project of which, according to this interpretation, he was an outstanding representative. This project conceives the task of philosophy of science *as showing the value of science by developing a normative theory of science, a methodology.* Its chief representatives were the logical positivists and Popper. Currently, Larry Laudan is the leading representative of this approach. The turn of the century conventionalists Poincaré and Duhem do not belong here: they were interested in the epistemological status of scientific theories, rather than in laying down universal norms for scientific practice.[3] It is for the same reason that later-day scientific realists like Richard Boyd or William Newton-Smith do not fit in here either. Their project is very much like that of the conventionalists, even though they reach practically the opposite conclusions. Kuhn and the other historically-minded authors also stand apart, for their approach is emphatically descriptive, not normative: they wish to study what happens in science rather than what should happen.

This project is often called "rationalism" both by its advocates and its critics. Lakatos uses a different term, "*demarcationism.*" He describes it as one of the three possible answers to what he considers the central problem of philosophy of science: that of being the normative appraisal of theories. He summarizes it in this way:

> In the demarcationist tradition, philosophy of science is a scientific standards watchdog. Demarcationists reconstruct *universal* criteria which explain the appraisals which great scientists have made of *particular* theories or research programmes. But medieval 'science,' contemporary elementary particle physics, and environmentalist theories of intelligence might turn out not to meet these criteria. In such cases philosophy of science attempts to overrule the apologetic efforts of degenerating programmes.

G. Kampis, L. Kvasz and M. Stöltzner (eds.),
Appraising Lakatos: Mathematics, Methodology and the Man, 73–83.

> Demarcationists differ over precisely what the universal criteria of scientific progress are, but they share several important characteristics. First, they all believe in the third world of Frege's and Popper's three worlds. The 'first world' is the physical world; the 'second world' is the world of consciousness, of mental states and, in particular, beliefs; the 'third world' is the Platonic world of objective spirit, the world of ideas. Demarcationists appraise the *products* of knowledge: propositions, theories, problems, research programmes, all of which live and grow in the 'third world' (whereas the producers of knowledge live in the first and second worlds). In line with this, demarcationists also share a critical respect for the articulated. They readily agree that articulated knowledge is only the tip of the iceberg: but it is exactly this small tip of the human enterprise in which rationality resides. Finally, demarcationists share a democratic respect for the layman. They lay down *statute law* of rational appraisal which can direct a lay jury in passing judgment. Of course, no statute law is either infallible or unequivocally interpretable. Both a particular ruling and the law itself can be contested. But a statute book – written by the 'demarcationist' philosopher of science – is there to guide the outsider's judgment. (1976, 226–227)

The other two answers are scepticism and elitism. *Scepticism*, as Lakatos understands it, denies that there is a distinction between good and bad theories. Every theory is as good as any other. Given his description (1976, 225; 1978c, 107–108), it would be better call this view relativism, but he likes to project contemporary positions into the past, and the label "scepticism" serves this purpose better. He does not seem to be bothered very much by this approach. He always describes it rather briefly and never takes the trouble to argue against it. This is somewhat surprising, since the only representative of this approach he mentions is Feyerabend, with whom he had been arguing for over ten years. Later on, I shall suggest that one reason for his not attacking it may be that he has no resources to deploy. This, in turn, may explain the label "scepticism": many people think that, when push comes to shove, one just cannot argue against scepticism.

His real enemy then, the enemy he does take on, is *elitism*. This is the view that even though there is a distinction between good and bad, the criteria cannot be articulated in terms of explicit rules. Some people, the scientific elite, just see what is good due to their exceptional talents or their long experience. So quality control is possible, but it is not amenable to public justification. If we want to find out the value of a theory, we cannot apply rules but have to turn to the scientific elite. Of course, we have to know who to turn to. So the elitist lays down sociological and psychological criteria for the identification of the elite.

The thinkers Lakatos identifies as elitists are those who are not relativists but do not traffic in universal rules: Kuhn, Polányi and Toulmin. They would probably object to being classified in this way, and they would have a point: their professed aim is not to solve the problem of normative appraisal but to analyse how science works in descriptive terms. However, Lakatos' regarding of them as elitists is not groundless. He could argue that, by rejecting the enterprise of rule-based normative appraisal, they imply that appraisal is to be left to the scientists, a group of people with certain privileges, i.e. an elite.

So their rejection of the problem amounts to an elitist answer. He could add that many things these thinkers say would very much fit in with the elitist position.

But why does Lakatos think that this approach is wrong? His first objection (1978c, 114) is that normative evaluation of products must precede the sociological or psychological identification of the producers. The whole point of identification is to find out whose opinions can be trusted in the future. Using purely descriptive means, it is possible to identify many groups – high energy physicists, astrologists, stamp collectors, etc. – but this alone will not answer the question of whom to trust. It seems that criteria are necessary in order to decide which opinions are trustworthy, after which one may then employ sociological criteria to identify the group whose members are likely to have trustworthy opinions. However, the criteria for selecting the trustworthy opinions are just the sort of explicit universal rules the demarcationists are looking for. So elitism either does not get us anywhere, or it is incoherent, for it builds on the results of the demarcationist program, which it rejects.[4] Alas, this objection is not particularly convincing. An elitist (and not just an elitist, for that matter!) may reply that one does not need explicit rules to start trusting some people. It is without rules that we trust parents and teachers. And these initial authorities may direct us to other authorities, say, scientists, and then we may start applying sociological techniques to identify that group.

Lakatos' second objection (1978c, 115) is that the elitist position implies that every change, in the elite's view, constitutes progress. If we ask the elite whether their current view is better than the earlier one, they will certainly answer "yes." So every change is improvement, which is absurd. However, this objection works only if the elitist is not fallibilist. The fallibilist elitist may say this: "Look, this is what the elite believes. Your best shot is to believe the same. It may happen later that they realize that they were wrong, and that their change for their current view was a step backwards. So from the vantage point of their future opinion, the change for the current view was not progress." So the elitist may re-evaluate his authority-based judgment when the authority's view changes, but this falls short of acknowledging that every change is progress.

The third objection is that elitism does not help if the elite is divided. That is true. Under these circumstances that elitist will not know what opinion to adopt. But is this a fatal shortcoming? Why should the elitist be able to provide an assessment at any time? Why cannot he suspend his judgment until the elite sorts out the questions?

I did not raise these difficulties to defend elitism, but to illuminate Lakatos' position. The fact that he does not consider these – not particularly ingenious – counterarguments suggest that he believes they are beside the point. And the reason why they are beside the point is that he believes that normative appraisal should be *objective,* in the sense that it should not rely on anything particular, local or personal. It should be made from the standpoint of someone with no specific features or determinations at all (with the exception of the minimal rationality all sane adults possess). If one insists on this sort of objectivity, then the reply to his first objection is irrelevant. It is just a biographical matter what authorities one first comes to trust, and to what further authorities one is directed by them. These personal factors cannot lead to objective value judgments. So the reply simply does not speak to the point. The reply to Lakatos' second objection can be dismissed on similar grounds. Since the elitist's judgments vary together with the views of the elite, such judgments cannot be objective, since objectivity requires independence of accidental matters. If we have two theories and a body of relevant evidence, and these are kept constant, one should always arrive at the same judgment concerning their comparative value, and the judgment should not depend on local conditions such as the elite's opinion. The judgment can be overruled only if the theory or the evidence changes. A change in the elite's opinion is not an objective factor. Roughly the same reasoning can be applied against the reply to Lakatos' third objection. It might indeed happen that we cannot give higher marks to one competitor than to the other, i.e. we have to suspend the judgment. But the suspension of judgment cannot be based on the local condition of the elite's being divided.

This conception of objectivity makes it easy to understand some of the things Lakatos says about rationalism, i.e. demarcationism. The normative rules should be *universal*, since local rules, by definition, are not objective. They must be *explicit* rather than tacit, for the uninitiated cannot apply but explicit rules. Whether someone has been trained to pass judgment in a certain way or not is again a personal matter, so an objective judgment should not depend on it.[5] Finally, the objects of objective judgment should belong to the *third world*, since second world objects, like beliefs, are not available for everyone. Newton's thoughts were available only for a few people at best, but his theory is there for everyone.

2. Finding The Rules

Hopefully, this analysis of Lakatos' demarcationism results in a clearer picture of the aim of the rationalist project: to show the value of science in terms of rules that are objective in the above sense. But how should the rationalist go about finding such rules? The only possibility seems to be to take some theories whose scientific credentials are impeccable, and compare them with those which do not qualify as being scientific. Popper actually tells such a story about how he arrived at his rules (1963, 33–37), and Lakatos repeats it (1971a, 123–124). Around 1919, when Popper first started to think about these issues, there were four widely discussed theories: Einstein's theory of relativity, Marx's theory of history, Freud's psychoanalysis and Adler's individual psychology. He was impressed by the first, but has come to question the scientific status of the other three. He was thinking in this way:

Einstein alone took a serious risk – Eddington's 1919 measurements could have refuted his view. The advocates of the other three theories, however, had claimed to have won without taking risks. Everywhere they looked, they found evidence. From here, he arrived easily at the criterion of falsifiability and the ban on ad hoc adjustments.

Now this way of finding objective rules may raise two sorts of concern. First, suppose Popper had adored Marx, Freud and Adler, but despised Einstein. Would not have he come up with a completely different set of rules then? Surely, the rules must be based on something more solid than personal taste. I will come back to this point, but only later in the paper. Let us now accept that his choice of good and bad examples was legitimate. The second worry is that he had helped himself to some particular examples, which became available only at a particular point in history. Suppose Popper had been a contemporary of Plato. Could he have *then* come up with the same criteria? Probably not. One's conception of what science should be like is certainly heavily dependent on what science is at the moment. Since her place in history is a specific fact about the prospective rationalist, it cannot be relied on in the formulation of the rules for science, since the rules must be objective. Fortunately, this worry can be put to rest by invoking the distinction between *discovery* and *justification*. What warrants the objectivity of the rules is that they can be justified in an objective manner. Their discovery can – and actually must – depend on particular facts, but that does not matter.

Thus, the objectivity of the methodological rules must be grounded in their justification. So how are methodological rules to be justified? The obvious strategy for this is the following. Find something suitable to serve as the ultimate goal of science, something that does not stand in need of further justification, and show that the consistent application of the rules leads towards the realization of this ultimate goal. The best (the only?) candidate for serving as the ultimate goal is truth. So you have to show that the theories your rules of appraisal favour are those which have a better chance to be true. If the logical positivists had been asked to defend their methodology, they would have said that the reason why scientists should prefer highly confirmed theories is that the better a theory is confirmed, the more likely it is to be true. However, justifications of this sort can be subjected to epistemological criticism. An example of this is Popper's attack on logical positivism. He pointed out first that the logical positivists' preference for highly confirmed theories has the disastrous effect of leading to theories poor in content. The richer a theory is, the less likely it is to be highly confirmed. In effect, the high confirmation rule favours truisms. Second, he repeated Hume's criticism of induction in order to sever the link between confirmation and truth. If induction is ruled out, there is no reason to believe that high confirmation is a mark of truth.

Popper's own rules suffered similar criticism at Lakatos' hands. He pointed out that Popper's requirement of high content and high falsifiability does not naturally hook up to anything that might make the game of science worth playing (1968, 181–191; 1974a, 154–167). Popper's rules do help us to free ourselves of mistakes, but the only thing we learn from falsification is that we were wrong. So Popper's injunction to learn from our mistakes will result in "learning without ever knowing" (1974a, 155). To be able to say that an eventually refuted theory which, however, had survived many tests before it failed did, in fact, get us closer to truth, Popper would have to accept some inductive principle. Therefore Lakatos submits a "plea to Popper for a whiff of 'inductivism'" (1974a, 159) which, of course, Popper cannot accept without abandoning what he takes to be his most treasured insight.

3. Coping With History

The search for abstract, epistemological justifications of methodologies could have continued. But, instead, the *Kuhnian revolution* broke out. At the beginning of the sixties Kuhn, Hanson, Feyerabend and Toulmin pointed out that the then available methodologies did not match the historical evidence. They produced many convincing historical examples which show that scientific research is not conducted according to the logical positivists' or Popper's rules. Indeed, it seems now that if scientists had consciously proceeded by these rules, science would not have become the flourishing enterprise it is. Of course, one might claim that the rules are fine and it is science which has gone wrong, but this would be difficult to take seriously. Lakatos puts this rather succinctly.

> Is it not then *hubris* to try to impose some *a priori* philosophy of science on the most advanced sciences? Is it not *hubris* to demand that if, say, Newtonian or Einsteinian science turns out to have violated Bacon's, Carnap's or Popper's *a priori* rules of the game, the business of science should be started anew? (1971a, 137)

It is in light of this development that Lakatos' distinction between scepticism, elitism and demarcationism can be understood. Given the historical criticism of the available rules, there are three options. To admit that there is no way to distinguish between science and non-science or between good science and bad science, is the sceptical option. To maintain that there are such distinctions, but that they cannot be expressed in terms of explicit rules, and to say that the decision should be left to the scientists, would be elitism, for in Lakatos' view, this is an "abject surrender to authority" (1974b, 324). The last option is to press on with the demarcationist or rationalist program. But once it has been admitted that history is relevant for the justification of methodological rules, this will have two important implications for the rationalist project. It is one of Lakatos' great achievements, that he recognized these implications.

First, the project has to be made more modest *by abandoning the ahistorical notion of objectivity* in favour of a historical one. The classical Enlightenment notion of objectivity, which is manifested both in the logical positivists' and Popper's works and which still haunts in Lakatos' arguments against elitism, was ahistorical. It held that the appraisal of a position on the basis of a given set of evidence is objective in the sense that it does not depend on historical contingencies. If we have the theory and the evidence, the judgment should be the same even if history had been completely different. If Einstein had put forward his theory in Plato's days, its merits should have been judged in the same way as they are judged today. Once history is considered relevant, this ahistorical conception of objectivity must go. The argument is as follows. History figures into the justification of methodological rules. If history, the path science took, had been different, that would have affected the extent to which various methodologies would have been justified. In other words, a different history could have justified different methodological rules. Therefore, the validity we attribute to certain methodological rules is relative to history. Now, particular appraisals of cognitive achievements can only be as objective as the justifications we have for the methodological rules we apply in making these appraisals. So if the justification of methodology is relative to history, so are the evaluative judgments the methodology issues.

This point can be understood in a pessimistic and in an optimistic way. Understood *pessimistically*, this is just historical relativism. Consider the Galileo contra Bellarmin case. Bellarmin challenged Galileo to deduce his physical principles from metaphysical principles. In light of the prevailing Aristotelian methodology of those days, this challenge was perfectly legitimate. Galileo tried to meet it and failed. Since then, we have abandoned Aristotelian methodology, so from our perspective Galileo's mistake was to take up the challenge, but his failure to meet the challenge does not speak against his theory. Bellarmin was right from his perspective, Galileo was right from our perspective, and the question of who was right simpliciter makes no sense. Bad enough. But does this indeed follow from allowing history to be relevant for the validity of methodological rules? I do not think so.[6] Why should we accept that Aristotelian methodology is just as good as the modern methodology Galileo was working towards? We do have more history at our disposal, and what happened afterwards suggests that Aristotelian methodology was wrong, so our methodology is superior, and so are the judgments it issues. The situation is not symmetrical: history lends evidence to the superiority of our standards. The relativist may perhaps come back and say: "The methodological change Galileo has initiated might be later reversed. Perhaps, two hundred years from now, we will be back with Aristotle and regard modern science as an unfortunate interlude." But this argument is weak. Future possibilities cannot overrule the evidence which is actually available.

This points towards a more *optimistic* reading. The historical relativity of the justification of methodology suggests that we learn from history. Methodology develops in time, just as science does. As we go along, methodology improves. This is exactly what Lakatos believes. He adds that methodological development may lag behind the development of science in the sense that the available methodologies are inappropriate to what goes on in science (1971a, 137). He sees his task as that of making methodology catch up with the development of science.

The relevance of history for the justification of methodology carries a second implication as well: it has to be worked out *how to use history to evaluate methodologies*. It is not obvious that history can be put to such use, for two reasons. First, history is factual, but methodology is normative. How can we use facts to evaluate norms? Genocide does happen, but this does not invalidate the moral judgment that it is evil. Second, we have seen that a radical mismatch between scientific practice and methodology is an argument against the latter. It seems, however, that a perfect match is equally problematic. Suppose you create an elaborate methodology which prescribes scientists to do exactly those things they actually do. In this case, we would suspect that it is rather a description of the scientists' behaviour, which has nothing normative about it except for the linguistic form of the sentences in which it is articulated. Alternatively, the perfectly matching methodology would issue the judgment that whatever scientists do is rational.

Lakatos' way out is elegant. He suggests that the historical test of methodology should consist in comparing its judgments about particular cases with the leading scientists' judgments. Since the scientists' judgments are normative, we compare the normative with the normative. It is not facts we use to evaluate norms. This is practically sufficient to solve the second problem as well. Methodology is not in the business of capturing what goes on in science. It is in the business of capturing the leading scientists' judgments. Since it is highly unlikely that the leading scientists judge every development rationally, there is little danger of methodology's degenerating into mere description or pseudo-normative rationalization of absolutely everything.

Elegant as this solution is, it invites certain objections. First, are the leading scientists *competent* judges of scientific developments? When we talk about relatively recent developments in their own specialty, this would be difficult to doubt. But the farther we go back in history and the farther we move away from their specialty, the less reliable their judgments become. For example, most scientists believe that Copernicus's system, as he himself developed it, is superior to Ptolemy's. But historical investigations have revealed that this superiority is at least questionable.[7] Lakatos would probably accept this criticism. He thinks that it is very important to teach the history of science to scientists as well, because otherwise, scientific "education" would degenerate into "training" (1963, 254). Historical erudition is indeed important for competent value judgments. But then he would probably add that the objection is not fatal. We cannot but use the best evidence we have, and we do not really have better evidence than the value judgments of the leading scientists. It is still better to measure with a not completely reliable device, than not to measure at all.

Secondly, there is the objection that Feyerabend keeps pressing: why should we turn to *scientists* rather than to, say, astrologists? Since scientists prefer science to astrology, their value judgments cannot be used to argue against astrology or other non-scientific ventures on a neutral basis. Of course, this is a relativistic, or, as Lakatos would put it, sceptical objection, and Lakatos never considers it. I do not see how he could answer. He only says that he views the position inherent in the objection with horror (1974b, 324). He can, at best, say that his methodology is meant for those who have already accepted the superiority of science. It serves to justify certain developments *within* science and is not intended to be the justification *of* science.

Thirdly, how do we *select* the scientists against whose value judgments we are to measure the adequacy of methodological rules? How do we choose the top scientists from the multitude of scientists? We cannot select them on the basis of their achievements, for those cannot be evaluated without resorting to methodological rules, which would result in a vicious circle.[8] The only option is to use sociological criteria, like the number of citations, evaluation by the in-group and the out-groups, etc. But this looks as if might were right: the use of sociological criteria would commit us to some sort of elitism. Lakatos does not consider this objection, and I believe that he could not rebut it. He could, however, point out that the elitism he is committed to is rather modest. First, because he does not demand that methodology capture all value judgments (1971a, 131–133), which would indeed be an unqualified "surrender to authority." Second, he goes even further than that. His view is not that methodology is subordinated to scientific authority, but may occasionally diverge from it. Rather, he believes that methodology possesses authority in its own right. The scientific elite and methodology are independent authorities, but each lends legitimacy to the other (1971a, 137). The scientific elite plays a vital role in the justification of methodology, and methodology provides a rationale for the intuitive value judgments of the scientific elite. So his elitism is rather modest, because of this exchange.

There is a final question to be asked. What is the relationship between the abstract *epistemological* justification and the *historical* justification of methodology? More specifically, can the values and standards justified in a historical way also be justified in abstract, epistemological terms? What a Lakatosian historically-based methodology achieves is the identification of certain values. In Lakatos' methodology, the all-important value is empirical growth. The scientific enterprise aims at increasing predictive and explanatory power. But can empirical growth be seen as progress towards truth? Lakatos is somewhat ambiguous about this. He clearly says that all current theories are, most likely, false. He does not believe with Popper that if we are lucky, we might just hit upon the truth (1961, 125–126; 1971b, 175). So present empirical success is not a mark of truth. But he also believes that growing empirical success is a sign of "ever-increasing verisimilitude" (1971b, 175). He approves Engels and Lenin saying that the "endless succession of human generations" may "draw closer and closer to objective truth (without ever reaching it)" (1961, 125). But as he showed in his criticism of Popper, one cannot link empirical success and verisimilitude without positing some inductive principle. So he admits that he needs to posit a principle of this kind (1971a, 113). In practice, however, he never tried to defend or describe an inductive principle. Perhaps he regarded this task as formidable – or he simply died before he could get down to it. But it is also possible that he just did not care about truth (Hacking 1983, 112–113, 119–121; Radványi 1998).

What does this all imply? What conclusions does the above analysis of Lakatos support? I would suggest the following:

1. The ahistorical conception of reason should be abandoned. Our conception of rationality is the product of history, and cannot be defended without relying on historical evidence.
2. Rationalism cannot provide arguments against relativism. It cannot triumph over the opponents of science. Rather, it helps those who already value science to state explicitly why they value it.
3. The theory of scientific rationality cannot be defended without relying on scientific authorities. This, however, is far short of blind submission.
4. The historical justification of a theory of scientific rationality does not in itself decide whether the procedures that are rational according to the theory lead us closer to truth. Here, there is room for epistemological arguments.

NOTES

1. The research for this paper was supported by an OTKA grant, no. F21229.
2. Different interpretations are suggested in Hacking (1979) and (1983, Chapter 8), Radványi (1989) and (1998), Forrai (1993). In Hacking's view Lakatos' aim was to give an account of the objectivity of science that does not rely on the correspondence theory of truth. The most probable explanation of Lakatos' rejection of the correspondence theory of truth is that he comes from the Hegelian – Marxian tradition where this notion is repudiated. Radványi's central idea is exactly that Lakatos pursues the Hegelian project of showing how scientific methodology is situated in history. However, his strongest arguments are based on Lakatos' philosophy of mathematics rather than his philosophy of science. I myself have argued that his philosophy of science was originally conceived as a continuation of the somewhat Hegelian project of his philosophy of mathematics, but he accepted certain doctrines which prevented him from doing the same to science as he did to mathematics. This is why his philosophy of science looks more Popperian than Hegelian. I still do not think the view I propounded is wrong. There are, however, still many things to be said for the sort of Popperian–rationalist interpretation the present paper assumes.
3. Of course, what they said can be reformulated as a normative theory, as Lakatos does (1971a, 105–107). This, however, is unjust: it draws them into a battle in which they never intended to participate and of which they might have been completely unaware.
4. The idea that the normative must precede the descriptive is a recurrent scheme in Lakatos' work. See his recommendations for historiography (1971a, especially 118–121).
5. Lakatos, of course, does not mean to license the ignorant to pass judgment on scientific matters. If you do not have enough math to calculate the consequences of a theory, you

cannot serve as a member of the jury. His point is rather that the knowledge required to pass judgment should be such that, in principle, it can be obtained by everybody.

6. A minor point which is relevant only for the Galileo contra Bellarmin case but nor for the general point is this. It is not at all certain that Aristotelian methodology was justified in those days. Deciding this, however, requires a good deal of historical scholarship which I do not possess.
7. Lakatos himself modifies his methodology to agree with the scientists' judgments about Copernicus. He has to regard as novel prediction the prediction of well-known facts which were not explained by earlier versions of the research programme (Lakatos and Zahar 1976, 184–185).
8. It seems reasonable to demand though that the methodology should present the scientists in terms whose value judgments it itself is justified as top achievers. If a methodology gives bad grades to these scientists, we will have little reason to trust their value judgments, and that would weaken the justification.

REFERENCES

Feyerabend, Paul K. (1975) *Against Method*, London: New Left Books.

Forrai, Gábor (1993) "From the Method of Proofs and Refutations to the Methodology of Scientific Research Programmes", *International Studies in the Philosophy of Science*, 7, 161–176.

Hacking, Ian (1979) "Imre Lakatos' Philosophy of Science", *British Journal for the Philosophy of Science*, 30, 381–410.

— (1983) *Representing and Intervening*, Cambridge: Cambridge University Press.

Lakatos, Imre (1961) "Necessity, Kneale and Popper", in (1978b) 121–127.

— (1963) "The Teaching of the History of Science", in (1978b) 254–255.

— (1968) "Changes in the Problem of Inductive Logic", in (1978b) 128–200.

— (1971a) "History of Science and Its Rational Reconstructions", in (1978a) 102–138.

— (1971b) "Replies to Critics", in: P.S.A. 1970, *Boston Studies in the Philosophy of Science*, 8, edited by R.C. Buck and R.S. Cohen, Dordrecht: Reidel, 174–182.

— (1974a) "Popper on Demarcation and Induction", in (1978a) 139–167.

— (1974b) "The Role of Crucial Experiments in Science", *Studies in the History and Philosophy of Science*, 4, 309–325.

— (1976) "Understanding Toulmin", in (1978b) 224–243.

— (1978a) *The Methodology of Scientific Research Programs, Philosophical Papers*, Vol. 1, edited by J. Worrall and G. Currie, Cambridge: Cambridge University Press.

— (1978b) *Mathematics, Science and Epistemology, Philosophical Papers* Vol. 2, edited by J. Worrall and G. Currie, Cambridge: Cambridge University Press.

— (1978c) "The Problem of Appraising Scientific Theories: Three Approaches", in (1978b) 107–120.

— and Zahar, Elie G. (1976) "Why Did Copernicus's Research Programme Supersede Ptolemy's?", in (1978a) 168–192.

Popper, Karl R. (1963) *Conjectures and Refutations*, London: Routledge and Kegan Paul

Radványi, János (1989) "A Mathematical Bildungsroman", *History and Theory*, 28, 26–42 (published under the pseudonym John Kadvany).

— (1998) "The Extraterritoriality of Imre Lakatos", in: *Late Editions*, edited by G. Marcus, Chicago: University of Chicago Press (forthcoming).

Department of History of Philosophy, University of Miskolc
H-3515 Miskolc-Egyetemváros
Hungary

JOHN WORRALL

"HEURISTIC POWER" AND THE "LOGIC OF SCIENTIFIC DISCOVERY": WHY THE METHODOLOGY OF SCIENTIFIC RESEARCH PROGRAMMES IS LESS THAN HALF THE STORY

1. AUTOBIOGRAPHICAL PREFACE

Since this paper is based on a presentation at a conference organised in honour of my late teacher and friend Imre Lakatos on the occasion of what would have been his 75th birthday, I hope I may be forgiven for beginning with a few autobiographical remarks, despite their irrelevance to the paper's intellectual content.

I went to the LSE as an undergraduate in 1965 intending to become a statistician. (I had been told by a careers advisor who briefly visited my school in the cotton and coal town in the North of England where I grew up that the professionals with the highest average pay in Britain were actuaries. And, although I had no idea what actuaries did but since I liked the idea of following a profession, and also, I am afraid to say, at that time liked the idea of becoming rich, I asked what I should study at University to become an actuary and was told "Statistics at the London School of Economics.") Fortunately, at the time, the LSE allowed first-year students to take options that were not closely connected with their intended specialisation and I chose, for no reason that I can now reconstruct, Alan Musgrave's course in deductive logic. Again fortunately, Musgrave included on the timetable alongside his own very fine basic lectures on propositional and predicate logic, Karl Popper's very non-basic lecture course on "Problems of Philosophy." Many logic students asked what Popper's lectures had to do with the course – by which they meant what had they to do with the examination; and Musgrave told them "nothing, just go if you enjoy them." Well I *did* enjoy them: it was heady stuff for an eighteen-year old working class lad, listening to the fascinating and wideranging thoughts of someone with real intellectual charisma (charisma largely based, I now think, on the fact that, whatever history may eventually decide, he clearly firmly believed himself to be a great philosopher – though in a way that somehow avoided overt arrogance).

I had been a happy apprentice statistician, but chi-squared tests, the analysis of variance and the rest paled by comparison with Popper's talk about everything from Einstein to the amoeba to Michaelangelo and back again (they were all operating by "conjectures and refutations"!), and I

G. Kampis, L. Kvasz and M. Stöltzner (eds.),
Appraising Lakatos: Mathematics, Methodology and the Man, 85–99.

obtained permission to switch to Philosophy, Logic and Scientific Method as my "special subject." So it was that Popper with assistance from Musgrave ensured that I would never be rich.

Until then I had only ever sighted Imre Lakatos at a distance – in the front row of the audience for Popper's lectures. But I was again fortunate. There was at that time only one option within the official Philosophy course at LSE: you chose either Moral and Political Philosophy or Mathematical Logic. Being already deeply committed to *im*morality and having been fascinated by Musgrave's dark hints of the esoteric splendours of Gödel's theorem, I chose Mathematical Logic. It turned out that this choice also determined the tutor to which you were assigned – those choosing Mathematical Logic had Imre Lakatos as tutor. Since this was a really tough course, it may have been Imre's way of ensuring that he was bothered by rather few undergraduates, and certainly I was his only one of my year. I loved Mathematical Logic and being assigned to Imre as tutor was the intellectual event of my life.

He frightened me out of my wits. He told me that, since I had a strong mathematics background, I must continue with the mathematics courses that had been part of the Statistics option and, without waiting to see if I was happy with that, telephoned someone in the Central Administration to get me special permission. (People in the Administration had long ago learned that it was always easier to give way to Imre.) He told me that he did not want to see me again until I had read and mastered a list of books – largely in mathematics, logic and physics (though I also remember some very welcome, light relief in the shape of Koestler's *The Sleepwalkers*). Perhaps he thought that I would never manage it – and certainly I remember often being close to tears over the difficulties of mastering transfinite set theory from books on my own. But I had believed him when he said that I was not to return until I had mastered all the material he had given me, and eventually I thought I had developed a reasonable grasp. When I went to see him again some months later he seemed to agree and my reward was to be called a "hopeful monster" (which, someone later explained to me, is a technical term in evolutionary theory and that his remark was probably intended to be complimentary). I was also rewarded with a copy of all the *BJPS* articles making up his "Proofs and Refutations" tied together with a green cotton tag.

This was the start of an intense eight-year relationship with him: as his undergraduate tutee, as his doctoral student, as his research assistant and then, for a sadly brief time, as his colleague. He could be infuriating; he could be a bully; he wanted to control every aspect of the life of his "apprentices" – vetting potential girlfriends for social acceptability, deliberately telephoning and keeping me talking for hours at times when he knew I wanted to be out partying. Some of the political opinions that he then held I now regard with horror (qualified, I like to think, with some understanding of how his espousal of them was connected with his earlier sufferings under Stalinism). But he had none of the "German professor" about him – he was genuinely delighted when I or Elie Zahar or Colin Howson found good criticisms of something he had written. He was always strongly encouraging about our own work and fiercely supportive over our careers. Imre could also be loveable, charming and of course quite astoundingly quick-witted. It is an honour for me to be honouring his memory on his 75th Anniversary.

2. Lakatos on the Logic of Mathematical and Scientific Discovery

In starting my Lakatos studies with "Proofs and Refutations," I began at the top. I have no doubt that this remains his chief claim to lasting intellectual fame. A leading idea of that work, one which of course owed much to the influence of another eminent Hungarian, George Polya, was that of a rationally reconstructible, genuine *logic of discovery.* Indeed the chief interest of "Proofs and Refutations" is not in how already articulated proofs of mathematical results are to be accredited as genuine proofs, but in the process of *developing* cogent proofs. Unlike Popper's notoriously misleading title *The Logic of Scientific Discovery,* Imre's Cambridge PhD from which "Proofs and Refutations" emerged had a title perfectly fitted to its subject matter – *Essays in the Logic of Mathematical Discovery.*

Lakatos himself saw – with, perhaps, varying degrees of clarity at different stages of his career – some of the leading ideas of his main contribution to the philosophy of science, his methodology of scientific research programmes (hereafter MSRP), as based on his work in philosophy of mathematics. This is in particular true of his notion of a programme's "positive heuristic."

It goes without saying of course that Popper was a major intellectual influence on Lakatos. Popper's official view on the logic of scientific discovery was straightforward: despite the English title of his most famous work, there is no such thing. I remember that Popper used to like to joke in his lectures about his somewhat anomalous situtation: his title was "Professor of Logic and Scientific Method," yet he had many times argued (and believed himself to have established) that there is no such thing as Scientific Method – at least not in its original meaning of a systematic way, specifiable in advance, of arriving at scientifically accredited results. Like Reichenbach, Popper held that logical considerations only come into play in the "context of justification" – once, that is, a theory had been articulated. But as for the way the theory is produced, Popper famously held that this process "neither call[s] for logical analysis nor [is] susceptible of it." More fully, Popper's view was as follows:

> The initial stage, the act of conceiving or inventing a theory, seems to me neither to call for logical analysis nor to be susceptible of it. The question how it happens that a new idea occurs to a man – whether it is a musical theme, a dramatic conflict, or a scientific theory – may be of great interest to empirical psychology; but it is irrelevant to the logical analysis of scientific knowledge. This latter is concerned not with *questions of fact* (Kant's *quid facti?*) but only with questions of *justification or validity*

(Kant's *quid juris*?). Its questions are of the following kind. Can a statement be justified? And if so, how? Is it testable? Is it logically dependent on certain other statements? Or does it perhaps contradict them? In order that a statement may be logically examined in this way, it must already have been presented to us. Someone must have formulated it, and submitted it to logical examination (Popper, 1958, pp. 31–2).

Lakatos is generally known for pushing hard the contrary line and for doing so in a rather more detailed and developed way than others – amongst them N.R. Hanson and Mary Hesse – who advertised somewhat similar views at around the same time. And certainly I remember long, frequent and richly-detailed three-cornered discussions on the logic of scientific discovery betweeen Imre, Elie Zahar and myself. But, on reviewing Imre's published work, it is difficult to see how this reputation could have been based soundly on anything that found its way into print. Even the – central – notion of 'positive heuristic' is very sketchily presented in his famous papers on MSRP and even this sketchy treatment is, I would argue, considerably over-influenced by what is an interesting but in several ways unrepresentative case (that of Newton's development of his theories of mechanics and gravitation).

Here is the sum total of Imre's general remarks about positive heuristic in his main MSRP paper:

Few theoretical scientists engaged in a research programme pay undue attention to "refutations." They have a long-term research policy which anticipates these refutations. This research policy, or order of research, is set out – in more or less detail – in the *positive heuristic* of the research programme. The negative heuristic specifies the 'hard core' of the programme which is 'irrefutable' by the methodological decision of its proponents; the positive heuristic consists of a partially articulated set of suggestions or hints on how to change, develop the 'refutable variants' of the research programme, how to modify, sophisticate [make more sophisticated], the 'refutable' protective belt.

... The positive heuristic sets out a programme which lists a chain of ever more complicated *models* simulating reality: the scientist's attention is riveted on building his models following instructions which are laid down in the positive part of his programme. He ignores the *actual* counterexamples, the available '*data*'. (1978, vol. 1, p. 50)

This is clearly a *very* sketchy (and, I would add, very one-sided) account. Notable strides towards a more detailed and better-rounded view of how scientists can be seen as *arguing to*, rather than simply "conjecturing," theories have been taken by Elie Zahar.[1] In this paper I shall try to take further strides by supplying analyses of three particular historical episodes, each of which, I claim, clearly illustrates an important facet of the logic of scientific discovery. All these episodes involve theoretical breakthroughs made in optics in the early 19th Century by Augustin Jean Fresnel. I shall show that each episode illustrates a general kind of case where the theoretical discovery at issue can be straightforwardly reconstructed as the result of a systematic *argument* from essentially uncontested – or, at any rate, widely accepted – premises.

3. THREE HISTORICAL CASES

3a. Fresnel's development of his theory of double refraction

This theoretical breakthrough made by Fresnel is, I shall show, an instance of the following sort of scheme:

(i) A *general* theory has already been accepted (because of its predictive success) and is taken as a premise.

(ii) Given that general theory as a premise, scientists systematically develop further specific theories from it *plus* data *plus* "background knowledge."

Here are the details. Fresnel's (1819) treatment of diffraction had proved to be very successful empirically.[2] (Diffraction effects are produced when an opaque body – a "straightedge," circular disk, narrow wire, etc. – is interposed in a beam of light.)

Fresnel's theory of diffraction was a specific theory based on the general ether or wave theory of light. Given the general idea that light consists of periodic disturbances transmitted through an elastic ether, Fresnel's account of the various diffraction patterns depended on working out the form of the wave surface in air (which closely approximated that in the "free ether") when various opaque (i.e. light absorbing) objects are interposed. This account of the wave surface was itself based on what was already known (that is, background knowledge) about the mechanics of continuous elastic media.

When Fresnel came to develop his account of the transmission of light, not in air, but within transparent crystals, he of course took it (relying on the general theory as a premise) that he was looking for an account of the form of the wave surface within such crystals – the form of that surface depending on the mechanical properties of the ether within the crystal. His account of "free" light waves involved the idea that any particle of free ether is subject to a restoring force when disturbed from its equilibrium position, a force that depends on the magnitude of the disturbance but not its direction. The idea that the value of the force is not, in this case, dependent on direction is itself not properly characterisable as an *assumption*, but was instead dictated by background knowledge plus observational results about the transmission of light in air/free ether. Although observational results dictated no dependence of the elastic restoring force on direction of disturbance for the particular case of light transmitted in air/free ether, "known" results (background knowledge again) about the mechanics of elastic media *in general* of course implied that the restoring force may well in fact depend on the direction of disturbance. Indeed the general form of the elastic restoring force was known to be specifiable in terms of three numbers: the "coefficients of elasticity" along three arbitrarily chosen, mutually orthogonal directions through the medium. Only in the special case where all three coefficients are equal does the absolute value of the restoring force fail to depend on the direction of disturbance.

Now, observational results known to Fresnel had already established that there are three types of transparent crystals: (i) unirefringent (ii) uniaxial birefringent and (iii) biaxial birefringent. (A unirefringent medium – ordinary glass is an example – is one in which an incident ray of light is simply refracted along a characteristic direction within the medium. A birefringent medium – such as calcite – is one in which such a ray is, in general, split into *two* refracted rays. In a uniaxial birefringent crystal, one of these two rays, the "ordinary ray," satisfies the Snell-Descartes "ordinary" law of refraction, while the other, "extraordinary" ray does not; in biaxial birefringent media neither refracted ray satisfies the "ordinary" law.)

Given this theoretical and evidential background, it was, to say the least, natural for Fresnel to infer that the three types of crystal are characterisable theoretically as crystals within which the light carrying ether is so constrained as to have

(i) all three coefficients of elasticity the same (unirefringent)

(ii) two coefficients the same, but third different (birefringent, uniaxial)

(iii) all three coefficients different (birefringent, biaxial).

Indeed this natural inference could readily be reconstructed as a fully-fledged deduction (using of course fairly substantial principles of background knowledge). This is not to say that Fresnel did not exercise startling ingenuity in seeing things so clearly and intuiting this argument. Mathematicians after all exercise great ingenuity in discovering what are in the end *purely deductive* proofs. (This in turn is not to say, of course, that the actual psychological process of mathematical discovery consists of the great mathematician laboriously working through the proof – certainly not consciously; rather the mathematician "intuits" or "sees" that a proof *can be* given, skipping over the details which can be filled in later. Indeed, at a much more mundane level, we all surely do this when we solve a mathematical or logical problem for the first time.)

This account of the mechanical properties of the ether within the three types of medium (plus of course a good deal of purely mathematical ingenuity) constitutes Fresnel's theory of double refraction – a theory which, like his theory of diffraction, turned out to be dramatically predictively successful. (Hamilton showed, in 1833, that the form of the wave surface within biaxial crystals dictated by the theory entails the existence within them of two hitherto entirely unknown phenomena – those of internal and external conical refraction. These phenomena were duly experimentally observed in the same year by Hamilton's friend Humphrey Lloyd.)

3b. Fresnel's development of his theory that light waves are transverse

My second example of a theoretical breakthrough was again made by Fresnel. This breakthrough is – I shall show – an illustration of the following sort of general case:

(i) The latest specific theory developed out of an already successful general idea has been empirically refuted; but

(ii) scientists develop (deduce!) a new specific theory taking as premises the same general idea plus background knowledge *plus the refuting data.*

Here are the details. Fresnel, as we already saw, had espoused the general theory of light as *some sort* of wave in *some sort* of elastic medium. Background knowledge (in the form of accepted theories of the mechanics of elastic media) allowed two fundamental types of waves in such media: longitudinal and transverse (plus combinations of the two). Longitudinal waves (sound waves in air are an example) are those in which the oscillating particles involved in the wave motion (in the case of sound, particles of air) oscillate *in the same direction* as the overall transmission of the wave. Transverse waves (the usual example is a rope attached to a wall at one end and being wiggled up and down at other) are those in which the oscillating particles (in this case, the particles of the rope) oscillate *at right angles* to the overall transmission of the wave motion (the wave-forms move *along* the rope toward the wall, but the individual parts of the rope move up and down).

Again background knowledge in the form of the mechanics of elastic media had much to say about the situation. In particular, it entails that all media, including fluids, can transmit longitudinal (or "pressure" waves) but only *solids* have the necessary resistance to shear to transmit transverse waves.

Fresnel, like everyone else, had thought of the luminiferous ether as a highly attentuated fluid: after all, the planets had to move with extreme freedom through the ether (gravitational theory explaining their motion entirely – or, at any rate, pretty well entirely –, so any frictional effect of the ether had to be negligible). It seemed much easier to conceive how the planets could move with, to all intents and purposes, complete freedom through a highly attenuated fluid than through even the most highly attenuated solid.

However the longitudinal wave theory had been *directly* refuted by Fresnel's and Arago's results on the interference of polarised light. Fresnel and Arago found that if the standard double-slit experiment is performed (yielding, initially, the standard interference pattern on the observation screen) but then polarising plates of crystal are interposed behind the two slits so that the light emanating from those two slits is polarised in mutually orthogonal directions, then the interference bands disappear completely. The general wave theory is committed to the idea, of course, that light consists of periodic disturbances. Near the centre of the two slit pattern, the disturbances emanating from each slit are travelling in very nearly parallel directions; hence if the oscillations making up the light operate in the same direction as the overall propagation, the oscillations produced by each slit separately must be nearly parallel and hence *must* "interfere" – in particular, at the two points on either side of the centre of the pattern where the distances from the two slits differ by exactly half a wavelength, the two series of disturbances must be consistently out of phase (what would have been a trough if only one slit were open always meeting what would have been a crest if only the other slit were open, and so on) – this should produce "destructive interference," that is, more or less complete darkness at those two points.

All this must remain true on the assumption that the waves are longitudinal no matter what (of course transparent) medium the beams from the two slits cross. But Arago and Fresnel found precisely that the interference pattern disappears entirely when the light from the two slits is oppositely polarised. Given the theoretical background that Fresnel accepted, this result established *directly* that at least the disturbances making up polarised light could not be longitudinal, nor could they have any longitudinal component that was relevant to the effects that are produced when two such (oppositely) polarised beams meet.

The only "pure" alternative that background knowledge allowed was that the waves are transverse. Fresnel could, of course, have assumed that the polarised light beams have both transverse *and* longitudinal components, but the experiments he had performed with Arago had shown that such longitudinal components (and their interference) had no effects. Hence it was clearly simpler to assume that the waves are purely transverse. (Simplicity surely plays a role in "deductions [of theories] from the phenomena"; but not in some vague general sense but rather in very precise and particular senses like this one.)

3c. Fresnel's (and Huygens') argument for the general wave theory of light

A predictable (and entirely reasonable) reaction to the above two examples is the following.

Maybe it can be shown in the sorts of cases you have analysed so far that the path to the theoretical discovery concerned is reconstructible as a logical argument, but that is *precisely because in those cases all the real theoretical invention has already occurred.* Let's concede that scientists *within* research programmes, or within paradigms, or whatever, can deduce specific theories "from the phenomena": very often, for example, a general theory (the "core" of a programme, or part of the general theoretical framework underpinning of a paradigm) may contain a free parameter whose value is then "read off" some experimental results – thus producing, by deduction from those results, a more specific theory. But to believe that this shows that all theoretical discovery is systematically reconstructible is to miss the site of the real scientific creativity, the real unanalysable-genius-stuff, which lies precisely in the invention of the core or paradigm-founding theories in thc first place. Logic may rule in "normal science," but the standard scepticism about logic of discovery, stemming from Reichenbach and Popper, is entirely justified in the case of "extraordinary," revolutionary science – exactly the site of the real scientific breakthroughs.

Lakatos seems clearly to have been inclined to go along with such an objector. For Lakatos, a "positive heuristic" gives guidance for the articulation of specific theories only within the context of a given scientific research programme. (The specific positive heuristic involved is in fact part of the characterisation of the corresponding research programme.) Similarly, the sort of relatively vague suggestions in Kuhn (about "puzzle solving traditions" and the guiding role of "exemplars") that point to much the same sort of idea are very clearly *intra*-paradigm notions. There is no suggestion in Kuhn that the path to a new paradigm itself can be reconstructed as a systematic argument. And there is no suggestion in Lakatos that the invention of "core" theories is anything other than a matter of logically unanalysable Popper-style conjecture.

But this objection is wrong. In conceding to it, both Lakatos and Kuhn made a consequential error that led them both vastly to overstate the extent of the discontinuities involved in "scientific revolutions." Or so I shall argue. I shall show that, not only did Fresnel infer specific theories from the data, plus the general wave theory, he *inferred the general wave theory* (i.e. the *hard core* of his programme itself) "*from the phenomena*."

Indeed a very similar argument had already been developed, with great clarity, at the end of the eighteenth century by Christiaan Huygens in his *Treatise of Light* (Huygens, 1690). Let me begin then by quoting Huygens at some length.

It is inconceivable to doubt that light consists in the motion of some sort of matter. For whether one considers its production, one sees that here upon the Earth it is chiefly engendered by fire and flame which contain without doubt bodies that are in rapid motion, since they dissolve and melt many other bodies, even the most solid; or whether one considers its effects, one sees that when light is collected, as by concave mirrors, it has the property of burning as a fire does, that is to say it disunites the particles of bodies. This is assuredly the mark of motion, at least in the true Philosophy, in which one conceives the causes of all natural effects in terms of mechanical motions. This, in my opinion, we must necessarily do, or else renounce all hopes of ever comprehending anything in Physics.

... Further when one considers the extreme speed with which light spreads on every side, and how, when it comes from different regions, even from those directly opposite, the rays traverse one another without hindrance, one may well understand that when we see a luminous object, it cannot be by any transport of matter coming to us from the object, in the way in which a shot or an arrow traverses the air; for assuredly that would too greatly impugn these two properties of light, especially the second of them [that is, the fact that two non-parallel beams can "traverse one another without hindrance"]. It is then in some other way that light spreads; and that which can lead us to comprehend it is the knowledge which we have of the spreading of Sound in air.

We know that by means of the air, which is an invisible and impalpable body, Sound spreads around the spot where it has been produced, by a movement which is passed on successively from one part of the air to another; and that the spreading of this movement, taking place equally rapidly on all sides, ought to form spherical surfaces ever enlarging and which strike our ears. Now there is no doubt at all that light also comes from the luminous body to our eyes by some movement impressed on the matter which is between the two; since as we have already seen it cannot be by the transport of a body which passes from one to the other. If, in addition, light takes time for its passage ... it will follow that this movement, impressed on the intervening matter is successive; and consequently it spreads, as Sound does, by spherical surfaces and waves ... (pp. 3–4)

Something like this is also found – clearly, if rather less explicitly – in Fresnel. By Fresnel's time, background knowledge had, of course, been modified and augmented – especially through the addition of some notable further experimental results. The inference that Fresnel used to argue for the wave theory of light can be reconstructed as follows (the similarities to Huygens' own explicit argument will be plain).

(i) Background knowledge (in the form of the "mechanical philosophy") entails that the physical world consists of matter in motion.

(ii) Hence light in particular consists of either matter in motion or motion through matter.

(iii) If light consisted of bits of matter in overall motion, then the emission of particles from a luminous source would form either (a) a more or less continuous stream or (b) a succession of discrete particles.

(iv) Possibility (a) is ruled out by the fact that two light beams can cross each other, at right angles say, without either being affected beyond the point of crossing (if the two beams were two streams there would surely be a good deal of interesting action where they crossed which would modify the beams in their further progress).

(v) No such problem need arise on possibility (b). This sort of "non-superposition" could be explained by assuming that the particles of light follow one another at great distances, hence making the probability of any collision between particles in beams that cross one another very small. However, at least in Fresnel's opinion, this possibility too was ruled out in a clear cut way by well-established experimental results – principally those concerning the diffraction of light.

(vi) It follows therefore from (i) to (v) that light must consist of motion through matter.

(vii) It is also part of background knowledge that light has a finite velocity (Huygens explicitly refers to Roemer as having established this); hence there must be a material medium intervening between source and receptor to carry the motion making up the light in the finite time-interval between emission and absorption. (The "luminiferous aether" is hence inferred not conjectured!)

(viii) All sorts of optical phenomena exhibit *periodicities* – properties that recur at regular spatial and temporal intervals: notably the phenomena of Newton's rings and various interference effects. (This premise, firmly emphasised by Fresnel, is missing from Huygens who really held a "disturbance," rather than a wave, theory of light.) Again the periodicity of light was part of commonly accepted background knowledge (accepted by Newton, for example, who, to explain this, conjectured that his "parts" of light revolve with given periods as they move along).

(ix) Hence light consists of regular, periodic oscillations transmitted from point to point in the ether.

Thus we have finally the classical wave theory of light: light consists of periodic disturbances transmitted through an all-pervading mechanical medium.

4. CONCLUSION: THE METHOD OF "DEDUCTION FROM THE PHENOMENA" AND ITS PROBLEMS

These three historical examples show, I hope with more clarity than has sometimes been achieved in this area, that substantive new theories can indeed be *argued to* on the basis of material that is taken to be already known rather than merely conjectured; and that such arguments are not restricted to ones whose conclusions are simply more specific versions of some general theories that were already around, but may instead be given even for general, "hard core" or "paradigm-forming" theories.

Let me be a little clearer about what I do, and especially do not, claim for this method.

First, I am not claiming that Fresnel (or any other great scientist) actually first discovered any of his theories (first arrived at any of them in his own mind) by consciously going through a detailed argument of the kind articulated here. Rather, I would claim that, just like the great mathematician "intuits" a proof (arrives at an often *very* sketchy proof-sketch in his mind) and only later fills in the details, so the great scientist "intuits" that some specific or general theory can be argued to in one of the ways described and may, if pressed, fill in the details later. But the fact that there is such a detailed argument to be articulated, the fact that its premises are widely known and fairly widely, if not always universally, accepted is crucial. It explains what would otherwise be the entirely mysterious process of theory-creation;[3] it explains what would otherwise be the entirely mysterious prevalence of simultaneous discovery or near simultaneous discovery in the history of science; *and* it explains what I think is the inescapable feeling for anyone who studies the history of science carefully that even the greatest scientists save science only a relatively few years – even where there was no simultaneous discoverer, there are generally others aside from the great genius who first articulated the new theory who clearly would have got to the same theoretical discovery within a few years. (It may also explain the much more mundane, linguistic fact that we tend to talk of theoretical *discovery*, when on the Reichenbach-Popper view "invention" would be the altogether more appropriate term.)

Secondly, I am not of course claiming that these "deductions from the phenomena" prove or establish the theories involved. As is made clear elsewhere,[4] I am not one of those inclined to "fuzz up" Fresnel's theory so as to be able to argue that it is still accepted in presentday science. On the contrary, the only clear-sighted view is surely that Fresnel's theory was *rejected* some fifty or so years after its birth in favour of Maxwell's electromagnetic theory. This obviously does not mean that it was rejected root and branch – much of its content (indeed, I would argue, all of its *structural* content) was preserved within Maxwell's theory. But the fundamental theoretical idea of an all-pervading elastic medium that carries the light waves was unambiguously rejected within (at any rate the "mature" form of) Maxwell's theory in favour of disturbances in a "disembodied" *sui generis* electromagnetic field.[5] Although the arguments to theories involved can be presented as genuine deductions and are the sort of thing Newton and others had in mind (if rather less clearly) in talking about "deductions from the phenomena," they clearly do not rely simply on phenomenal or observational premises: background knowledge of a theoretical (indeed sometimes highly theoretical) kind is involved as, in the case of Fresnel's argument to the general wave theory, is an obvious *judgment* (basically that the available evidence made the corpuscular theory hopeless). Whatever one might think about the observational premises (and my own view is that, if one goes "low" enough to the level of Poincaré's "crude facts," then they are incorrigible[6]), these further premises are clearly defeasible – which is just as well since clearly a deduction with a false conclusion must have at least one false premise.

Some of the premises – both the general wave theory in the case of the first two arguments, for example, and the "mechanical world picture'" in the case of the third argument for that general wave theory itself – were not only defeasible but eventually defeated. None the less they were as a matter of fact generally accepted at the time – Fresnel certainly did not have to "conjecture" them – and they clearly formed the basis for Fresnel's discoveries.

There are many problems about the method of "deduction from the phenomena" (more properly deduction from the phenomena *plus "background knowledge"*). Obvious problems, for example, about the status of background knowledge and how it itself gets established – at any rate, in the case of theoretical rather than merely observational background knowledge; epistemological problems about what exactly such a deduction establishes and why, if at all, its conclusion should be considered any more secure – any more worthy of scientific acceptance – than a theory that *had* somehow (pretty well *per impossibile,* I would say) been conjectured out of the blue and then (successfully) tested. Important problems also arise in cases more complicated than the ones I have discussed here about the role and status of the "correspondence principle." (Newton's famous deduction of his theory of universal gravitation from Kepler's "phenomena" can indeed be reconstructed as *largely* deductive, but since, as Duhem and Popper both pointed out, Newton's theory is strictly *inconsistent* with Kepler's laws, this can only be with the help of something like the correspondence principle.[7]) But, as Lakatos had the Teacher say at the end of the original "Proofs and Refutations" (quoting Popper), "a scientific inquiry begins and ends with problems." (To which Beta plaintively adds "But I had no problems at the beginning! And now I have nothing *but* problems!") I have aimed in this paper only to do enough to show that these problems about the logic of scientific discovery are worth pursuing further.[8]

Notes

1. See in particular his (1989). See also the very interesting (1989) paper by Alan Musgrave. Some valuable material is also contained in the recent literature aimed at rehabilitating Newton's idea that theories may be "deduced from the phenomena." See for example Glymour (1980), chapter VI.
2. For the details (and the true story of the famous 'white spot' episode) see my (1989).
3. Popper used often to repeat the story of how he once went into a classroom and instructed his students to 'observe!' Popper reports that they were "of course" non-plussed and asked "what I *wanted* them to observe" (Popper, 1963, p. 46). I have always thought that Popper's (Viennese!) students must have been a rather dull lot. Certainly when I repeated the experiment nearly all my students simply arbitarily chose some feature of the "bloomin', buzzin' confusion" to observe without asking me anything (some looked out of the window

concentrating on passers-by; embarrassingly many concentrated on the viusal peculiarities of the maniac at the front of the classroom who was giving them weird instructions). Popper's point, however – that *interesting* observations demand a previously given point of view, that you will not, on the whole, observe anything interesting unless an interesting theory tells you what is interesting to observe (and very often tells you how to observe it) – is of course valid. But what would have happened if Popper had gone into the class and instructed its members "conjecture!". An equally, or even more, confused reaction is likely (and actually occurred in the admittedly small, unsystematic and entirely uncontrolled experiment I performed) – more are genuinely non-plussed in this case than in the "observe!" case, but the general reaction is again to choose more or less arbitrarily some topic to conjecture about. The results are mostly pretty idle ("I conjecture Manchester United will lose tomorrow") and unconnected with one another. I think an analogous (but entirely contra-Popperian) conclusion can be drawn here to the one drawn by Popper in the "observe!" case. It is essentially the conclusion drawn by Newton in his attack on "hypotheses" – namely that without *some already given, more general theoretical background*, any specific conjectures are likely to be either completely uninteresting or entirely speculative and untestable or both. (Indeed the marginally most frequent, and least uninteresting, type of response in my little experiment *was* clearly itself based – if rather loosely – on background knowledge. It was along the lines of "I conjecture he's going to use our responses to make some point about the philosophy of science.") Of course there are places where Popper shows himself to be implicity aware of this – especially when he stresses how important it is for a scientist to be "immersed in the problem background" before she can hope to produce a sensible innovation. But he never seems to have realised that this correct intuition is quite at odds with his official articulated view on scientific discovery.

4. See my (1989a) and (1994).
5. I say the "mature" version of Maxwell's theory since, as is well known, Maxwell himself and many others (notably Kelvin) strove mightily to explain the field further in terms of the contortions of some highly complex material ether. It was only after repeated failure to provide such explanations (or rather to provide any that was both coherent and independently testable) that the accepted view became that the field must be accepted as a separate, independent primitive constituent of the universe alongside matter.
6. See my (1991).
7. See Zahar (1989).
8. Since this paper was completed in 1998, I have published an extended treatment of some parts of it in Spanish in my (2001) and (2001a). I have also published a paper detailing my own view of the issues raised by the recent revival of Newton's method of "deduction from the phenomena" (see footnote 1 and section 4) in my (2000).

REFERENCES

Glymour, C. (1980) *Theory and Evidence*. Princeton University Press.

Huygens, C. (1690) *Treatise on Light*. All references to the Dover Reprint, (1962).

Lakatos, I (1978) *The Methodology of Scientific Research Programmes: Philosophical Papers*, Vol. 1, edited by J. Worrall and G. Currie, Cambridge University Press.

Musgrave, A.E. (1989) "Deductive Heuristics" in: Gavroglu, Goudaroulis and Nicolacopoulos (eds.) *Imre Lakatos and Theories of Scientific Change*. Kluwer.

Popper, K.R. (1958) *The Logic of Scientific Discovery*. Hutchison.

Popper, K.R.(1963) *Conjectures and Refutations*. Routledge.

Worrall, J. (1989) "Fresnel, Poisson and the White Spot: The Role of Successful Predictions in the Acceptance of Scientific Theories" in: D. Gooding, T. Pinch and S. Schaffer (eds.) *The Uses of Experiment – Studies of Experimentation in Natural Science*. Cambridge University Press, pp. 135–57.

Worrall, J. (1989a) "Structural Realism: the Best of Both Worlds?", *Dialectica*, 43/1–2, 99–124. (reprinted in D. Papineau (ed.) *Philosophy of Science*, Oxford Readings in Philosophy, 1996).

Worrall, J (1991) "Feyerabend and the Facts" in: Gonzalo Munevar (ed.) *Beyond Reason*, Kluwer, pp. 329–353.

Worrall, J (1994) "How to Remain (Reasonably) Optimistic: Scientific Realism and the 'Luminiferous Ether'" in: M. Forbes and D. Hull (eds.), *PSA*, Vol. 1, Philosophy of Science Association, pp. 334–342.

Worrall, J (2000) "The Scope, Limits and Distinctiveness of the Method of 'Deduction from the Phenomena': Some Lessons from Newton's 'Demonstrations' in Optics", *British Journal for the Philosophy of Science*, 51, pp. 45–80

Worrall, J (2001) "De la Matematica a la Cience: Continuidad y Discontinuidad en el Pensamiento de Imre Lakatos" in: Gonzalez, W.J. (ed.), *La Filosofía de Imre Lakatos: Evaluacion de sus propuestas*, UNED, Madrid 2001.

Worrall, J (2001a) "Programas de investigacion y heurística positiva: Avance respecto de Lakatos", in: Gonzalez, W.J. (ed.), *La Filosofía de Imre Lakatos: Evaluacion de sus propuestas*, UNED, Madrid 2001.

Zahar, E.G. (1989) *Einstein's Revolution – A Study in Heuristic*. Open Court.

Department of Philosophy, Logic and Scientific Method
LSE
Houghton Street
London, WC2A 2AE
UK

PÉTER SZEGEDI

LAKATOS ON CRUCIAL EXPERIMENTS AND THE HISTORY OF INTERPRETATIONS OF QUANTUM MECHANICS

1. "EXPERIMENTUM CRUCIS"

Let us begin with a few words on the notion of crucial experiment, or experimentum crucis.[1] The idea appears in the seventeenth century in Francis Bacon's *Novum Organum*, as the *instantia crucis* – that is, a crucial instance, or in the English edition, "instance of the fingerpost." One of Bacon's examples of crucial instances is that of the decision between two theories of tides: according to the first, the tides are due to a to-and-fro motion of the waters, as in a basin, while the second theory asserts that the tides are a periodic lifting and falling of the waters. Bacon's crucial question was: Are there high tides at the same times at the coasts of Spain and Florida as well as at those of China and Peru? These coasts are at opposite sides of the imaginary basins, so the basin theory does not allow a positive answer to this question.

Galileo also used this method – without mentioning the Baconian name – referring to the phenomenon of the tides that as a way of deciding between the two world systems, the Ptolemaic and the Copernican. The term "experimentum crucis" was first written down by Descartes, then Boyle and Newton, as well. The notion was thus introduced to science, and it gained growing importance on up to the present. The first philosopher of science who was able to deny the method was Duhem, at the turn of the century. In his view, experimental observation is theory-laden, because of the network of hypotheses necessary to the use of a measuring instrument. This fact excludes the conclusive role of experiment. Later, however, Popper wanted to rehabilitate crucial experiments – in his case, only falsifying ones. In the expanded English version of his *Logic of Scientific Discovery,* he writes that "in most cases we have, before falsifying a hypothesis, another one up our sleeves; for the falsifying experiment is usually a *crucial experiment* designed to decide between the two."[2] And at another place:

> It should be noted that I mean by crucial experiment one that is designed to refute a theory (if possible) and more especially one which is designed to bring about a decision between two competing theories by refuting (at least) one of them – without, of course, proving the other.[3]

G. Kampis, L. Kvasz and M. Stöltzner (eds.),
Appraising Lakatos: Mathematics, Methodology and the Man, 101–111.

2. Lakatos on Crucial Experiments

The originally Popperian Lakatos took a radical turn away from this point of view. On the basis of his methodology of scientific research programmes, he takes a new look at crucial experiments, too. By his opinion, "*the history of science has been and should be a history of competing research programmes* ...,"[4] and so the problem arises: "*how are research programmes eliminated*?"[5] This problem leads Lakatos to the crucial experiments. He distinguishes the "*minor crucial experiments*" and the "*major crucial experiments*." The first appear *within* a research program, that is, within a series of theories with a common hard core. We can often find such kinds of crucial experiments. As Lakatos writes on the subsequent theories,

> Experiments easily 'decide' between the *n*th and (*n*+1)th scientific version, since the (*n*+1)th is not only inconsistent with the *n*th, but also supersedes it. If the (*n*+1)th version has more corroborated content in the light of the *same* programme and in the light of the *same* well corroborated observational theories elimination is a relatively routine affair.[6]

"*Major crucial experiments*" are necessary for deciding between two research programmes. In the following, these will be our subjects. According to Lakatos' reconstruction of the battles and the war between competing research programmes, and according to his examples from the history of physics (the corpuscular and wave models of light, Mercury's perihelion precession, Brownian motion, the Michelson-Morley experiment, etc.), the experiment is seen – only *with hindsight,* occasionally several decades later – to have been crucial, and only in the light of some superseding theory. The conclusion of his considerations is that "*there are no such things as crucial experiments*, at least not if these are meant to be experiments which can *instantly* overthrow a research programme."[7] Two years later, Lakatos' conviction in this area seems to be strengthened. "'Crucial experiments' in Popper's sense do not exist: at best they are honorific titles conferred on certain anomalies *long after the event*, when one programme has been defeated by another one."[8] – he writes. In his survey paper, "The Role of Crucial Experiments in Science," he repeats his statements in more explicit form: "*Neither the logician's proof of inconsistency nor the experimental scientist's verdict of anomaly can defeat a research program at one blow. ... 'Crucial experiments' in the falsificationist sense do not exist* ...".[9]

3. Interpretational Schools as Research Programmes in Quantum Mechanics

Now we turn to the interpretations of quantum mechanics. At first glance, the history of these interpretations, or schools fits better with Lakatos' methodology of rival scientific research programmes than, for example, with

Kuhn's theory of absolute dominant paradigms, for Lakatos allows the simultaneous existence of different programmes.

Let us remind the reader of the criteria of progressivity and degenerativity of programmes in Lakatos' methodology:

Let us say that such a series of theories is *theoretically progressive (or 'constitutes a theoretically progressive problemshift')* if each new theory has some excess empirical content as the unrefuted content of its predecessor, that is, if it predicts some novel, hitherto unexpected fact. Let us say that a theoretically progressive series of theories is also *empirically progressive (or 'constitutes an empirically progressive problemshift')* if some of this excess empirical content is also corroborated, that is, if each new theory leads us to the actual discovery of some *new fact*. Finally, let us call a problemshift progressive if it is both theoretically and empirically progressive, and degenerating if it is not.[10]

By the Lakatosian criteria, the rise of quantum mechanics is itself a progressive problemshift (perhaps not a revolution), for it was both theoretically and empirically progressive, as it predicted novel, unexpected and later corroborated facts. But if we arrange the interpretations of quantum mechanics into series of theories, then we can perhaps discern theoretical progressivity. There is, for example, the Aharonov-Bohm effect, which was discovered by Bohm in searching for hidden parameters: it had general importance for physics – but hardly possessed empirical progressivity, for no effects different from the ones usual in quantum mechanics were measured. So by these criteria, the interpretational programmes can be scientific but degenerative; in other words, they are characterised by stagnation. Of course, the evaluation can change in the future, and Lakatos himself wanted programmes to be evaluated over long run: "Moreover, it occasionally happens that when a research programme gets into a degenerating phase, a little revolution or a *creative shift* in its positive heuristic may push it forward again."[11]

So we can hope that his recommendation to repress the degenerate programmes is not valid for the interpretations of quantum mechanics. To say more, Lakatos himself gives us a more sophisticated description – for example, of the Michelson-Morley experiment – than one of this progressive-degenerative sort. Referring to just a quantum mechanical problem, he writes:

But one should not forget that *two specific theories, while being mathematically (and observationally) equivalent, may still embedded into different rival research programmes, and the power of the positive heuristic of these programmes may well be different.* This point has been overlooked by proposers of such equivalence proofs (a good example is the equivalence proof between Schrödinger's and Heisenberg's approach to quantum physics).[12]

Here, as well, he speaks on the possibility of a degenerated programme's coming back, and he mentions again a moment of the history of quantum mechanical interpretations, namely the Dirac-ether used by Jean-Pierre

Vigier. Some followers of Lakatos later pointed out that, without essential modifications of the Lakatosian conception, the programmes could be classified in a more sophisticated manner (e.g. cumulative progression or heuristic progression), and that the degenerated programmes (like the nuclear magnetic resonance programme in physics) can actually become successful ones if, say, they provide results useful to other programmes.[13]

I was speaking of different interpretations of quantum mechanics as research programmes, but another question, that of how to identify the structural elements of the Lakatosian research programmes in these series of theories, also arises. It is a rather difficult task – not only in quantum mechanics, but in other physical disciplines, as well. One might mention James Cushing's works on high-energy physics, for example; Cushing tried to reconstruct the history of particle physics in a Lakatosian sense.[14] In his opinion,

> It might be better to speak of a central core of a research program as a combination of the hard core and the heuristic since it is not clear that these two latter elements exist as separate entities. For purposes of general discussion it may be schematically simplest to present these components as a fixed hard core to which auxiliary assumptions are successively joined by a creative application of a guiding heuristic to generate a sequence of theories, thus guaranteeing a continuity from one theory to the next within one research program.[15]

But, after all, one can distinguish in these schools a permanent hard core and a changeable element, which can be described as able to accommodate problems that arise. In Louis de Broglie's programme, for instance, the effort to obtain a synthesis of wave and particle pictures is always present as a hard core, while the actual forms of the realisation are the following, in order of appearance: the principle of double solution, the theory of pilot waves, the non-linearity hypothesis, and hidden thermodynamics. In the case of David Bohm, the hard core is the hidden parameter theory and quantum potential; later on he added the stochasticity hypothesis. Vigier generally followed the same hard core, but with the additional assumption of the subquantal level, or of the hypothesis on hidden degrees of freedom, later of Dirac-ether and also stochasticity. There are other programmes in the interpretation of quantum mechanics, where stochasticity is not an additional hypothesis, but the hard core. This is the Fényes-Nelson-de la Peña line. In its first form, it uses the diffusion processes, then Brownian motion and finally stochastic electrodynamics. The same structure can be found in the series of hydrodynamic models, where the characterisation of the fluid becomes increasingly difficult. The positive heuristic of these programmes is changing to a certain extent, but generally contains the relativistic approach, the principle of determinism or causality and the principle of the unity of nature (or, concretely, the unity of quantum mechanical processes – namely time evolution and the measurement process). One can say that the orthodox interpretation is also a programme; in this case we can identify – partly following Cushing – the hard core as the canonical commutation relations and Hamilton's equations of motion. The guiding hypotheses in the positive heuristic are the applicability of the classical forms of Hamiltonians for specific systems, the correspondence principle and the principle of observables, while the operators-observables relation is an example for the auxiliary assumption. Lakatos' evaluation of this programme is rather negative:

> In the new, post-1925 quantum theory the 'anarchist' position became dominant and modern quantum physics, in its 'Copenhagen interpretation,' became one of the main standard bearers of philosophical obscurantism. In the *new* theory Bohr's notorious 'complementarity principle' enthroned [weak] inconsistency as a basic ultimate feature of nature, and merged subjectivist positivism and antilogical dialectic and even ordinary language philosophy into one unholy alliance. After 1925 Bohr and his associates introduced a new and unprecedented lowering of critical standards for scientific theories. This led to a defeat of reason within modern physics and to an anarchist cult of incomprehensible chaos.[16]

In light of the history of interpretations of quantum mechanics, therefore, the Lakatosian methodology can be justified; the following section will attempt to justify his standpoint on crucial experiments, as well.

4. "CRUCIAL EXPERIMENTS" IN THE FOUNDATIONS OF QUANTUM MECHANICS

4.1. The EPR experiment

The series of experiments that I will mention begins with a gedankenexperiment, specifically, the Einstein-Podolsky-Rosen experiment.[17] I do not want to go into the details of this well-known experiment; I would like only to stress that, although – as a gedankenexperiment and not a real experiment with real instruments – it has nothing to do with the above-mentioned Duhem-thesis, despite the fact that later commentators discern (at a minimum) five hypotheses in it: the principle of realism, the validity of quantum mechanical formalism, the completeness hypothesis, the principle of separability and, finally, the validity of classical logic.

According to the EPR argument, at least one of the five hypotheses is false. But this gedankenexperiment was quite far from being a practically manageable one. The next step was taken by Bohm, who reformulated the gedankenexperiment with spins.[18] In this way, the experiment got closer to being a real one, but it did not yet seem to be a crucial experiment, for nobody reported the different theories producing different measurement results.

4.2. Bell-inequality

John Bell's work in the mid-sixties raised the hope that it might be possible to test interpretations experimentally. Bell studied the EPR experiment or, rather, Bohm's version of it.[19] He pointed out that there must be differences between the quantum mechanical and hidden parameter predictions. He supposed that in a real experiment, we would be able to measure probabilities in the manner depicted in Figure 1.

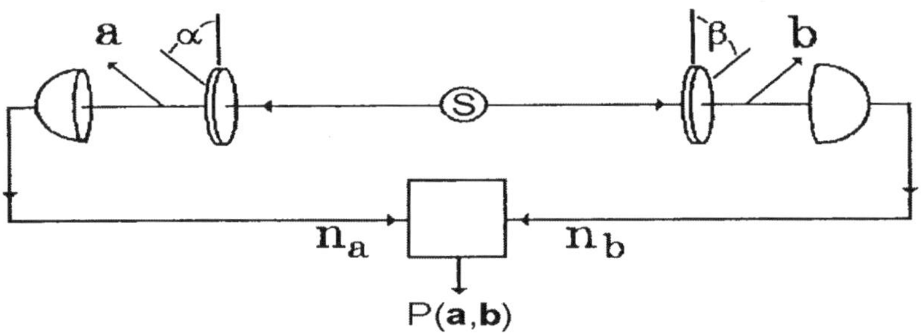

Figure 1. The measurement scheme of the Bell-inequality using photons. The photons emerging from the source S are measured by polarisation filters (not by Stern-Gerlach magnets as in the case of electrons) with given orientations (angles). After the filters there are the counters (hemispheres in the figure). From the counters the signals – corresponding to the photon number – go to the coincidence circuit (square in the figure), which gives us only the simultaneous beats.

From certain assumptions, he derived his famous inequality for the expectation values of the product of outcomes of spin measurements on the two particles:

$$|\, P(\mathbf{a},\mathbf{b}) - P(\mathbf{a},\mathbf{c})\, | \leq 1 + P(\mathbf{b},\mathbf{c})$$

Bell explicitly presupposed reality, locality or separability, determinism, and hidden parameters – and not so explicitly, formalism and classical logic, as well. Towards the beginning of the seventies, the Bell-inequality drew nearer and nearer to the real conditions of a manageable experiment.

4.3. Realizations and auxiliary hypotheses

In the seventies, there were essentially three types of experiments to measure the Bell-inequality. In the first type, the photons were gained from an atomic cascade process[20] (Figure 2).

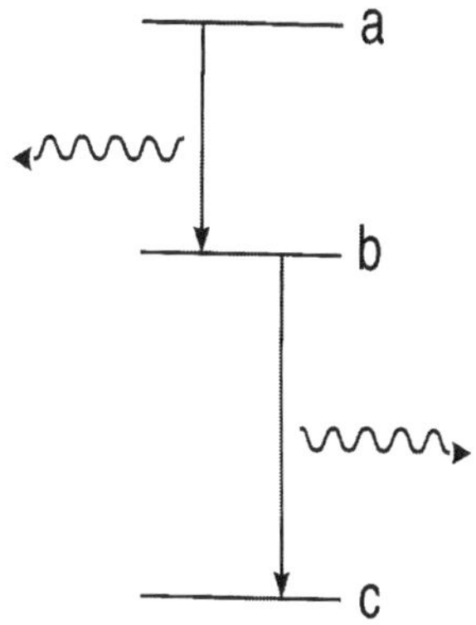

Figure 2. Cascade of photons. The electron in the atom is excited from the energy level *c* to *a* by two laser beams, for example, and when the electron returns, the atom emits in rapid succession two photons which have opposite polarisations (spins), because of the features of the atomic energy levels.

In most of the experiments, the quantum mechanical predictions were justified. In the meantime, however, theoreticians continuously analysed the presuppositions for the derivation of Bell-inequalities. First Bell, then Clauser and Horne[21] showed that determinism is not necessary for their derivation, since the local theories with non-dynamical stochastic hidden parameters are also sufficient for it. Later Ballantine, Bohm[22] and Bell pointed out that even the hidden parameters are not necessary. So only the reality and locality hypotheses remained. The Aspect experiment intended to measure just the separability.[23]

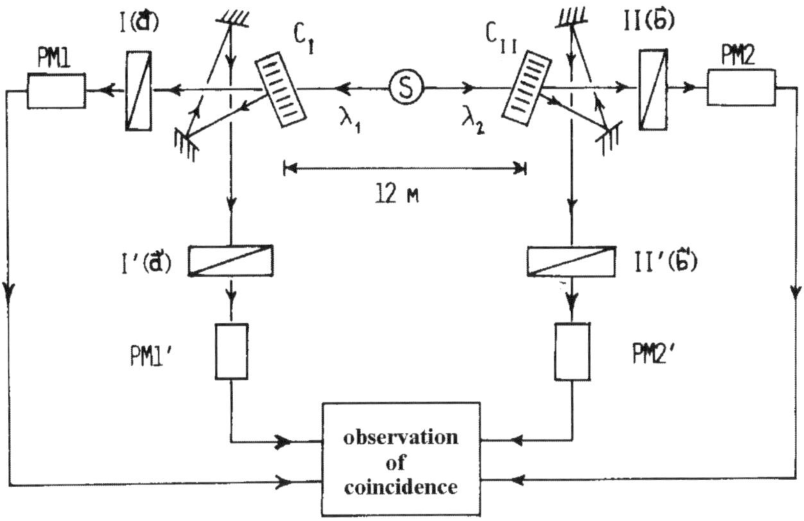

Figure 3. The arrangement of the Aspect experiment. S is the photon source. C_I and C_{II} are the optical switches, which are followed by the differently oriented I, I′ and II, II′ polarization filters, respectively. Behind the filters are the PM photon counters. The coincidence circuit allows the observation of the beats by photon pairs.

For the measurements, however, some further additional presuppositions were inevitable; Bell's derivation essentially assumed certain ideal conditions, where we can measure, for example, the polarization of photons in any direction. In the real process, the photons emerging from the source fall to a polarization filter, which, depending on the polarization of the photons (and perhaps on other factors), either does or does not absorb them. In a real experiment, only those photons, which arrive at the detectors in pairs, are important to us – and the efficiency of detectors is only 10–20%. For the interpretation of measurement, we have to make some assumptions. Clauser-Horne-Shimony-Holt, for instance, assume in their paper that the probability of photon pair detection does not depend on the presence and orientation of polarisation filters.[24] It is a plausible presupposition, but one that is hard to verify. And this assumption restricts the interval of values of the critical observable to its one hundredth part. The other possible assumption was the no-enhancement principle,[25] which has similar features. Because of these assumptions – and because of low efficiency – the different interpretational schools were able to develop models that saved their respective programmes.

The second type of experiments measured the Bell-inequality with high energy photon pairs obtained from electron-positron annihilation.[26] In this second case, the presuppositions are even stronger than in the previous one; one must, for instance, use quantum mechanics itself, and so on. The results were in good agreement with quantum mechanics, but Bell and others worked out *ad hoc,* hidden parameter theories, which were unfalsifiable under the given experimental conditions, because their predictions were the same as those of quantum mechanics.

The third type of experiment was conducted with low energy proton-proton scattering, and the experimenters measured the Bell inequality of spin correlations.[27] Of course, they needed auxiliary hypotheses as well, so the loophole problem remained, and one can say that Lakatos' negative heuristic works well in all the cases, and that it is easy to verify his thesis that there is no "experimentum crucis" in the falsificationist sense.

In the nineties, new experiments were suggested, which were intended to overcome the loophole problem by measuring other observables rather than spins, and by measuring a far larger part of the emerging useful objects, not merely a fraction of them. With the help of photons and interferometers it is possible, according to J.D. Franson's suggestion, to test Bell-type inequalities with position and time co-ordinates, rather than with spins.[28] For such experiments one can use the so-called "parametric downconversion" (see Figure 4).

In short, the effect is the following: a non-linear crystal splits the laser beam into two beams, each with double the original wavelength (in the symmetric case). It is possible, as we shall have two uniform photons, to make interference with them in a Mach-Zender apparatus; and changing the optical path, a Bell type inequality can be measured for the phase and momentum of the wave. The efficiency of the device was much greater than

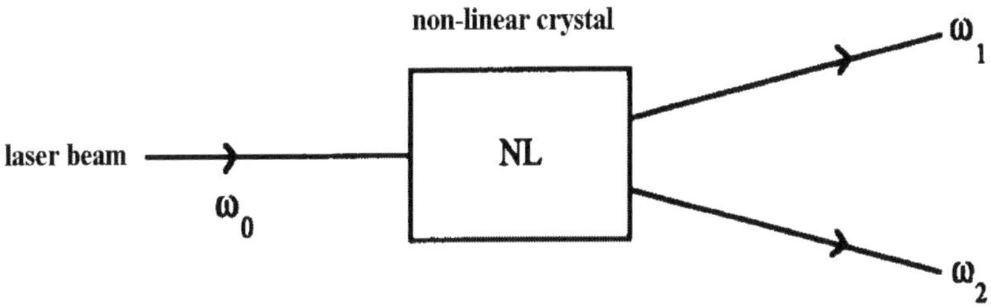

Figure 4. Parametric downconversion. The photon (coming from the laser) with frequency ω_0, after the interaction with the non-linear crystal NL becomes two photons where $\omega_1 + \omega_2 = \omega_0$.

in the earlier experiments,[29] so there was hope of avoiding the auxiliary hypothesis and the *ad hoc* explanations.

Unfortunately, though, some theoreticians worked out different models based on a local realistic hidden parameter theory, and these models are able to reproduce – under different conditions – the old, as well as the new, experimental results.

Further suggestions arrived at, for instance, in the paper by Greenberger, Horne, and Zeilinger, urge that the measurement in the original EPR experiment be undertaken with three particles, rather than two.[30] They hope that the difference between the quantum mechanical and hidden parameter prediction will grow with this solution, and that it will be more difficult to make *ad hoc* models for three particles. But the measurement will also be much more difficult, we think.

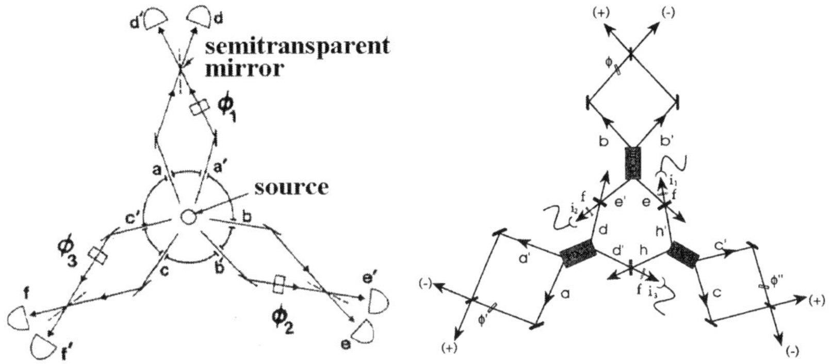

Figure 5. GHZ's original suggestion for measurement with three particles is to the left. To the right is their current concept with the use of parametric downconversion.

5. MORAL

If we believe in Lakatos and in the skill of theoreticians, then we may not expect radically new results, but must await an empirically progressive new programme; with hindsight, we shall see which experiment was crucial. Despite this, I do not want to claim that these experiments are useless. These experiments and their interpretations brought us somewhat closer to the understanding of quantum mechanics and, in some cases, they were useful for practical purposes, as well. Over the course of this history of these interpretations, theoreticians and experimental physicists developed technologies that could be applied in other fields of physics, for example in quantum correlations. Some people hope that the final results will include quantum cryptography, quantum communication,[31] and quantum computers.[32] In conclusion we can say that, although there are no crucial experiments in science, there do exist useful experiments.

NOTES

1. For a detailed discussion, please turn to the paper by Márta Fehér, "The Rise and Fall of Experimentum Crucis", in: *Doxa – Philosophical Studies* 6, 1985, pp. 59–97.
2. Karl R. Popper, *The Logic of Scientific Discovery.* Section 22, note 1. London: Routledge, 1992, p. 87.
3. *Ibid.*, Section 85, note *2. p. 277.
4. Imre Lakatos, "Falsification and the Methodology of Scientific Research Programmes", in: John Worall and Gregory Currie (eds.), *The Methodology of Scientific Research Programmes, Philosophical Papers by Imre Lakatos Vol. I*, Cambridge: Cambridge University Press 1978, p. 69.
5. *Ibid.*
6. *Ibid.*, p. 71.
7. *Ibid.*, p. 86.
8. Imre Lakatos, "Popper on Demarcation and Induction", *ibid.*, p. 150.
9. Imre Lakatos, "The Role of Crucial Experiments in Science", in: *Studies in History and Philosophy of Science* 4, 4, 1974, pp. 319–320.
10. Lakatos, "Falsification and the Methodology of Scientific Research Programmes", *op.cit.*, pp. 33–34.
11. *loc.cit.*, p. 51.
12. *loc.cit.*, p. 77, note 2.
13. See for instance H. Zandvoort, "Intrinsic Success and Extrinsic Success of Research Programs", in: *7th International Congress of Logic, Methodology and Philosophy of Science, Vol. 3 Abstracts of Section 6*. Salzburg 1983, pp. 289–292.
14. James T. Cushing, "Models and Methodologies in Current Theoretical High-Energy Physics", in: *Synthese* 50, 1982, pp. 5–101.
15. *Ibid.*, p. 73.
16. Lakatos, "Falsification and the Methodology of Scientific Research Programmes", *op.cit.*, pp. 59–60.
17. Albert Einstein, Boris Podolsky and Nathan Rosen, "Can Quantum-Mechanical Description of Physical Reality Be Considered Complete?", in: *The Physical Review* 47, 1935, pp. 770–780.
18. David Bohm, *Quantum Theory.* Englewood Cliffs: Prentice Hall, 1951.
19. John Bell, "On the Einstein, Podolsky Rosen paradox", in: *Physics*, 1, 1964, pp. 195–200, and "On the Problem of Hidden Variables in Quantum Mechanics", in: *Reviews of Modern Physics* 38, 1966, p. 447.

20. See a detailed review e.g. in D. Home and F. Selleri, "Bell's Theorem and the EPR Paradox", in: *La Rivista del Nuovo Cimento* 9, 1991, pp. 1–95.
21. John F. Clauser and Michael Allan Horne, "Experimental Consequences of Objective Local Theories", in: *Physical Review D* 10, 1974, pp. 526–535.
22. A. Baracca, D.J. Bohm, B.J. Hiley, A.E.G. Stuart, "On Some New Notions Concerning Locality and Nonlocality in the Quantum Theory", in: *Il Nuovo Cimento B* 28, 2, 1975, pp. 453–465.
23. A. Aspect, "Proposed Experiment to Test the Nonseparability of Quantum Mechanics", in: *Physical Review D* 14, 1976, pp. 1944–1951; Alain Aspect, Philippe Grangier, and Gérard Roger, "Experimental Tests of Realistic Local Theories Via Bell's Theorem", in: *Physical Review Letters* 47, 1981, pp. 460–462; "Experimental Realization of Einstein-Podolsky-Rosen-Bohm Gedankenexperiment: A New Violation of Bell's Inequalities", in: *Physical Review Letters* 49, 1982, pp. 91–94; Alain Aspect, Jean Dalibard and Gérard Roger, "Experimental test of Bell's inequalities using time-varying analyzers", *Physical Review Letters* 49, 1982, pp. 1804–1807.
24. John F. Clauser, Michael A. Horne, Abner Shimony, and Richard A. Holt, "Proposed Experiment to test local hidden-variable theories", in: *Physical Review Letters* 23, 1969, pp. 880–884.
25. See Clauser and Horne, ref. 21.
26. Among others A.R. Wilson, J. Lowe and D.K. Butt, "Measurement of the Relative Planes of Polarization of Annihilation Quanta as a Function of Separation Distance", in: *Journal of Physics G: Nuclear Physics* 2, 1976, pp. 613–624. Its interpretation in D. Bohm and B. Hiley, "Nonlocality and Polarization Correlations of Annihilation Quanta", in: *Nuovo Cimento B* 35, 1976, pp. 137–144.
27. M. Lamechi-Rachti and W. Mittig, "Quantum Mechanics and Hidden Variables: A Test of Bell's Inequality by the Measurement of the Spin Correlation in Low-Energy Proton-Proton Scattering", in: *Physical Review D* 14, 1976, pp. 2543–2555.
28. J.D. Franson, "Extension of the Einstein-Podolsky-Rosen Paradox and Bell's Theorem", in: *Physical Review D* 26, 4, 1982, pp. 787–800; James D. Franson and Kenneth A. Potocki. "An Experimental Test of the Quantum Theory", in: *Johns Hopkins APL Technical Digest* 5, 3, 1984, pp. 305–308; J. D. Franson, "Bell Inequality for Position and Time", in: *Physical Review Letters* 62, 1989, pp. 2205–2208.
29. Z.Y. Ou, X.Y. Zou, L.J. Wang and L. Mandel, "Observation of Non-Local Interference in Separated Photon Channels", in: *Physical Review Letters* 65, 1990, p. 321; P.G. Kwiat, W.A. Vareka, C.K. Hong, H. Nathel and R.Y. Chiao, "Observation of Nonlocal Two-Photon Interference in a Dual-Beam Michelson Interferometer", in: *Physical Review A* 41, 1990, pp. 2910–13; J.G. Rarity, P.R. Tapster, E. Jakeman, T. Larchuk, R.A. Campos, M.C. Teich and B.E.A. Saleh, "Two-Photon Interference in a Mach-Zender Interferometer", in: *Physical Review Letters* 65, 1990, pp. 1348–1351; Paul G. Kwiat, Klaus Mattle, Harald Weinfurter, Anton Zeilinger, Alexander V. Sergienko and Yanhua Shih, "New High-Intensity Source of Polarization-Entangled Photon Pairs", in: *Physical Review Letters* 75, 24, 1995, pp. 4337–4341.
30. D.M. Greenberger, M.A. Horne and A. Zeilinger, "Going beyond Bell's Theorem", in: M. Kafatos (ed), *Bell's Theorem, Quantum Theory and Conceptions of the Universe*. Dordrecht: Kluwer 1989, pp. 73–6; Anton Zeilinger, Michael A. Horne, Harald Weinfurter, and Marek Žukowski, "Three-Particle Entanglements from Two Entangled Pairs", *Physical Review Letters* 78, 16, 1997, pp. 3031–3034.
31. Y.H. Shih, A.V. Sergienko, T.B. Pittman, D.V. Strekalov and D.N. Klyshko, "Two-Photon 'Ghost' Image and Interference-Diffraction", in: Daniel M. Greenberger-Anton Zeilinger (eds.), *Fundamental Problems in Quantum Theory*. New York: The New York Academy of Sciences 1995, pp. 121–132.
32. Tycho Sleator and Harald Weinfurter, "Quantum Teleportation and Quantum Computation Based on Cavity QED", *Ibid.*, pp. 715–725; I.L. Chuang, R. Laflamme, P.W. Shor, W.H. Zurek, "Quantum Computers, Factoring, and Decoherence", in: *Science* 270, 5242, 1995, pp. 1633–1635.

Department of History and Philosophy of Science
Faculty of Sciences, Eötvös University, Budapest
Pázmány P. Sétány 1/A, H-1117 Hungary

II.

Underway to a Philosophy of Contemporary Mathematics

Lakatos, Tamás Kövári and Tamás Lipták at a Colloquium on Measure, Theory, Balatonvilágos, 19 September 1955
(By kind permission of Éva Pap)

Lakatos with his wife Éva Pap, at Porto di Brenzone, 10 September 1957
(By kind permission of Éva Pap)

DAVID CORFIELD

ARGUMENTATION AND THE MATHEMATICAL PROCESS

INTRODUCTION

We have reached a position today from which to evaluate judiciously Lakatos' contribution to the philosophy of mathematics. We should applaud unreservedly his decision to take the development of mathematics as a topic worthy of philosophical consideration and admire the first steps he took to elucidate patterns of theory change via his identification of certain mechanisms of concept-stretching. We must recognise, however, that the *method of proofs and refutations* accounts only for the modification of a concept's definition,[1] omitting more fundamental varieties of theory development, and consequently that much remains to be done. The two tasks before us today are to describe an expanded range of varieties of theory production, especially radically innovative mathematical conceptualisations, and to ponder the philosophical issues relating to such a description. Regarding the former, see Corfield (1998a) and Kvasz (1998). Through this paper I shall largely confine myself to the latter, endeavouring to convey a sense of the vast array of possibilities now open to us.

In the first part of this paper I shall survey some current ideas about how the development of mathematics should be studied and show that, despite the wild variations in the approaches adopted by commentators and the absence of any serious engagement with each other's positions, there is still plenty of common ground. I shall attempt to forge a path through some of the literature by constructing links between the contributions to be considered. Some of these links take the form of questions which I raise here not to answer, but to provide markers for the construction of an arena in which the mathematical process may be studied. This is to be seen as suggestive of a much larger project to weave together the very partial and disjointed insights gained to date. In the second part I shall propose some measures for bringing the parties together to allow this weaving to take place.

G. Kampis, L. Kvasz and M. Stöltzner (eds.),
Appraising Lakatos: Mathematics, Methodology and the Man, 115–138.

1. THE PRESENT

Mathematical judgement

For all the attention he devoted to informal mathematics and its heuristics, there still remained in Lakatos a residual streak of logical empiricism, in the sense that the scientific and mathematical knowledge of greatest concern to him, philosophically speaking, was embodied in theoretical statements. Hidden lemmas, he tells us, emerging during the 'proofs and refutations' stage of discovery, turn into the axiomatic contents of the hard core (1978, p. 96) which must be defended at all costs. We also see this stance displayed in the following quotation, where, while criticising Toulmin, Lakatos posits the 'knowing that' component of scientific knowledge as philosophically fundamental and adopts a correspondence theory of truth:

> This stress [on Toulmin's part] on the inarticulable shifts the problem of 'knowing that' to the problem of 'knowing how'; from knowledge expressed in propositions to knowledge expressed in skills and activities. This in turn leads from the classical conception of truth – a proposition is true if it corresponds with the facts – to pragmatism – a belief is true if it gives rise to useful or effective action ... Propositions, and thus the 'third world,' are redundant (1978, p. 228).

Moreover, Lakatos maintained that axiomatisation marks the end of the creative process. He indicated that readjustments might be required if the informal material had not been satisfactorily captured, but gave no hint that an axiomatisation could act as a springboard for further theoretical development. This, coupled with a recognition of the achievement of a considerable stabilisation in standards of rigour, left him with something close to Francis Fukuyama's vision in political theory of the 'End of History' (Fukuyama, 1989): at the level of ideas, all battles are over, the idea of liberal democracy having won. This parallel, no doubt attributable to their Hegelianism, extends to the fact that they bemoan the lack of excitement to be found in the realms of current political and mathematical thought. For Lakatos, once a mathematical theory has entered the axiomatic stage "imagination is tied down to a poor recursive set of axioms and some scanty rules" (1978, p. 68).

However, Lakatos is wrong on both these counts. He should have known that in the axiomatic era mathematics is much larger than its accepted statements. This excess is not only non-linguistic know-how, although essential, but may be expressed in statements: 'This definition captures the concept at the right level of generality'; 'Construction A is a better way of generalising construction B than is construction C'; 'Theory F can be used to produce simple/short/generalisable proofs of results in area G'; 'This notation is extremely efficient'; 'Proofs often prove more than was intended and as such should be sifted through to extract their extra power'; 'That line of research has been disappointing because it has provided no new insights.' Even had he later come to recognise this kind of knowledge as belonging to the heuristic of a mathematical research programme, he still would have had difficulty coupling it to his rigid conception of the axiomatic process.

Axiomatisation is not the end of the road. It acts to support a more refined kind of informal thought, allowing plenty of room for further disagreement. The good news is that there are still arguments being conducted, often unnoticed by outsiders, which concern the direction of research, what constitutes a breakthrough, which style of research to encourage, which way to think about a theory, which way to generalise a concept, and so on. Mathematicians have not relocated to the realm of logic, but are as ever stationed within the realm of dialectic, where there is room for inventive concept-formation and concept-stretching. There is still excitement. What are invoked in these arguments are selection criteria which for the most part are unrelated to issues concerning the correctness or validity of mathematical results. Indeed, you will hear mathematicians claiming about another group of researchers that they have 'gone too far,' with no hint of an accusation of lack of rigour.

In a heuristic-filled textbook for aspiring topologists (Gilbert and Porter, 1994), the authors, having earlier provided the means for the reader to construct a formal proof of Van Kampen's theorem, invite her to do so as an exercise (p. 222). Their next piece of advice is to

> ... go to your library and search out books which contain a formal proof of the Van Kampen theorem. Compare their methods with yours. *Criticize* both your proof and theirs. Try to improve your proof with ideas, phrases, or notation from their version. Do not think their version will necessarily be fundamentally better than yours, but as the authors tend to have had years' more experience of writing proofs, you should expect to gain something, at least in style (Ibid., pp. 22–3, my emphasis).

What is at stake here is not the kind of proof criticism by counterexample that we find in *Proofs and Refutations*, but one concerning conceptual organisation, notation and style. The truth or correctness of the result is not at issue, nor is the rigour of the proof. It is all about mathematical taste and understanding.

To move quickly to the central point I wish to make here, what have typically been ignored by philosophers of mathematics are the judgement skills of mathematicians, their *bons sens* or *finesse*, to return to Pierre Duhem's expressions. In effect, what I have been indicating above is the fact that Lakatos' appreciation of this *bon sens* was somewhat restricted. This is also revealed by his decision to elaborate on the first of George Polya's principal brands of 'plausible reasoning' (Polya, 1954), *induction*, when, as I shall discuss elsewhere, it is the second, *analogy*, which counts for more in the estimation of mathematicians. Yet Lakatos' efforts to learn from mathematicians, historians of mathematics and mathematics educationalists gave him a far greater sense of mathematical culture than those who can think of nothing more interesting to wonder about mathematics than whether numbers exist. Our task then is to persuade sufficiently many people that it is towards an analysis of the varieties of mathematical judgement that our resources should be devoted. We must recognise, however, that we are up against a deep-seated ignorance of phenomena in this realm.[2] Indeed, we appear to be dealing with a widespread form of agnosia and, as such, the difficulty in persuading the sufferers of their deficit should not be underestimated.

For those of us who have long known that there was more to be said about mathematics, our passage has appeared to be blocked by barricades erected at one end by the logical empiricists – mathematics is just a bunch of contentless tautologies and is only of interest as directly applied in science or everyday life, and at the other by Platonists – mathematics is about objects dwelling in some ethereal realm. If these obstructions are confronted on their own terms, one ends up confined to a kind of philosophical ghetto on a diet of Gödel's results, constructivism, or the Benacerraf paradox. Small wonder that some of those wishing to break out have turned to category theory. It possesses the twin qualities of being arguably as 'foundationally' significant as set theory (e.g. topos models for Cohen's forcing and for constructive logic), and yet also of revealing its tighter connections to mainstream mathematics, arising as it did out of concrete problems in algebraic topology studied in the 1940s. But once it is accepted that there is some philosophical juice to be extracted from the development of the central branches of mathematics, we may also be permitted to talk about the careers of the remarkable ideas of Gauss, Cauchy, Abel, Dirichlet, Galois, Riemann, Kronecker, Dedekind, Hilbert, Poincaré, Weyl and Noether, not to mention a host of theorists of the past sixty years. Their achievements should be our bread and butter, just as anyone studying the development of physics is expected to have the essential ideas of Faraday, Maxwell, Helmholtz, Einstein, Bohr and Heisenberg at their fingertips.

Today there is particular cause for hope that after decades in which little interest has been shown in what constitutes mathematics 'done well' the pendulum may be swinging in our direction, since we now begin to find our services required by the practitioners of two disciplines, namely, science studies and mathematics education.

Sociologists, historians and historically sensitive philosophers of physics have until now largely been concerned with the struggle to capture and question the various kinds of *best practice* of the theoretician and of the experimenter, and have treated the mathematics as something already there, implicit within the physics. They may justify their neglect of mathematical best practice by pointing to the fact that in the past physics and mathematics had not separated. As late as 1915, the world's leading mathematician, David Hilbert, was near enough to the cutting edge of theoretical physics to come a close second to Einstein in the race to publish a theory of general relativity. Then they will remind us of the likes of Weyl, von Neumann and Wiener. Even that quintessential 'pure' mathematician Karl Weierstrass did physics.

We should respond by saying that while there is a considerable overlap between the physicists' and the mathematicians' sense of judgement, it is just an overlap, one which may vary over the passage of time as a proportion of the whole. Certainly, von Neumann worried that this overlap can become too small, leading to degeneracy on the part of mathematics, but he was far too good a mathematician to wish to see mathematical research swamped by scientific considerations. Mathematical and scientific conceptual development are not locked together in a tight embrace; their course resembles more a piece of modern dance than it does a waltz. So even if valid objections are raised to the presentation of their encounters as being inexplicably fortuitous, there ought to be just a little more surprise displayed at the fact that Appolonius' work on conics, and then, more efficiently, the differential calculus, could allow Newton and his followers to model the motions of the solar system, or that Einstein did not have to look far for an appropriate calculus for general relativity, given the reasons behind the invention of the mathematics.

Given this non-coincidence, to put things a little naïvely, the more a piece of mathematics has developed due to internal considerations of a mathematical nature, the more there is to explain about its later applicability.

The striking thing is that the non-coincidence is today coming to the fore in that credit is accruing to physical theories if they make *mathematical*, rather than *empirical* predictions:

> ... quantum field theory has had its credibility enhanced by its success in making correct mathematical predictions. Given the lack of rigorous foundations for quantum field theory, these successes provide great encouragement to physicists that their ideas are fundamentally sound (Atiyah, 1995, p. 6070).

If mathematical statements are merely tautologies, it is hard to see the source of this enhancement of credibility.

Another reason the mathematical physics of the past twenty or so years calls for attention is that the recent interaction between mathematics and physics comes after a period when their grip on each other has been at its weakest.[3] Central to the state-of-the-art mathematics which is now intriguing physicists are the ideas formulated by the differential and algebraic topologists, practitioners of branches which have such a good century that "... by the 1970's, the whole of mathematics was saturated with topological ideas" (Atiyah, 1995, p. 6069). While it will be possible to find common considerations governing the development of gauge potentials in gauge field theory, on the one hand, and of connections on fibre bundles, on the other, there will still be more work to be done to think through this type of theoretical convergence.

Some less sanguine about the concept of 'mathematical' predictions will, no doubt, claim that theories such as topological quantum field theory are just mathematics dressed up as physics and not proper empirical science. Be that as it may, there are still good grounds to anticipate a useful partnership between those who study less recent aspects of the scientific and mathematical processes.[4]

Let us turn now briefly to a second group of prospective clients – the mathematics educationalists. Mathematics has always had an intimate relationship with the way it is taught, reflecting its etymology. The Babylonians bequeathed us tablets containing worked solutions for the benefit of their students. In the *Meno* Plato writes about a mathematics lesson Socrates gave to a slave. The pedagogic format of *Proofs and Refutations* may be accounted for by the fact that Lakatos himself worked in the Ministry of Education in the post-war communist government of Hungary and was well versed in the didactic literature.

If we turn the spotlight on the mathematical process as a whole, the training of future mathematicians will certainly need to be investigated. Teaching practices reflect cultural conceptions of the nature of mathematics and have a marked impact on research practices.

The division of labour between those studying the mathematical process and the mathematics educationalist may be put as one between the study of the 'phylogenetic' and 'ontogenetic' development of mathematical understanding. These, of course, are not isomorphic, but as one educationalist tells us:

> The difficulties they [the students] encountered, the tentative understandings of a still very unclear situation were often quite close to those experienced by mathematicians in the past (Sierpinska, 1994, p. xi).

Calls have been made by educationalists to introduce historical information into the classroom. When teaching the quadratic equation, for example, one could contrast the Babylonian formulaic method with the Greek ruler and compass construction and the Arabs' much closer geometric mirroring of the relevant algebra. More recent episodes could be useful for university education. How many mathematics students have turned up to the first session of a course on rings and modules to be told without further ado that "A ring is a set with two binary operations ..."? How can a concept such as this be grasped without a sense of the problem situations which gave rise to the emergence of rings of algebraic integers, polynomials, entire functions, p-adic numbers, etc.?

As with science studies, the relationship with mathematics education should be a dialogue. Ideas concerning the mathematical understanding of students at all levels are germane to the study of the mathematical process. Here, it is noteworthy that many educationalists have turned away from Anglophone thought to other traditions, for example, to Piaget, Vygotski and Bachelard, in their attempts to think about the nature of mathematics and the implications for the way it is taught. This inclination towards psychology is something to which I shall return below.

Phenomenology versus social history

So far I have not said much about the status of this *mathematical judgement.* The first point a sociologist of knowledge might make on hearing the term is that its transmission sounds a lot like indoctrination, the passing on of the values of the élite. Certainly, I am claiming that it has much to do with the production and selection of mathematical concepts and strategies, which taken together form a process governed by comparatively few people. The question is whether, having shifted from a dichotomous discovery-justification model, a rationality may be posited as underlying this process, and, if so, where this rationality resides. The range of options is wide. It runs from seeing the mathematical process largely as a power struggle between groups of mathematicians to control the purse strings and so the direction of research, to seeing it as a messy realisation of the unfolding of disembodied thought in some other realm. Between these extremes are to be found some intriguing options. A further point for those who take rationality to be at stake when choices are to be made in specific problem situations is whether such piecemeal rationality can be patched together to provide a long-term version.

Lakatos' writings have attracted appreciative glances, but comparatively little effort has been expended in building on his work. As I argue elsewhere (Corfield, 1998), his ideas require some radical recasting to have a hope of working. Let us then consider how the ideas of Lakatos' rival, Thomas Kuhn, are faring in the mathematical domain. We can find a dozen or so appraisals of the relevance of Kuhnian concepts to mathematics contained in the volume *Revolutions in Mathematics* (Gillies, 1992). Many interesting points are raised here, yet a failing of this book was a reluctance on the part of many of the contributors to consider whether the rationale for Kuhn's work in science was relevant to mathematics, revolutions being judged for the most part as descriptive devices. Of those who did go beyond the question of whether or not revolutions occur, two contrasting developments are due to Luciano Boi (1992) and Herbert Mehrtens (1976, 1992), a phenomenologist and a social historian.

Their papers do reveal considerable agreement when it comes to their suggestions as to the best devices to describe the mathematical process. Mehrtens gives a subtle descriptive framework of a disciplinary matrix in terms of beliefs in particular models, values, exemplars, concepts, and standard problems (pp. 31–5), and, as we can see from the first two of Boi's "most significant points" (pp. 206–7), they are in accord over this.

1. Mathematical thought undergoes conceptual transformations which embrace at the same time its methods, concepts, 'objects,' symbolism, and techniques; all these elements contribute to the development of mathematical knowledge ...

2. In mathematics there are traditions, problems, and traditions of problems which are just so many ways of giving a reply and a solution to these problems.

That this amount of agreement may be found is impressive. From here, however, they part company, their paths diverging towards a social history and a phenomenology. Boi clearly has no time for the sociological direction of Mehrtens[5]:

> No sociological or extra-mathematical reasons could help in understanding the nature of mathematical knowledge and the intrinsic reasons for its development and changes (Boi, 1992, p. 197).

Boi's preference is for the phenomenological construction of a 'genealogy of ideal forms':

> 3. There is an intrinsic hermeneutics of mathematics which, as we have seen, is characterized by the following two aspects: the inter-translation and auto-interpretation of theories.
>
> 4. Mathematical knowledge is indissociable from the establishment of a theory of concepts and a theory of structural analogies which bind the theories together.
>
> 5. There are certain fundamental intuitions which inspire and guide mathematical discovery and development; they are alinguistic and cannot be completely formalized. An example is the spatial continuum, which cannot be reduced to any axiomatic construction (Boi, 1992, p. 207).

Returning to Lakatos, each side could claim him as their own. He could be steered down the social history route by taking research programmes in the direction of social representations theory and concentrating on the practical realities of adhering to a programme, or down the phenomenological route by emphasising his Hegelian idealism and by focusing on what gets left out by formalization (cf. Boi's point 5). Unlike Boi, I have no objection to those who choose the former route, particularly if Sal Restivo's point is taken that, at times, the influence due to the external social milieu is much weaker than that due to the internal social milieu, that is, to factors operating within the relevant mathematical research community (Restivo, 1992, pp. 139–141). Indeed, not being a fan of a 'World 3' cast in stark opposition to the two other Worlds, I rather hope that the two approaches may interact creatively.

What appears to be at stake between the approaches is whether the stress should be placed on the development of mathematics as a necessary or as a contingent process. On the side of necessity, we often find, alongside the phenomenologists, mathematicians taking the time to reflect on their discipline as a whole. What they appear to be doing is arguing for a picture of mathematics which reflects the kinds of qualities they find attractive in particular pieces of mathematical theorising. In their attempts to promote the latter they will use the language of necessity: "This concept has been formulated independently in many fields. Its definition appeared natural to several mathematicians. It acts to unify these fields and allows the fruitful transfer of further concepts. There was an inevitability to its discovery."

In view of the complexity of the necessity/contingency question, it is all too easy to discuss aspects of it without meeting head on treatments arising from the opposite orientation. Hence the need to find specific issues to debate. A good place to commence a dialogue concerns what is perhaps the largest bone of contention between the opposing factions, namely, the difference between their respective presentations of modern mathematics as fragmented or as unified.

The unity of mathematics

The key question here is whether this dispute has a basis or whether it is simply a matter of a difference in emphasis. The sociologists and social historians tend to stress disunity: Eduard Glas (1989) outlines the incommensurability between the programmes of Monge and of Laplace and Lagrange; Mehrtens (1990) portrays the demise of the grand programme and the fragmentation into specific concrete problems; Restivo (1992) concentrates on the organisational structure of research over very long time scales where naturally variation will be easy to find.

Opponents will accuse them of a lack of feel for the structure of mathematics and its interacting branches (algebraic geometry, differential topology, analytic number theory, etc.) and a lack of feel for the flow of mathematics and its continuity. The Babylonians posed problems such as that of finding the side of a square field given that eleven times its area added to seven times its side amounts to $6\frac{1}{4}$. It generally comes as a surprise to find them presenting a solution which amounts to the calculation of

$$\{\sqrt{[(7/2)^2 + 11(6\tfrac{1}{4})]} - (7/2)\}/11 = 1/2,$$

demonstrating that quadratics were solved 4000 years ago in a very similar fashion to the way we teach our teenagers today. Then, why should the unpublished ideas found in Gauss' *Nachlass* contain so many of the major developments of the first half of the nineteenth century? Furthermore, much current research is hard to classify as belonging to a single branch, blending as it does ideas originating in diverse fields. To pluck out a single example, William Thurston's work in three-dimensional geometry has points of contact with a vast array of branches.

But by this search for continuity and unity, we might inaccurately be presenting today's régime as inevitable. Perhaps the sociologists are justifiably avoiding acting to legitimise the ways things are. Might Boi be seen in the same light as Kant, the legitimator of Euclidean geometry, someone who inhibited the development of geometry by arguing for its necessity?

At times mathematicians certainly do sound like legitimators. André Weil, for example, claims of one of his historical works that

[t]he main thesis will be the continuity of number theory for the last three hundred years and the fact that what we are doing now is in direct continuation of what has been done by the greatest number-theorists since Fermat started it all in the seventeenth century (Weil, 1974, p. 279).

Michael Atiyah goes so far as to make the claim that:

The difference [between modern and traditional mathematics] is more in the manner than the substance and if Newton or Gauss were to reappear in our midst only a short refresher course would be necessary before they could understand the problems being tackled by the present generation of mathematicians (Atiyah, 1977, p. 275).

Elsewhere Atiyah (1984, p. 307) aligns himself, in terms of mathematical taste, with Hermann Weyl. Is this self-aggrandisement or is there a genuine continuity of (successful) personal heuristic? Again, why did Emmy Noether admire Richard Dedekind so greatly?

Perputed continuity is used to legitimise, but so are claims about the cohesiveness of mathematics. Here, as elsewhere, we might learn some lessons from our colleagues studying the natural sciences. In an important contribution, Ian Hacking in his paper *The Disunities of the Sciences* (Hacking, 1996) distinguishes two senses of the term 'unity': unity as *singleness* and unity as *harmonious integration*. We might say that the various foundationalist enterprises and Bourbaki's monolithic *Elements of Mathematics* have aimed at the former, whereas mathematicians today are promoting the second by conveying their sense that when any of their kind worth his or her salt is working on a problem in one area, they cannot help but stumble across ideas central to another, as when Vaughan Jones' studies of von Neumann algebras allowed him to discover a new link polynomial in knot theory. This, it appears, is what Atiyah means when he talks of the 'unity of mathematics' (Atiyah, 1978). Saunders MacLane (1986), meanwhile, prefers the term 'connectivity.' We may then wonder whether it is just part of their attempt to impose their preferred mode of doing mathematics or whether it is in some sense *rational* to aim for harmonious integration?

MacLane also talks of the 'Protean' appearance of ideas, that is, the same idea cropping up in many guises in different branches and epochs. What is this notion of the 'same' idea? It suggests a stability of form, dissociated from the particular temporal instantiation – anathema to the sociologists – closer to the phenomenologists.

In the same paper, Hacking discusses *methodological* disunity while renewing his advocacy of the following classification of scientific styles proposed by the historian A.C. Crombie:

(a) postulation in the axiomatic mathematical sciences, (b) experimental exploration and measurement of complex detectable relations, (c) hypothetical modelling, (d) ordering of variety by comparison and taxonomy, (e) statistical analysis of populations, and (f) historical derivation of genetic development (Hacking, 1996, p. 65).

To these Hacking adds "laboratory science ... characterized by the construction of apparatus intended to isolate and purify existing phenomena and to create new ones" (ibid.). Hacking applauds Crombie's inclusion of (a) as "restoring mathematics to the sciences" (ibid.) after the logic positivists' separation, and extends the number of its styles to two by admitting the algorithmic style of Indian and Arabic mathematics. I am happy with this line of argument, especially if it prevents mathematics being seen as activity totally unlike no other, just so long as this restoration does not lead to its annexation by science studies. Indeed, mathematicians do more than postulate axioms and devise algorithms; it would hardly be figurative to say that mathematicians also engage in styles (b), (c) and (d). Jean-Pierre Marquis (1997) has even made a start on an analysis of the notion that some mathematical constructions are used as apparatus to explore the features of other mathematical entities, for instance, K-theory to probe topological spaces, although he seems inclined to see this as a metaphorical usage (Ibid., p. 254). The fact that mathematicians employ such a wide range of styles suggests we might learn from current studies of scientific argumentation. However, annexation might occur were we to ignore the differences between, say, classifying finite simple groups and tabulating their properties, and doing similarly for the chemical elements, fundamental particles, or zoological phyla.

In sum, we could expect the theme of 'the unity of mathematics' to provide an enlightening topic for debate.

The inevitability of the mathematical process and the applicability of mathematics

Returning to the social historians, their investigations certainly throw up philosophical questions. Consider Andrew Pickering's recent attempt to extend his 'mangle of practice' to mathematics via his analysis of William Hamilton's invention of quaternions (Pickering, 1995, chap. 4). After relating how Hamilton aimed to create a three dimensional number system by transporting certain features of the complex numbers to a new setting, Pickering arrives at the question whether "Hamilton's search for triplets was doomed in advance (or fated to arrive at quaternions)" (p. 141n). Two quick answers are possible here, the second favoured by Pickering:

(a) Yes, as Frobenius later showed there are just three associative division algebras over the reals: the reals, the complex numbers and the quaternions;

(b) No, the acceptance of the notion of an associative division algebra over the reals was a product of sequences of practices taking place after, and largely arising from, Hamilton's work. "Since this concept [associative division algebra] was not available to Hamilton, he cannot have been looking for new instances of it." (Ibid.)

Pickering's question would seem to bring us directly to what I would take to be philosophical questions. These are not so much about the realism of mathematical objects, but more along the lines of 'Are some concepts (here the notion of a division algebra) inevitable?' If along with Mehrtens we hold that "mathematics is about something that offers resistance to the mathematician and calls for treatment" (1976, p. 24), we see that Hamilton did meet resistance. What is its nature?

This episode may also be viewed as illustrating an answer to a historical question with sociological overtones and philosophical implications, along the lines of Peter Galison's (1987) 'How Experiments End': How mathematical experiments to find analogue constructions end. The physicists of the nineteenth century were not nearly so impressed by the quaternions as were Hamilton and Tait, and so hacked off their non-scalar part to form a three dimensional vector calculus. This raises the question as to whether the formation of these vectors could count as the proper end of the experiment Hamilton had begun, despite their lack of resemblance to the complex numbers: two multiplications, each allowing products of non-zero vectors to be zero. This argument could not be resolved purely by logical means, but is a matter of justification and criticism.

The quaternions have gone on to have other uses relevant to modern physics, for instance, via the isomorphisms between the unit quaternions, the three sphere and the the Lie group SU(2), double covers of the Lie group SO(3). This brings me to the question of applicability. MacLane (1986) has briefly sketched an explanation for the applicability of mathematics which begins with our activities in the world (measuring, counting, shaping) and our psychological capacities (generalising, abstracting, analogising). The mathematics thus generated then finds application because the world happens to have structure similarities at different levels (very small and very large). By contrast, David Ruelle pictures mathematics as formed by periodic injections from science coupled to a filling-in process, which is governed by the mind according to conceptual structures inherent in language, since "our 'logical thinking' is linked with visual intuition and tied to illogical 'natural languages'" (1988, p. 266).[6] Thus, for Ruelle, our mathematics is not so very inevitable. Different physical problems might have inspired us, and a different psychology would have led to a different filling-in process.

In that the products of this filling-in process have a human flavour to them, even with Ruelle's picture of scientific injections, we might still wonder at the later applicability of mathematics produced in this way. Atiyah, for instance, claims that he had no idea of the future applicability of the Atiyah-Singer index theorem, and that he even had to be taught the significance of Fourier analysis (1988, p. 2). This kind of comment subtly suggests the theorem's importance. If a piece of theory has unexpected applicational uses, any thought of its contingency vanishes.

In response one might allow an implicit 'intentionality' implanted by earlier workers which adheres to the mathematics as it is transformed. After all, the course on which your dancing partner sets off is devised for later reunion. Remember that in Atiyah's case part of the ground was prepared by those trained in the Soviet Union, where those aspiring to become pure mathematicians were given a thorough grounding in mathematical physics.

Psychology and phenomenology

Notice a common use being made of psychology to explain mathematical development. Is it an accident that mathematicians tend to prefer psychological to sociological explanations? Given our situation as biological and social beings, emphasis on the former would favour the *continuity of mathematics* picture. Sociologists might be happier with the idea that changing mathematical conceptions of space reflect and produce alterations in, say, the means of production, modes of transport and communication, styles of architecture and urban planning, and our lifespace in general, to include the construction of the body image and selfhood. Mathematicians, on the other hand, seem to prefer the idea that they are extracting principles employed by us in very everyday perceptual and motor activity. Just think of the complex visual and proprioceptive feat involved in walking towards a box and picking up a ball contained within it. Can we imagine a culture in which a healthy five-year-old could not do this? Yet researchers constructing machines to do similarly need to pack in a huge amount of differential geometry. Might we say that the cultural environment is what provides the requisite frames of mind for the extraction and development of the principles employed by the mind in its everyday activities?

The idea that mathematical understanding is rooted in our embodied condition has been touted by various cognitive linguists (Jackendoff, Lakoff, Johnson, Talmy), who would have mathematics seen as an elaboration of certain modes of thought, e.g. image schemas arising in us through our embodiment. For instance, Talmy (1988) argues that the experience of the world's resistance to our intended actions lies at the base of our modal and logical concepts. Here we are not so far from Poincaré and Einstein's conception of thinking with the body image or musculature, nor from the variety of phenomenology put forward by a philosopher such as Ferdinand Gonseth.[7]

Phenomenology is something of a blanket term, but generally commendable as an approach which requires an immersion in the subject matter with which it is dealing: to discuss the possible, you need to have engaged with the actual. However, as Gonseth argued, phenomenology can find itself ending up in some rather barren dead-ends, especially when it aspires to new forms of foundationalism and essentialism. A position developed by Joseph Margolis over the course of his trilogy, *The Persistence of Reality*, is that, foregoing foundationalism, we should move towards a naturalised phenemonology, which he sees as equivalent to a phenomenological naturalism.[8]

I can think of no better illustration of the potential for a naturalised phenomenology than the improbable convergence between a recent discovery in cognitive neuropsychology and Martin Heidegger's idea that prior to any propositional knowledge of objects, tools for example, the tool is *ready-at-hand* and will only emerge as *present-at-hand*, i.e. as a 'thing' with properties, when there is a disruption to the action of using the tool. Compare this to the theory of Melvyn Goodale (1995) which distinguishes two streams of visual processing, a ventral stream concerned with visually guided motor behaviour, and a dorsal stream for "identifying objects in the visual world and attaching meaning and significance to them" (p. 176). Subjects with a lesion interrupting the dorsal stream cannot recognise even the most familiar objects, yet are able "to grasp that object under visual control as accurately and as proficiently as people with normal vision" (p. 169). For instance, they don't recognise a pencil when they see one, yet they can still pick it up and draw with it; it does not exist as an identifiable object for them, yet it is ready-at-hand.[9]

To what extent, we may wonder, do mathematicians use theories instrumentally, without propositional awareness? Certainly, much of their activity goes on with tacit understanding, only surfacing occasionally when things go wrong, during foundational crises, for instance, or, more prosaically, when students fail to learn the game. Other elements of their understanding will only need to be made explicit for the purposes of conventional exposition. William Thurston (1994, p. 167) explains how he can skip sections of a paper when he knows "how it will go," since he would find it quicker to construct his own proof rather than wade through someone else's. This he likens to the way you don't bother reading the 16-page instruction booklet for a new toaster when you already know how to use one.

Possibly this kind of linkage between mathematical thinking and phenomenologically sensitive psychology is something new that those studying the mathematical process could bring to science studies. Psychologists themselves have made attempts to offer their services to science studies, but they have encountered a great deal of resistance due largely to the persistent stand-off between sociology and psychology.

Summary

To sum up where we have reached so far, researchers adopting a variety of approaches to the study of the mathematical process are currently addressing interrelated questions. But, whereas science studies provides a meeting ground for sociologists, historians and philosophers to compare notes, there is no satisfactory arena at present for analogous serious and sustained dialogue concerning mathematics to occur. Or, if you prefer, the mathematical wing of science studies is extraordinarily weak. Either way, a community of researchers must learn to engage with each other's points of view and through participation in dialogue adopt a more explicit sense of identity. Since the correlation between the viability of a discipline and its worthiness is not particularly high, strategic planning is not to be decried, even for so good a cause as ours.

2. The Future

Studying the mathematical process

What I am proposing is the creation of an analogue to science studies. For this we shall first need a name. *Mathematics studies* would be a natural choice, although such titles are now seen by many in a bad light. Another possibility is to wrestle the term *metamathematics* from the formalists. The term is, however, weighed down by a history of foundationalism. A third option would have us stay with *philosophy of mathematics*, while moving away from the standard fare offered up in the English-speaking world under this description.

Does this field not already exist? Certainly many studies could be included under its banner, but there is a lack of self-conscious identity that would help promote conferences and the creation of journals. Aspray and Kitcher (1988, p. 17) have dubbed as belonging to the 'Maverick Tradition' those philosophers of mathematics who pose such questions as

> How does mathematical knowledge grow? What is mathematical progress? What makes some mathematical ideas (or theories) better than others? What is mathematical explanation?

Three points may be made here:

(1) The term 'Maverick' I find unhelpful. Samuel Maverick was a Texan cattle-raiser, famous for not branding his calves. His name has been transformed into a word with two meanings: (a) dishonestly obtained animals, anything dishonestly obtained; (b) unbranded by its owner, not belonging to a school. The latter sense is intended by Aspray and Kitcher, rendering the term 'Maverick Tradition' as something of an oxymoron. At a time when co-operation is needed, it portrays isolated scholar-mavericks united only in their common disillusion with the traditional school's questions of the existence of mathematical objects and the truth of mathematical statements.

(2) The list of questions seems to require us to find some ahistorical criteria for judging growth and progress. We hardly need brand ourselves 'post-modernist' to wish to query this outlook. While at first glance this century would seem to present itself as one of enormous growth and progress, listen to mathematicians such as Vladimir Arnold, Réné Thom and Morris Kline for doubts on these scores.

(3) They position this approach solely within philosophy, which is fine just so long as this allows room for contributions to be made by historical writing with a point: Kline (1972) fears for the degeneracy of a mathematics that veers away from applications; Mehrtens (1990) claims that the image of a unity of mathematics put forward by modernist thinkers like Hilbert and Bourbaki has been punctured, leaving us with a welcome multiplicity of practices; McLarty (1990) proposes that the history of topos theory be written in a way to give its roots in algebraic geometry their due and so to challenge set theoretic reductionism.

But we can look beyond philosophy and history to take insights from sociology, psychology and anthropology. In this way, if you wanted to study the relevance of psychological theories of analogy for the development of mathematics, and later do a piece of interpretative history, you would not be accused of drifting out of your field in the form of a 'But-that's-not-philosophy' objection. This, after all, is asking for less of a reversal of the division of labour than did Karl Marx when he held out a vision of our being able to milk cows in the morning, and do philosophy in the afternoon.

Mathematics must be allowed to speak to us through its history and not act as a screen on which to project our philosophical or sociological fantasies. This is what Lakatos intended with the first half of his warning that "Philosophy without history is empty, history without philosophy is blind." Rather than its being empty, instead we see philosophers readily filling the void with their flights of fancy. Lakatos himself courted such dangers with his rational reconstructions of developments in World 3. Witness how the large jump in *Proofs and Refutations* from the nascent algebraic topology of the 1830s to Poincaré's *analysis situs* of the 1890s, made without a mention of the names of Riemann and Betti, induced him to rely too heavily on the method of proofs and refutations as the engine of mathematical progress.

There is a need for a deep engagement with mathematics; a large part of our research time should be spent learning and reading about mathematics and soaking ourselves in its culture. Above our door we might erect a plaque bearing the inscription 'Let no man (or women) enter who is ignorant of Riemann surfaces or Galois theory.' We shall certainly need to attract the interest of mathematicians to help us mathematically, but in a more practical sense we shall also need them to provide the resources to support our work by allowing us to teach, say, a module per year to their students. That, after all, is how most history and philosophy of science departments earn their keep. But to do this we had better not make what to them will appear as laughable mistakes. No doubt errors have crept into my papers, but I hope they including nothing comparable to claims, which may found in the literature, such as that the mean curvature at a point on a surface is an intrinsic property of that surface, or that a group is a product of a normal subgroup and the corresponding quotient group.[10] If we carry on in this vein, mathematicians will not wish to let us loose on their students and we shall remain unemployable.

What needs to be done?

Regarding the present weaknesses of our fledgling discipline, foremost is the pitiful state of our knowledge of twentieth century mainstream mathematics. It is a lamentable fact of modern academia that 'Star Trek: The Next Generation' has considerably more academic research time, conference time, and journal space devoted to it than has the recent development of the central branches of mathematics. The foundationalist legacy has caused far too much time to be spent on mathematical logic and set theory, at the expense of geometry in all its forms. This concern with foundations even carries over into a book such as *Revolutions in Mathematics* (Gillies, 1992) where the only two twentieth century studies concern logic and set theory.

In the history of mathematics, we find a complete imbalance as regards time, as though historians were wondering whether enough time has elapsed to discuss the Crimean War; twentieth century historical work is all too often done by mathematicians. Typically, when Kitcher and Lakatos turned their attention to the history of mathematics of the nineteenth century it was to the foundations of real analysis rather than the glorious nexus of concepts relating to Riemann surfaces (elliptic curves, holomorphic differentials, harmonic functions, topological genus, uniformisation).

The selection process occurs on different time scales and at different levels of commitment: the qualities sought by journal editors, what to teach and how, which problem to recommend to a doctoral student, whom to employ, which line of research to take, whom to be awarded a prize. There is, therefore, a danger similar to one treated by Kuhn in his paper *Mathematical versus Experimental Traditions in the Development of Physical Science* (Kuhn, 1976). Here Kuhn discusses how two 'noncommunicating' historiographical modes have emerged in the history of science: one which describes the development of a particular science, while the other describes the evolution of science taken as a whole. The danger run by the first approach is of artificially isolating a speciality based on present day understanding. For example, to see the antecedants of the modern theory of electricity as comprising a unified endeavour is historically inaccurate. The second approach, by taking on the whole of science, runs the risk of not delving sufficiently into the technicalities of any specific area thus lacking the resolution to explain the development of particular scientific theories. So both in their way underplay the subtle and complex dynamics of scientific evolution whose depiction requires the historian to occupy a 'difficult middle ground' where he or she may enter sufficiently into the technicalities within a field while being aware of the influence of other fields. In mathematics we find a similar situation obtains: *Proofs and Refutations* could be accused of artificially isolating algebraic topology and certain sociological works could be accused of bypassing technicalities.

Questions raised by the applicability of mathematics must feature prominently in our discipline, requiring very strong links to be made to science studies. The latter has made some important advances and we should feel free to try out some of their ideas in our patch, while maintaining the necessary distinctions. This has already been done for Lakatos, Kuhn and Laudan, but there should be insights to be taken from the writings of people like Peter Galison, Ian Hacking and Thomas Nickles. We should remember that applications although deepest in physics do occur elsewhere: coding, control theory, optimization, theoretical computing, etc.

I mentioned above my belief that some strands of psychology should prove important. Besides the suggestions of the cognitive linguists, there have been important debates concerning the nature of mental imagery. Might we not use these to interrogate the claims one hears to the effect that the education system is stacked in favour of the algebraic manipulators over the geometric intuitionists, and, even worse, that our imagery-producing faculties are atrophying due to neglect? A knowledge of styles of mathematical thinking would be useful in this context.

An enormous field is opened up, some of which should be accorded greater priority. In other words, we must worry about our own selection process. One strategy is to follow Lakatos' advice to study episodes in accordance with the basic value judgements of the élite (1978, p. 124). We might start by listening to mathematicians, so long as we do so with caution. Already there has appeared some important work arising from attempts to think through their comments. The treatment by Marquis (1997), which I mentioned above, arose from Alain Connes' claim that some mathematical objects act as machinery, allowing the finer examination of other objects. Leo Corry (1996) has investigated the claim that mathematics is the study of structure, demonstrating the problematic nature of specific assertions to this effect. The claim that mathematics is the 'science of analogy' has emerged as a *leitmotiv* in the writings of mathematicians throughout this century and also needs interrogating.

It should be possible for people from a wide range of backgrounds to contribute to our understanding of phenomena occurring within the mathematical process. Along with any of the questions raised in Part 1, we might pose more specific problems, such as evaluating the significance of the emergence through this century of algebraic topology as a vital core subject. We might hear responses begin as follows:

(1) The extension of the study of equations to several variables yielded solution spaces of dimension higher than two. Whereas the closed Riemann surfaces require a single topological invariant for their classification, viz. their genus, higher dimensional spaces require more subtle topological invariants. These are necessary for the better understanding of the differential, holomorphic and algebraic structures borne by manifolds.

(2) Its course was held up during the nineteenth century by an epistemological obstacle preventing the investigation of higher-dimensional spaces.

(3) Generally speaking, Poincaré's work was taken up slowly due to his unwillingness to form a school. While Birkhoff later took up his dynamical systems theory, three other American residents, Alexander, Veblen and Lefschetz, developed his algebraic topology strongly from 1920 onwards, allowing it to act as a symbol of the vigour of a new mathematical nation.

(4) It marks a new chapter in the centuries long elaboration of the relationship between the two classes of our sense modalities, the visual-kinaesthetic-proprioceptive and the linguistic-syntactic-propositional. The decomposition of spatial objects into subparts and a registering of their intersections is one of the strategies employed by the visual systems of the brain.

(5) Its recent applicability in physics may be related to its role in mediating between the continuous and the discrete, in the guise of the cohomology and homotopy groups of spaces, etc.

Along with the successes, we ought also to study the 'failures.' Of special interest are ideas neglected at the time but later revived: Grassman's exterior algebra and the exterior differential calculus, Kronecker's work on ring extensions and scheme theory, etc.

Studying the mathematical process as useful to mathematics itself

We might wonder what became of Lakatos' legacy. He himself abandoned mathematics for science and none of his students continued his first interest. Philip Kitcher (1983) later attempted to reconcile the two philosophies of mathematics, but the parts of his book were better received than the whole. Much could be learned from this failure to succeed in the impossible task of bridging the gap between a historically sensitive treatment of mathematical development and a philosophy embracing a correspondence theory of truth and knowledge as warranted true belief. Kitcher also later turned to science.

Even within the study of the scientific process, there are signs that the divisions of History and Philosophy of Science departments are moving apart. The historians view the philosophers' historiography as unsophisticated, and are turning towards sociological, postmodernist and phenomenological ideas for inspiration. Meanwhile, even historically-aware philosophers are using logical formalisms to encapsulate an ahistorical vision of the scientific process of model formation.

This divergence was made abundantly clear to me in a recent seminar in which a presentation was made of certain rival positions in the philosophy of mathematics. There were expressions of incredulity and murmurs of 'Scholasticism' from the historians when *modal fictionalism* was discussed. I cannot claim a profound grasp of this curious hybrid, but I take this position to be proposing that there exists some possible world in which mathematical objects exist as fictions. What is so disturbing about our own world that makes highly intelligent adults seek refuge in this style of thought? What is to stop them from going on to consider the existence in a possible world of a organism imagining a possible world in which mathematical objects exist (as fictions) ...? Are we so far from Star Trek studies (or is that *fictional modalism*)?

At least the Scholastics were overt about what they were doing, rather than engaging in a surreptious form of theology which reveals more of the inner phantasy life of the philosopher than anything about mathematics. Surely there is enough work to be done understanding the intricacies of this world:

> To suppose ... that we possess criteria of rationality which are independent of our understanding of the essentials of the scientific process is to open the door to cloud-cuckoo land (Kuhn, 1962/70, p. 264).

As I have remarked earlier, I do not believe we have much a grasp on the essentials of the *mathematical* process.

We may pose ourselves the question as to whether ahistorical philosophies of mathematics have anything to offer us. Consider a fictionalist's attempt to talk about algebraic topology's century. Given their sole concern with applied mathematics, the best we could hope for is that they know that this branch has spent much of its life as a provider for other branches. But what do we gain from the notion that mathematicians have devised some useful fictions, useful to other mathematicians in understanding their fictions, which have helped other mathematicians to produce fictions useful in physics?[11]

One slender indication of there being a potential for dialogue comes in a portion of Hintikka's recent book, *The Principles of Mathematics Revisited* (Hintikka, 1996),[12] where despite dismissing interest in mathematical practice as a 'fetish' (p. 102), he does note that mathematicians do not spend the majority of their time proving theorems, but engage in what amounts to concept-stretching. This I can agree with: the neglect of the act of mathematical description in favour of a near exclusive concern with mathematical proof has been a large mistake.

What should be a cause for concern for philosophers of mathematics is the fact that mathematicians show not the slightest degree of interest in their work. An article (Jaffe and Quinn, 1993), recently appearing in a mathematics journal, argues for the necessity of rigorous proof in the mathematics now being devised for the latest physical theories and expresses the fear that the ungrounded speculation of those romping through a field and attaching their name to all the big conjectured results may stunt the growth of promising new ideas, and hence lead to the loss of potentially important new theory. Their claims may thus be seen as the flip-side of Lakatos' concern with premature axiomatisation. The next issue of the journal included a flurry of replies to what was considered a provocative article. What was immediately noticeable was that not only did no philosopher enter into this discussion, none was even mentioned.[13] Here, mathematicians are carrying on their own debates about production and selection practices,when there is surely a place for critical input. Indeed, the only non-mathematician to contribute to the debate was Jeremy Gray, a historian, who corrected some erroneous historical claims made by Jaffe and Quinn.

So, we do find spontaneous outbursts of debate among the mathematical community concerning ideas of best mathematical practice, but their claims often suffer from a lack of precision. Ours has the potential to be not only an intellectually challenging field, revealing something of the human condition, it could also usefully allow a more richly structured setting for debates within mathematics and provide the resources for issues to be explored from a variety of angles. The findings could then be fed back into the process via modifications in educational methods.

Conclusion

These are critical times for the future course of a discipline which takes as its object the mathematical process. The numbers of researchers and their level of commitment have reached a level at which such a discipline can now become viable, but this will only happen in the event of sufficient accommodation on all sides to allow for meaningful dialogue. In this paper, I hope to have given some indication of the way our efforts might be better co-ordinated. Even those from opposite ends of the spectrum, radical sociologists of knowledge and Husserlian phenomenologists, should be able to find a sufficiently specific issue in which to engage each other in constructive debate.

A third of a century after *Proofs and Refutations* first appeared, the emergence of this discipline is by now long overdue. In view of his interest in the historical flow of mathematics, it can be no accident that we find Lakatos to be one of the few philosophers about whom mathematicians speak well. I have the impression, but no proof, that his ideas had a real effect on textbook writers, encouraging them to include more of the informal side of the mathematics they were discussing, a little of its history and its motivation. Ultimately, the source of the fascination held by *Proofs and Refutations* rests in the fact that it presented a brief, albeit distorted, snatch of a real mathematical argument. What we need do today is carry out a wide-ranging review of other snatches of mathematical debate. By doing so we shall prepare ourselves to be able to provide support for present-day mathematical conversations, and, who knows, we might even be able to join in.

NOTES

1. This relates to the point made by Feferman (1981) that Lakatos' concept-stretching may be represented by the predicate calculus. Lakatos himself attempted to use logical symbolism to represent the method of proofs and refutations in *The Method of Analysis-Synthesis* (1978, p. 70). On the limitations of Lakatos' philosophy of mathematics see Corfield (1997) and Kvasz (this volume).
2. Witness the fact that Duhem discussed *bons sens* and *finesse* in the field of physics, contrasting them to *géométrie* as an automatic mode of thought to which the mathematician is restricted (cf. Crowe, 1990).
3. David Ruelle, a mathematician as receptive as any to issues raised by physics, claims that "... it is clear that 20th century mathematics now largely produces its own problems, and that physics is only a secondary source of inspiration" (1988, p. 260).
4. Without a doubt, we should allow include all fields of engineering here. The computer scientist Nicholas Metropolis (1993) argues forcefully for more work to be done on the history of computer science. He provokatively maintains that:

   > The relationship between computer science and mathematics scarcely resembles that which exists between physics and mathematics. The latter may best be described as an unsuccessful marriage, with no possibility of divorce. Physicists internalize whatever mathematics they require, and eventually claim priority for whatever mathematical theory they become acquainted with. Mathematicians see to it that every physical theory, sooner or later, is freed from all shackles of reality and liberated to fly in the thin air of pure reason.
   >
   > Computer science, in a very different mode, turns to mathematics in much the same way that engineering always has. It borrows freely from already-existing mathematics, developed for altogether different purposes or, more likely, for no purpose at all. Computer scientists raid the coffers of mathematical logic, probability, statistics, the theory of algorithms, and even geometry. Far from resenting the raid, each of the disciplines is buoyed by the incursion (Metropolis, 1993, p. 123).

5. Cf. in particular section 2.3 of Mehrtens (1976) and his (1992).
6. Contrast with Boi's point 5 above.
7. Jean Petitot (1992) argues for the relevance of Gonseth's philosophical outlook to cognitive science, and in particular to the work of the cognitive linguists I mentioned above. These latter I have discussed in my unpublished PhD thesis.
8. See in particular Margolis (1989). He also advocates some deconstructionist input to stop the relationship becoming too cosy.
9. Cf. Heidegger in *Being and Time*: "The less we just stare at the hammer-thing, and the more we seize hold of it and use it, the more primordial does our relationship to it become, and the more unveiledly is it encountered as that which it is – as equipment" (1962, p. 98).
10. This last claim is true in a trivial sense, but is badly wrong in what I take to be its intended sense.
11. In recent years homotopy theory has been applied directly to account for faults in nematic crystals, so presumably requiring Fieldian nominalisation, the smuggling of structure from one formalism to the next.
12. Cf. I have review this work in Corfield (1998b)
13. While Lakatos was not mentioned in the debate, it is hard to think of any other philosopher with any comparable claim to inclusion. Cf. Stöltzner (this volume) for a detailed interrogation of this debate.

BIBLIOGRAPHY

Aspray W. and Kitcher P. (eds.) (1988) *History and Philosophy of Modern Mathematics.* University of Minnesota Press.

Atiyah M. (1977) *Trends in Pure Mathematics.* Printed in Atiyah (1987), pp. 263–276.

Atiyah M. (1978) 'The Unity of Mathematics', *Bull. Lond. Math. Soc.* 10, pp. 69–76. Reprinted in Atiyah (1987), pp. 277–286.

Atiyah M. (1984) 'An Interview with Michael Atiyah', *Mathematical Intelligencer* 6, No. 1. Reprinted in Atiyah (1987), pp. 297–307.

Atiyah M. (1987) *Collected Works, Vol. 1: Early Papers, General Papers.* Oxford University Press.

Atiyah M. (1988) *Collected Works, Vol. 4: Index Theorem 2.* Oxford University Press.

Atiyah M. (1995) 'Quantum Theory and Geometry', *Journal of Mathematical Physics* 36 (11), pp. 6069–72.

Boi L. (1992) *The 'Revolution' in the Geometrical Vision of Space in the Nineteenth Century, and the Hermeneutical Epistemology of Mathematics.* Chapter 10 of Gillies (1992).

Corfield D. (1997) 'Assaying Lakatos's Philosophy of Mathematics', *Studies in History and Philosophy of Science* 28(1), pp. 99–121.

Corfield D. (1998a) 'Beyond the Methodology of Mathematics Research Programmes', *Philosophia Mathematica*, pp. 272–301.

Corfield D. (1998b) 'Come the Revolution ...', a critical notice of Hintikka (1996), *Philosophical Books*, 39(3), pp. 150–156.

Corry, L. (1996) *Modern Algebra and the Rise of Mathematical Structures.* Birkhaüser Verlag, Boston.

Crowe M. (1990) 'Duhem and The History and Philosophy of Mathematics', *Synthese* 83, pp. 431–47.

Feferman S. (1981) *The Logic of Mathematical Discovery vs. the Logical Structure of Mathematics.* PSA 1978, Vol. 2, pp. 309–327.

Fukuyama F. (1989) 'The End of History?', *The National Interest*, Summer 1989, pp. 3–18.

Galison P. (1987) *How Experiments End.* University of Chicago Press.

Gilbert, N. and Porter, T. (1994) *Knots and Surfaces.* Oxford University Press.

Gillies D. (ed.) (1992) *Revolutions in Mathematics.* Oxford University Press.

Glas E. (1989) 'Testing the Philosophy of Mathematics in the History of Mathematics. (a) The Socio-cognitive Process of Conceptual Change, *Studies in History and Philosophy of Science* 20, pp. 115–131.

Goodale M. (1995) 'The Cortical Organisation of Visual Perception' in: *Visual Cognition, An Invitation to Cognitive Science*, 2nd edn. Vol. 2, S. Kosslyn and D. Osherson (eds.), MIT Press, Cambridge MA.

Hacking I. (1996) 'The Disunities of the Sciences', in: *The Disunity of Science: Boundaries, Contexts and Power*, P. Galison and D. Stump (eds.). Stanford University Press.

Heidegger M. (1962) *Being and Time.* Translated by J. Macquarrie and E. Robinson, Harper and Row, New York.

Hintikka J. (1996) *The Principles of Mathematics Revisited.* Cambridge University Press.

Jaffe A. and Quinn F. (1993) '"Theoretical Mathematics": Towards a Cultural Synthesis of Mathematics and Theoretical Physics', *Bull. Am. Math. Soc.* 29, pp. 1–13.

Kitcher P. (1983) *The Nature of Mathematical Knowledge.* Oxford University Press.

Kline M. (1972) *Mathematical Thought from Ancient to Modern Times.* Oxford University Press.

Kuhn T. (1962/70) *The Structure of Scientific Revolutions.* University of Chicago Press.

Kuhn T. (1976) 'Mathematical Versus Experimental Traditions', *The Journal of Interdisciplinary History* 7. Reprinted as Chapter 3 of *The Essential Tension.* University of Chicago Press, 1977.

Kvasz L. (1998) 'History of Geometry and the Development of the Form of Language'. *Synthese* 116, pp. 141–186.

Lakatos I. (1976) *Proofs and Refutations, The Logic of Mathematical Discovery*, J. Worrall and E. Zahar (eds.), Cambridge University Press.

Lakatos I. (1978) *Mathematics, Science and Epistemology, Philosophical Papers*, Vol. 2, J. Worrall and G. Currie (eds.), Cambridge University Press.

MacLane S. (1986) *Mathematics: Form and Function.* Springer Verlag, New York.

Margolis J. (1989) *The Persistence of Reality III: Texts without Referents.* Basil Blackwell, Oxford.

Marquis J.-P. (1997) 'Abstract Mathematical Tools and Machines for Mathematics', *Philosophia Mathematica* (3), Vol. 5, pp. 250–272.

McLarty C. (1990) 'The Uses and Abuses of the History of Topos Theory', *British Journal for the Philosophy of Science* 41, pp. 351–375.
Mehrtens H. (1976) 'T.S. Kuhn's Theories and Mathematics', *Historia Mathematica* 3, pp. 297–320. Reprinted as Chapter 2 of Gillies (1992).
Mehrtens H. (1990) *Moderne-Sprache-Mathematik*. Suhrkamp.
Mehrtens H. (1992) *Appendix (1992): Revolutions Reconsidered.* Chapter 3 of Gillies (1992).
Metropolis N. (1993) 'The Age of Computing: A Personal Memoir', in: A New Era in Computing, N. Metropolis and G.-C. Rota (eds.), MIT Press, pp. 119–130.
Petitot J. (1992) 'Schématisme, sciences cognitives et physiques mathématiques: hommage à Ferdinand Gonseth', in: *Espace et Horizon de Réalité*, M. Panza and J.-C. Pont (eds.), Masson, Paris.
Pickering A. (1995) *The Mangle of Practice: Time, Agency and Science.* The University of Chicago Press.
Polya G. (1954) *Mathematics and Plausible Reasoning, Vol.1: Induction and Analogy in Mathematics*, Princeton Universtiy Press.
Restivo, S. (1992) *Mathematics in Society and History: Sociological Inquiries.* Kluwer, Dordrecht.
Ruelle D. (1988) 'Is our Mathematics Natural? The Case of Equilibrium Statistics', *Bull. Am.. Math. Soc.* 19, pp. 259–268.
Sierpinska, A. (1994) *Understanding in Mathematics.* The Falmer Press, London.
Talmy L. (1985) 'Force Dynamics in Language and Cognition', *Cognitive Science* 12, pp. 49–100.
Thurston W. (1994) 'On Proof and Progress in Mathematics', *Bull. Am. Math. Soc.* 30, No. 2.
Weil A. (1974) *Two Lectures On Number Theory, Past and Present.* Collected Papers, Vol. 3, pp. 279–302.

Department of History and Philosophy of Science
University of Cambridge
Free School Lane
Cambridge, CB2 3RH
UK

Thomas Mormann

Towards an Evolutionary Account of Conceptual Change in Mathematics: *Proofs and Refutations* and the Axiomatic Variation of Concepts

1. Introduction

Although Lakatos' *Proofs and Refutations* (henceforth PR) is one of the classics of twentieth century philosophy of mathematics, few attempts have been made to build on its ideas.[1] Rather, quite often PR has become an object of respectful reference and detailed exegetical efforts. Of course, Lakatos is credited for having taught philosophers that philosophy of mathematics has to take into account the history and practice of mathematics. This orientation towards the history and practice of the discipline certainly represents progress beyond the ahistorical traditions of logicism, formalism, platonism, and intuitionism. It should help to get rid of the most pernicious vice with which the philosophy of mathematics is plagued; to wit, elementarism, which considers elementary arithmetics of natural numbers or Euclidean geometry as typical for all of mathematics. I think, however, that there is more to learn from Lakatos than just a general orientation towards the history and practice of real-life mathematics. As I want to show in this paper, these insights concern the role of inventing and varying mathematical concepts. More precisely, I want to use Lakatos' ideas of "concept-formation" and "concept-stretching" to sketch an evolutionary theory of mathematical knowledge, which takes axiomatic variation of concepts as the fundamental driving force of the ongoing evolution of mathematics (PR, 83ff.).[2]

Although Lakatos considers concept-stretching only for informal, and not for axiomatized mathematics, I contend that it is not restricted to this realm. Indeed, the role of conceptual variation (which I take to be a "neutral" term for Lakatosian concept-formation, concept-stretching and related conceptual activities) becomes even more important in axiomatic mathematics.

In the following, I propose to generalize Lakatos' approach of concept-stretching in such a way that the history and practice of modern axiomatic mathematics can be conceived of as an evolution of mathematical concepts based on axiomatic concept variation. In other words, I contend that a philosophical understanding of mathematics may be provided by perceiving it as a *conceptual evolution*. This implies that, in some sense, mathematical

G. Kampis, L. Kvasz and M. Stöltzner (eds.), Appraising Lakatos: Mathematics, Methodology and the Man, 139–156.

concepts are to be considered as more basic than mathematical theorems.[3] At first glance, this contention may be considered to be absurd. According to time-honored philosophical wisdom, mathematical knowledge is deductive knowledge, deposited in proven theorems. We know what deductive proofs and theorems are – at least, much better than what is meant by mathematical concepts. Thus, it may appear implausible, to say the least, to base a philosophical understanding of mathematical knowledge on the elusive notion of a mathematical concept. Hence, I hasten to add that the concept-based approach outlined in this paper does not dismiss the notions of theorem and proof at all. Rather, these provide the necessary environment for mathematical concepts, which never occur in isolation, but always in the context of mathematical theorems. The importance of concepts depends on the importance of the theorems in which they occur.

Mathematical concepts may be conceived of as entities, on which one may conduct trials or experiments. The trials or experiments are the attempts at proving theorems, in which the concept in question occupies a central position. Then, a successful experiment, i.e. a valid proof, may be considered as an outcome that is positive for the concept in question. To take Lakatos' example: the "Euler-experiment" on some variant of the polyhedron concept is the attempt to prove a version of the Euler conjecture for it. Usually, in such an experiment, several conceptual variants compete against each other. If several attempts at proving a conjecture fail for a conceptual variant, or only lead to an uninteresting contrived version of the original conjecture, that variant may be considered to be a failure. In this way, conjectures provide objective challenges, or survival tasks, for a group of competing variants of concepts, which have to cope with these. This does not mean, however, that there is a one-way relationship between concepts and theorems, in the sense that concepts have to strictly respond to the rules defined by the theorems. Often, new conceptual variants give rise to new theorems which cannot be proved or even formulated for the older conceptual variants. These new theorems alter the "theorematical" environment, and change the terms of competition for future conceptual variants. Thereby, even the assessment of past conceptual achievements may change.

The main thesis of this paper claims that this competition among conceptual variants may be described within the framework of an evolutionary theory, which conceptualizes the evolution of mathematical knowledge as a selection process of conceptual variants taking place in a "world" whose challenges are determinded by varying theorematical environments.

The outline of this paper is as follows: in Section 2, we take as our point of departure Lakatos' notion of concept-stretching. It may be considered as an informal precursor to the more powerful notion of axiomatic concept variation, which is characteristic for the conceptual practice of modern mathematics. We consider Hamilton's construction of the quaternions, as an early historical example of an axiomatic concept variation. In Section 3, the basic principles of an elementary theory of evolution are proposed. In

Section 4, these principles are applied to the evolution of mathematical concepts. A sketch of the evolution of some conceptual variants of the integral is offered as an example of an evolutionary account in Section 5. In Section 6, we conclude with some remarks on the specific features of the evolution of mathematical concepts, which are in need of further investigation.

2. CONCEPTUAL CHANGE: CONCEPT-STRETCHING AND AXIOMATIC VARIATION

Traditionally, mathematical concepts have been conceived of as citizens of some Platonic realm. This conception rendered them immutable and eternal entities. Lakatos was one of the first philosophers not to conceive of mathematical concepts in this manner. For him, mathematical concepts might undergo substantial alterations and changes – to wit, concept-stretching, or the reversal, concept-contracting. This is to say that mathematical concepts do not exhibit the immutability and eternity that platonistic accounts used to ascribe to them; mathematical concepts vary. Lakatos develops a sort of dialectic of proof and counterexamples, which is intended to shed some light on the process of the variation of concepts. All this has been the subject of detailed discussion and interpretation (cf. Koetsier, 1991; Larvor, 1997), which need not be recounted here. It is sufficient to note that a large part of mathematical practice may be described as an activity dealing with mathematical concepts, be it concept-stretching in the sense of Lakatos, or generalizing, specializing, adapting, or any sort of concept variation whatsoever. As a result of this complex concept-trafficking, typical mathematical concepts come in a multiplicity of variants. For instance, Lakatos gives five or six different variants of the key-concept "polyhedron" (cf. PR, pp. 14–22) which, according to his notorious footnotes, have actually occurred in the mathematical history of that concept.[4]

Lakatos restricts the practice of concept-stretching to the realm of informal, non-axiomatized mathematics. As soon the subject becomes formal axiomatized mathematics, he assumes that concept-stretching comes to an end, which is, according to him, a reason why modern axiomatic mathematics lacks much of the freedom and the excitement of earlier times. I think that this is quite wrong. Rather, as Corfield has emphasized:

> [t]he appropriate use of rigorous definition and axiomatization has not acted as a hobble on the creativity of mathematicians, but rather an invaluable tool in the forging of new mathematical theories and the extension of old ones. (Corfield, 1997, 100).

For established modern mathematical theories, as Corfield points out, the concepts involved are usually defined by systems of axioms. And it is precisely these systems of axioms which enable us to vary mathematical concepts in a very effective and conscious way not available in an informally formulated "intuitive" mathematics. Although I fully endorse Corfield's enthusiasm for the creative role of the axiomatic method, I have reservations with regard to his usage of the notion of "forging." When one "forges" an object – e.g. an instrument or, metaphorically, a theory – one assumes to have total control over the product one is forging. One knows what will be the result of one's doing. This, however, is most often not the case with mathematical concepts. The inventors of new mathematical concepts often don't know what exactly will be the outcome, and how their "brainchildren" will score in the competition. Of course, they are not completely in the dark. After all, they have designed their new concepts to serve some definitely useful purposes. But often, they have to modify their original intentions, and reshape them in the presence of resistance and obstacles, which they had not foreseen and are unable to overcome.

At first glance, such resistance in the conceptual realm of mathematics may look strange, since philosophers have repeatedly emphasized the freedom of mathematics: anything that is not inconsistent might be possible. Furthermore, the axiomatic method has been hailed as having fully brought to the fore this unique freedom of mathematics. A closer look at the practice of mathematics, however, reveals that the alleged freedom of mathematics is far from being a total one. Quite to the contrary, the conceptual practice of mathematics is severely restricted by the rules we have to follow in our conceptual practices. As every mathematician knows, the conceptual practice of mathematics tends to soon become difficult. The workings of the mind may lead the mind itself into difficult and sometimes unsolvable problems.[5] Thus, a primary task of the philosophy of mathematics is to explain this somewhat paradoxical fact.

Recently, Pickering contended that this restricted freedom of conceptual practices, such as mathematics, may be conceived of as a special case of what he calls the "mangle of practice" in scientific practices in general (Pickering, 1995). According to him, the essential ingredient for getting a non-distorted, "non-abstract" picture of science (mathematics included) is to conceive of this endeavour as being a temporarily-situated practice, in which a human and a non-human agency interact. The contours of both agencies are not fixed from the outset, but emerge in the temporality of practice. That is to say, they are never to be considered as fully determined and complete. Seen from the perspective of the human role, this means that the goals and intentions of the scientist are continuously under revision. Scientists do not simply fix their goals ("prove Euler's conjecture") once and for all and stick to them, come what may. Rather, they find themselves engaged in the "mangle of practice" in which their intentions undergo a permanent goal-revising dialectic of resistance and accommodation in such a way that the outcome only remotely resembles the original plannings (cf. Pickering, 1995, 22–23).[6]

As an example of the mangle of conceptual practice, which is characteristic for the restricted freedom of the axiomatic variation of concepts, let us consider the episode of Hamilton's invention of the quaternions (cf. Pickering, 1995). Originally, Hamilton intended to generalize the geometric representation of complex numbers in the 2-dimensional Euclidean plane, in order to get a three-dimensional analogue:

> $\sqrt{-1}$ [being] perpendicular to 1, it is natural to conceive that there may be another sort of $\sqrt{-1}$, perpendicular to the plane itself. Let this new imaginary be j; so that $j^2 = -1$, as well as $i^2 = -1$. A point in space may suggest the triplet $x + ij + jz$ (Hamilton, 1967, 103).

At this stage, Hamilton aimed at a rather straight-forward variation of complex numbers: instead of couples (x, y) he considered the triples (x, y, z), and attempted to generalize the arithmetics of complex numbers to their envisaged 3-dimensional counterpart.

The conceptual environment, with which this variant had to cope, was determined by the arithmetic laws of complex numbers. These may be succinctly described by the following formulas:

$$(x, y) + (x^*, y^*) = (x + x^*, y + y^*), (x, y) \bullet (x^*, y^*) = (xx^* - yy^*, xy^* + x^*y) \quad (2.1)$$

That is to say, Hamilton was looking for 3-dimensional analogues of these laws. Of course, the extension of the additive laws (i.e. associativity, commutativity etc.) posed no problems. The multiplicative law turned out to be the crucial test the envisaged triple numbers had to pass.[7] He tried for more than fifteen years to find one, but never succeeded. Today, it is well-known that such a 3-dimensional extension of real numbers cannot exist.[8] Rather, as Hamilton discovered in 1843, one must take a further step towards a 4-dimensional generalization of the real numbers leading to quadruplets (x, y, z, w), as a viable variant of the complex numbers (x, y). This variant of complex numbers (or, more remotely, of real numbers) are called the hypercomplex numbers, or quaternions H. They do *not* satisfy commutative multiplication. Rather, if $1, i, j$ and k denote the standard basis of H, then the multiplicative behaviour of quaternions is determined by the following law:

$$i^2 = j^2 = k^2 = i \bullet j \bullet k = -1 \quad (2.2)$$

As can easily be seen, (2.2) leads to an *anti*-commutative multiplication for i, j, and k, that is to say, $i \bullet j = -j \bullet i = k$ etc. Thus, Hamilton's new generalized complex numbers H, obtained by an axiomatic concept-variation concerning the multiplicative law, do not satisfy the original arithmetic laws required for "honest" numbers. Hamilton invented a variant of complex numbers with rather different features than those he originally had looked for.

It should be noted that the axiomatic variation of concepts, as exhibited in Hamilton's invention, shows a feature which defies a strictly Lakatosian account: namely, it does not leave intact the theorems or laws which were originally considered to be the touchstones of its respectability. In PR, the concepts of the polyhedron may be stretched or even torn; Euler's conjecture, however, never underwent any variation.[9] As he started, Hamilton seems to have considered the arithmetic laws of the complex numbers in a similar vein as Lakatos understood the Euler Theorem, to wit, as a sort of experimentum crucis. It must be considered as a sign of Hamilton's genius that he was prepared to revise the touchstone laws, thereby opening the gate to a "new world," the riches of which he himself only dimly perceived. In modern terms, this "new world" may be described as the theory of division algebras and related topics. In quite the same manner, most of the newly generated variants of mathematical concepts do not satisfy *all* theorems which hold for their ancestors. The dialectics of resistance and accommodations, which steers the axiomatic variation of concepts, are more flexible than Lakatos' Popperian account which is fixated on the original "bold conjecture."

Hamilton believed that the quaternions would open a new era for mathematics. This turned out to be an exaggeration. Modern mathematicians see Hamilton's quaternions as not being so important, in and of themselves. After all, H has not turned out to be a particularly interesting variant of a division algebra. Rather, the importance of Hamilton's construction resides in the methodological fact that it may be conceived of as one of the first conscious and deliberate cases of an axiomatic variation of a mathematical concept or, in Lakatos' terms, as a bold stretching of the number concept.

In the following, I would like to expose the thesis that the dialectical method of axiomatic concept variation may be considered as being the fundamental driving force behind the evolution of modern mathematical knowledge. It is the axiomatic method, which enables us to vary mathematical concepts in a systematic and effective way. If axiomatic variation of mathematical concepts is the basic mechanism underlying the progress of mathematical knowledge, then it is of utmost importance that a philosophical understanding of mathematics describe this mechanism as precisely as possible.

Obviously, the process of axiomatic variation is not arbitrary. Not anything goes. On the other hand, the process of axiomatic variation need not yield unique results, which are unanimously considered to be optimal. The evolution of mathematical concepts is not aimed at predetermined ends. Rather, axiomatic variation leads to a plurality of variants, which compete among each other. Even the leading mathematicians of their eras often err in predicting which conceptual variation in the long run will score best.[10] Most of the conceptual variants produced by mathematicians over the course of time are short-lived. They are confined to a rather small temporal interval, and do not find a fruitful continuation. Most are forgotten; i.e. in evolutionary terms, they die without offspring. Sucessful variants, on the other hand, spread. They are taken up and developed by other authors, leading to important new theorems and conceptual connections with other fields of mathematical knowledge. As it seems, many mathematicians are seduced by this fact into occupying a Whiggish perspective, which only remembers the winner's approach and forgets the other conceptual variants of the past, which may someday be revived.[11]

3. TOWARDS A SELECTION THEORY OF MATHEMATICAL CONCEPTS

If one conceives of the history of mathematics as being the evolution of mathematical concepts, one has to explain what is to be understood by evolution, and what its mechanisms are. One should not expect that any existing theory of evolution – say, Darwin's or Lamarck's account of biological evolution, or some of their modern successors – be directly applicable to the evolution of mathematical concepts. Rather, the evolution of mathematical concepts may have its own idiosyncratic features not occurring in any other evolution. For instance, time seems to play a different role in the evolution of mathematical concepts than in, say, the evolution of biological species. To some extent, the past is always present in mathematics. Hence, past forms of mathematical creatures may be revived, in a way which has no counterpart in biological evolution. The best that a general theory of evolution can do, is to offer a framework or format for a evolutionary understanding of mathematics. Without discussing the pros and cons of other evolutionary accounts, I propose that an elementary evolutionary theory be characterized by the following four principles:

3.1. The principle of variation

At any stage in the history of a given species, there will be variation among the members of this species. Different variants will have different properties.

3.2. The principle of the struggle for existence

At any stage in the history of a species, more variants come into existence than can survive to reproduce.

3.3. The principle of variation of fitness

At any stage in the history of a species, some of its variants have characteristics that better dispose them to survive and reproduce.

3.4. The principle of inheritance

Most properties of a variant are inherited by its descendants.

From these principles, one may infer that the history of a species will show the modification of that species in the direction of those characteristics which better dispose their bearers to survive and reproduce.

The principles (3.1)–(3.4) may all be traced back to Darwin's evolutionary theory. A theory obeying them, however, is not restricted to the biological realm, provided we interpret the key terms such as survival, reproduction and so on, in an appropriate, non-biological way. Hence, these principles may be considered as a general and abstract scheme for various sorts of evolutionary theories.

If we want to cast the development of mathematical knowledge into the framework of an evolutionary theory, the first question we have to answer is what should be considered as the units of selection, i.e. what entities we take as species, which evolve over time and show one or more different variations. This is not a trivial question, and it may not have a unique answer.[12] In this paper, I propose as units of evolution "medium-sized" mathematical concepts, such as *function, continuity, differentiability, integral, measure, topology* etc. This choice is not the only possible one. For instance, it may be convenient to chose larger conceptual entities such as categories, in the sense of Mac Lane and Eilenberg, as units of selection.[13] In the following, however, I would like to concentrate on a more pedestrian account, taking concepts of the above-mentioned type as units of selection. The main reason for this procedure is simply that otherwise one would have to introduce a lot of conceptual machinery, which would exceed the scope of this paper. In fact, I believe that categories would even better fit the evolutionary approach presented here.

Although the principles of natural selection have no empirical content, the presuppositions of their applicability are empirical. These presuppositions form the empirical core of the evolutionary theory of mathematics. In the next section, these general principles are reformulated, in order to give them empirical content for the evolution of mathematical concepts.

4. PRINCIPLES OF THE EVOLUTION OF MATHEMATICAL CONCEPTS

In order to get an evolutionary theory of mathematical knowledge started, one has to reformulate the above-mentioned four principles in such a way that they apply to mathematical concepts as units of selection. To do this, one needs to answer the following questions:

1. What is the "world," in which the evolution takes place?

2. What are the competing units of selection, which are the protagonists of the ongoing "struggle for existence" in that world?

In the preceding sections, these questions have already been answered as follows: The protagonists of the evolutionary selection process are variants of mathematical concepts. The world in which the competition takes place is a world of proof-problems, and conceptual growth in depth and coherence. An elementary example of such an evolution is Lakatos' account of Euler's conjecture E: For the various variants p_i of the polyhedron concept p, E defines the crucial task p_i has to cope with. A variant p_1 will supersede a variant p_2 iff p_1 gets along better with E than p_2. Admittedly, the world Lakatos designs for the species of polyhedra concepts is a rather primitive and poorly structured world, and one may well suspect that the worlds in which mathematical concepts live are actually much more complicated. This is indeed the case, as can be seen from the case of the integral concept. There is not simply one theorematical demand which defines a concept's fitness ("enable a proof of Euler's conjecture") – there are many.

Before we go into this, let us give a brief sketch of how the four general principles of evolution given in the last section may be applied to the case of the evolution of mathematical concepts.

4.1. Principle of variation

At any stage in the history of mathematics, there will be several different variants of a mathematical concept; these different variants, belonging to the same species, will have different properties but have to cope with the same, or at least similar, challenges.

Lakatos' case studies in PR may be read as evolutionary stories about competing variants of mathematical concepts such as polyhedron, continuity, continuum, etc. The actual history of mathematics is more complicated. In particular, many more competitors attend the race. This will be discussed in some detail in the next section, using the concept of the integral. I would like to contend that these examples exhibit the generic case – in other words, that they may be considered as species, which occur in several different variants differing in "mathematical fitness." Conceptual species having only a single and unique variant throughout the history of mathematics are extremely rare, if, in fact, there are any at all.[14]

In order to apply the Principle of the Struggle for Existence, one has to explain what it could mean when one variant of a mathematical concept "survives" and another one "becomes extinct." I propose the following interpretation:

4.2. The principle of competition

At any stage in the history of a mathematics, there is competition between variants of concepts. This competition results in a ranking of the competing variants. The high ranking of a conceptual variant means that it becomes acknowledged as a useful and fruitful device in proving important theorems, and/or giving rise to new and promising theoretical contexts. A variant loses the competition when it falls into oblivion and does not find interesting applications.

Today, for instance, Cauchy's variant of the integral may be considered as uninteresting and irrelevant to the further evolution of mathematics. It has been outscored extensionally by Riemann's and Lebesgue's variants, and moreover, it is to be considered a conceptual deadlock. Metaphorically, this may be expressed by saying that Cauchy's variant has died out. In the course of time, most conceptual variants suffer this fate. In contrast to biological extinction, however, mathematical concepts have, at least in principle, the chance of being resuscitated. For instance, the Riemann integral has undergone some new conceptual developments and has found new applications (cf. Edalat, 1995; Henstock, 1991). The Principle of Competition opens the door for sociological considerations. The environment of mathematical concepts, i.e. the theorems that are considered to be important, and the conceptual relationships that are considered to be fruitful, is a man-made world. It may depend on changing tastes and preferences. For instance, if the social pressure on the mathematical community is such that only those mathematical achievements are highly esteemed which can be put directly to practical use, then the resulting criteria for successful concepts may be quite different from those which reign if other "more theoretical" preferences are expressed. In some sense, then, the evolution of mathematical concepts is relative; it cannot be understood as the unfolding of some absolute conceptual platonic realm.[15]

4.3. The principle of variation of fitness

The key-term of this principle ("fitness") is not easily determined in a non-tautological way (as is the case for its biological counterpart). For mathematical concepts, one, and only one, measure of their degree of fitness is the generality of theorems they make it possible to prove. A concept for which a more general theorem can be proved is, ceteris paribus, better than one having a more restricted domain of applicability.

Some not overly convincing examples may be found in PR. A better example is provided by comparing some variants of the concept of the integral to be discussed in the next section. Beyond extensional generality, there are other criteria of fitness such as "fruitfulness" or "conceptual depth." Admittedly, these notions are difficult to nail down. The dependence of an evolutionary account of mathematical knowledge on such "difficult" notions should be considered, however, as virtue, rendering it a demanding and non-trivial acount. This shows, once again, that the standard elementarist accounts do not even come close to addressing the real problems of a philosophy of mathematics.

4.4. The principle of inheritance

This principle ensures a sort of historical continuity for mathematical concepts. The new variants have to resemble their ancestors, in order to meet at least some of the environmental challenges defined by the latter. It is a matter of the history of mathematics to elucidate the lineage of mathematical concepts, from their humble beginnings to their sophisticated most recent descendents.

The principle of inheritance ensures some sort of cumulative growth for mathematical knowledge, at least to a certain extent. Its cumulative character, however, should not be overestimated. It may well be the case that the modern descendants of mathematical concepts have lost some of the problem-solving capacities of their ancestors, since these problems have become obsolete in modern contexts.

In the next section, some episodes in the evolution of the concept of the integral are sketched according to these evolutionary principles.

5. ON THE EVOLUTION OF THE CONCEPT OF THE INTEGRAL

As an example of the evolution of a mathematical concept, let us now consider the evolution of the concept of the integral. It goes without saying, that only some aspects of the evolution of this concept can be dealt with, and even this only in a quite superficial way.[16]

In common usage, one speaks of "the" integral, as though this concept were a uniquely determined concept once and for all. This is, of course, wrong. In high-school mathematics, the office of the integral is usually held by Cauchy's variant of the integral. Most mathematicians probably understand as an "integral" Lebesgue's variant of this concept. Specialists in integration theory know a lot other variants, e.g. Denjoy's, Burkill's, Perron's, Taylor's, etc. (cf. Henstock, 1991). As exemplified by the recent work of Henstock, Edalat and others, the evolution of the concept of the integral can in no way be described as having reached its terminal point (cf. Henstock, 1991; Edalat, 1997). Rather, a huge and complex family of subspecies has evolved, whose members have adapted to a variety of quite different contexts. It goes without saying, that I do not intend to describe this complex evolution in its barest outlines. Rather, the rest of this section is devoted to a brief sketch of some aspects of this evolution, starting with Cauchy's variant and ending with Lebesgue's.

The relationships between the different variants of the concept of the integral are quite intricate. In particular, it cannot be described as a story of linear progress. Of course, it is known that the Lebesgue integral is more general than the Riemann integral (to say nothing of Cauchy's variant), in the sense that any function that is Riemann integrable on an interval $[a, b]$, is Lebesgue integrable, but the converse does not hold. For most purposes,

Lebesgue's variant turned out to be the more versatile and fruitful one, clearly outscoring the more traditional variants.[17]

Let us have a closer look how this happened, by starting with Cauchy's variant of the integral. He defined the definite integral of a uniformly continuous[18] function f as follows: Let f(x) be a uniformly continuous function defined on the closed interval [a, b]. Consider a partition of this interval

$$a = x_0 < x_1 < \ldots < x_{n-1} < x_n = b \text{ and the sum } S(P, f, a, b) = \sum (x_i - x_{i-1}) f(x_{i-1})$$

Using the uniform continuity of f, he was able to show that for finer and finer partitions, the values of S(P, f, a, b) approached a limit. This limit he defined as the definite integral of f on [a, b], to be denoted by $\int(f, a, b)$. Now define a function F by $F(x) = \int(f, a, x)$. For F, Cauchy established the following results, known as the Fundamental Theorem of Calculus:

(1) F is a primitive function for f, i.e. $F' = f$.

(2) If F is a primitive function for f, all primitive functions for f are of the form $G = F + c$, where c is a constant. That is, if G is a function with a continuous derivative G′, then $\int(G', a, x) = G(x) - G(a)$.

(3) If G is a function such that $G'(x) = o$ for all x in [a, b] then $G(x) = c$ for some constant c.

Since the times of Leibniz and Newton, any viable variant of the integral concept had to satisfy some version of the Fundamental Theorem. In the course of time, especially in the 19th century, however, the authority of the Fundamental Theorem as the crucial proof-experiment for an admissible variant of the integral concept diminished.[19] Other challenges for the integral concept came to the fore; for instance, various types of convergence theorems, which described the successful interplay of integral and various limit operations. One reason for this change in environmental challenges was the gradual expansion of the domain of functions, for which the Fundamental Theorem was assumed to hold.

Fulfilling the demands of the Fundamental Theorem became more and more difficult, as the domain of functions for which the Fundamental Theorem was designed to be applicable got more and more expanded. In other words, the features of the environment of the concept of the integral gradually changed. As is well known, one of the main protagonists of this change was Dirichlet, who for the first time introduced the concept of a truly "general" function, to be conceived of as a general rule correlating the elements of the domain and the range of the function. Dirichlet was the first mathematician to call attention to the existence of functions that are discontinuous on an infinite set of points in a finite interval, as well as to the problem of extending the concept of the integral to functions of this nature. For some time, the best response to this new challenge was Riemann's variant of the integral: He interpreted the meaning of integrability in terms of the Cauchy sums of the function f(x) to be integrated: Under what conditions on the function f(x) do these sums approach a unique limit as the norm of the corresponding partitions approaches zero? In order to answer this question, Riemann formulated two equivalent conditions. Then, he showed that his definition was extensionally superior to Cauchy's (and Dirichlet's) by constructing functions that were discontinuous on a dense set of real numbers, but nevertheless integrable according to his integral – but not in the sense of Dirichlet and Cauchy. As is pointed out by Hawkins (1970, 18), the advantage of Riemann's variant of the integral is not exhausted by its larger domain of applicability. In hindsight, one may see in Riemann's integrability conditions the germs of the concepts of Jordan measurability and outer content, which were bound to play a crucial role in modern integration theory, as it was to be developed much later by Lebesgue. Conceptual progress cannot be characterized by extensional considerations alone. If the competition among the various integral concepts could be fully described in this extensional way, it would essentially follow the lines of Lakatos' discussion of the concept-stretching of the concept of the polyhedron carried out in PR (cf. also Fine, 1981). Actually, things are much more complex. Conceptual mathematical progress is not measured exclusively by the expansion of the domain of intended application. Rather, the development of new conceptual relationships plays an essential role. The superiority of a new conceptual variant is also – and one may even say, mainly – demonstrated by the new lines of research and by the new perspectives it opens. Thus, the evolutionary value of a new variant of a mathematical concept is also assessed by the new conceptual relationships it opens up. In the case of the evolution of the integral, this is most impressively demonstrated by that of Lebesgue. As is well-known, Lebesgue's variant is extensionally superior to Riemann's. This advantage, however, is not the most important one. Lebesgue's new variant of the integral turned out to be a real mutation, so to speak. It opened up a completely new field and defined a new framework for the concept of the integral. For the first time, it elucidated the intimate relationship between the concepts of integral, measure, and topology.

From Lebesgue (and Borel) onwards, any variant of the concept of the integral had to consider relations with the conceptual field of measure theory and topology, which were quite alien to Cauchy's concept of the integral and only dimly visible for Riemannan's variant. Hence, the transition from the Riemann variant to the Lebesgue variant cannot be fully understood simply as an extensional expansion. Rather, it also involves an intensional, that is to say, conceptual progress insofar as the Lebesgue variant is able to build up connections and interrelations to many more other conceptual fields.[20]

6. CONCLUDING REMARKS

As is shown by the ongoing evolution of mathematical knowledge, the "essence" of mathematics is not to be found in some hidden nature of mathematical objects, but rather is revealed in the specifics of the ongoing evolutionary process of mathematical knowledge or, as Neo-Kantians such as Cohen and Natorp would have formulated it some one hundred years ago: the fact of mathematics (science) is a fact of becoming ("Werdefaktum").[21] If this is true, then it is of the utmost importance for a philosophical understanding of mathematics to include some insight into the mechanisms of this ongoing process. The best device for an insightful description of this process is provided by some sort of evolutionary theory.

In order to apply such a theory, a necessary prerequisite is to understand that mathematical concepts are not eternal, unalterable "Platonic" entities. Rather, they tend to occur in several different variants, which develop over the course of history. These variants differ in fitness, in terms of the wealth and value of the theorems which can be proved for them. Variation may occur on different levels: for instance, on the level of concepts, on the level of propositions, or on the level of theories. This ubiquity of variation renders modern mathematical knowledge a definitely un-Platonic epistemic endeavor, making variation (on which level whatsoever) the very essence of mathematics. Variation is not a 'noise' that may be ignored or filtered out by clever philosophical devices, in order to grasp the "essence" of mathematics. I consider it to be the great achievement of Lakatos' philosophy of mathematics to have paved the way for this insight.

The acceptance of a general evolutionary framework for a philosophical understanding of mathematics may not be too instructive. The evolution of mathematical knowledge exhibits some very special features, which it does not share with other evolutions. Hence, it is of crucial importance to explicate these special features:

(1) The environment in which the evolution takes place may change in a rather quick and drastic manner. This change may be brought about, at least partially, by the conceptual innovations themselves.

(2) Temporarily unsuccessful variants may survive in a "frozen state," and may be resuscitated.[22] Hence, compared with other evolutions, the struggle of existence in the conceptual evolution of mathematics is less cruel. Although there is strong competition among conceptual variants, this need not lead to the total extinction of the inferior variants. They may survive in some niche or, perhaps, may be resuscitated after the terms of competition have changed.[23]

(3) Conceptual evolution in mathematics is a group affair. That is to say, not a single concept but, rather, a group of connected concepts evolves

simultaneously. For instance, the evolution of the concepts of the integral is intimately related to the evolution of the concept of measure.

These features show that an evolutionary conceptualization cannot rely on some already-told evolutionary stories which occurred elsewhere. Rather, philosophers and historians of mathematics have yet to invent the language for telling such an evolutionary story in the case of mathematics. The first steps for this endeavour have been taken by Lakatos, with his theory of concept-formation and concept-stretching.

NOTES

1. There are, of course, exceptions. For instance, Fine (1981) attempts to embed Lakatos' account of concept-formation and concept-stretching, as pursued in PR, in a more general theory of concept-refinement, not restricted only to mathematical concepts, but embracing the conceptual development of all kinds of scientific concepts. As it seems, his sketchy outline has not been taken up by other authors.
2. Of course, an evolutionary theory of (scientific) concepts is not new at all. For instance, Toulmin offers an evolutionary description of the historical development of theories, and, more generally, intellectual disciplines, in the framework of a population model (Toulmin, 1972, Part II). Toulmin's account, however, is deliberately general. He is interested in the general mechanisms of the evolution of ideas, not in the specific mechanisms of a special discipline, such as mathematics. The point I want to make in the following is that the mechanism of axiomatic concept variation is characteristic of the development of mathematical concepts, and deserves a closer study than has hitherto been carried out.
3. The key-role of concepts for the development of mathematical knowledge is clearly acknowledged in PR: As Lakatos asserts in a footnote: "The key to Euler's result (i.e. the Euler conjecture, T.M.) was just the invention of the concepts of *vertex* and *edge* ...." In contrast, Descartes did not find the key for the formulation of Euler's conjecture, since he "did not make a conscious revolutionary change to the concepts of 0-dimensional vertices, 1-dimensional egdes and 2-dimensional faces as a necessary and sufficient basis for the full characterization of polyhedra" (PR, p. 6, footnote 1).
4. As we shall see in Section 5, this number is not at all an exaggeration. For instance, the number of variants of the concept of the integral can only be guessed. Even if we restrict counting to "modern" variants, one may distinguish more than one hundred from Cauchy onwards (cf. Hawkins, 1970; Henstock, 1991).
5. This has turned out not to be a coincidence. One may interpret the incompleteness theorems as stating that the *non-trivial* workings of the mind *necessarily* lead to unsolvable problems.
6. Pickering takes PR as a case of the mangle in *conceptual* practice: "The exhibition of novel counterexamples to specific proofs of the theorem counts, in my terminology, as the emergences of resistances, and Lakatos describes very nicely the revision of proof precedures as open-ended accommodation to such resistances, with interactive stabilization amounting to the reconciliation of such precedures to given counterexamples." (Pickering 1995, 119, Footnote) To a large extent, this may be correct. But I think that Pickering overestimates the open-endedness of Lakatos' account. After all, even in the second chapter of PR, in which he deals with Poincaré's proof of Euler's theorem, he concentrates on that theorem and nothing else. The reason may be that he is still in the grip of the Popperian "statement view" of (mathematical and scientific) theories, according to which single statements occupy the centre stage of scientific theorizing.
7. This was clearly recognized by Hamilton. After he had found the 4-dimensional quaternions, he wrote: "But I considered it essential to try whether my equations were consistent with the law of moduli (i.e. the law of multiplication), ... without which consistence being verified, I should have regarded the whole speculation as a failure" (Hamilton, 1967, 108).

8. Hamilton could have already proved this (cf. Ebbinghaus et al., 1983, 131).
9. This claim contradicts Koetsier's reconstruction of Lakatos' argument. According to Koetsier, PR "shows us a quite dramatic successions of versions of Euler's theorem for polyhedra, V – E + F = 2, ... , each version being temporarily the best until it is refuted and has to be replaced by its successor." (Koetsier, 1991, 24) All the "different" versions of Euler's theorem have the following form: "For all X-polyhedra V – E + F = 2" Here, "X-polyhedra" is a variable for "convex polyhedra," "ordinary polyhedra" (whatever this means), etc. I propose to take these different versions of Euler's theorem as one and the same theorem concerning different variants of the concept of polyhedron. This may be seem to be only a verbal quibble, but I believe it is not. As we shall see in Section 5, in contrast to the Lakatosian account, the theorems considered as the challenges with which conceptual variants have to cope, change in a much more dramatic way than Lakatos would have been prepared to admit.
10. For instance, even an expert in the field, such as Fubini, was rather skeptical whether Lebesgue's new variant of the integral was useful or not.
11. How dangerous such an attitude may be is nicely described in Rota's "On Reading Collected Papers" (Rota, 1997, 248): "In our ahistorical age, it is a fool's paradise to believe that the reading (even the casual scanning) of collected papers is more likely to enrich our knowledge than a feverish plunge into the latest periodicals ... The masters had a variety of ideas that are missing in later acounts of their work." According to Rota, in order to solve an outstanding problem, a mathematician has to seriously study the history of that problem.
12. Even for biological theories this, question has not found an unanimously accepted answer. One may consider, as Darwin did, individuals as the unit of selection. Other fashionable answers are to consider species – or even suborganismal parts – as units of selection, i.e. as agents of evolution.
13. Meanwhile, mathematicians are well aware of the selectional character of their research. This holds true, in particular, for fields such as lattice theory or topology. There, some quite strict "fitness criteria" are generally accepted: for instance, a "good" topological category should possess some nice closure properties; i.e. topological spaces should have products and coproducts, an exponential law for topological spaces should hold, etc.
14. A candidate for such an exceptional species may be the concept of natural numbers. It seems hard to find different variants of it. A closer look, however, reveals that even for this apparently undifferentiated species, several different variants exist. In axiomatic mathematics, a natural number concept is defined by some version of Peano axioms. Hence, the different variants of the Peano axioms may be considered as defining different concceptual variants of *natural numbers.* Moreover, if one takes into account that in different toposes, different natural number objects exist which may enjoy strikingly different structural properties, the principle of variation seems to be secured even for the *natural number* concept. This basic pluralistic character of mathematical concepts can only be ignored by a narrow elementarist perspective on mathematics, which takes into consideration only a small fragment of mathematical knowledge.
15. This opens up the possibility of "alternative mathematics" (cf. Bloor, 1991). Actually, Bloor quite heavily relies on Lakatos' account of mathematical knowledge in order to show that even in mathematical knowledge, there may be some sort of "cultural relativity."
16. For a detailed presentation of the origins and the developments of Lebesgue's theory of integration, see (Hawkins, 1970); for the modern post-Lebesgue evolution, see (Henstock, 1991). Judging by a very rough count of the different definitions occurring in Hawkins and Henstock, several dozens of different variants of the integral have been proposed in the last one hundred and fifty years. This shows that Lakatos' five or six conceptual variants of the polyhedron as exhibited in PR form a rather poor collection.
17. This outcome wasn't that clear in the first quarter of the century (cf. Hahn, 1914, 1915). Several attempts were made to modify Riemann's definition of the integral so that it would accommodate unbounded integrands and encompass conditionally convergent integrals. This took some time. Lebesgue's account changed the terms of competition in such a way that the superiority of Lebesgue's variant was generally acknowledged, at least for some time and some domain of applications.

18. Of course, the concept of a uniformly continuous function was explicitly defined only after Cauchy. However, Cauchy often implicitly assumed a function to be uniformly continuous when he allegedly dealt with continuous functions.
19. For instance, discoveries of Dini, Volterra and others show that the second part of Fundamental Theorem (5.1(2)) is not generally valid for Riemann's variant of the integral, since there are functions with bounded, non-integrable derivatives. Lebesgue's variant in some sense restores this part of the Fundamental Theorem, at lest for bounded derivatives. On the other hand, for Lebesgue's variant, the first part of the Fundamental Theorem (5.1(1)) does not hold in its full generality. It only holds in a version "almost everywhere" (cf. Hawkins, 1970, 137).
20. This points to the fact that the evolution of mathematical concepts should not be thought of as the evolution of a single concept, but rather as a group-evolution of many concepts simultaneously.
21. See Cohen's *Logik der reinen Erkenntnis* (Cohen, 1977 (1902)). The process character of scientific knowledge is even starker emphasized in Natorp's *Die logischen Grundlagen der exakten Wissenschaften* (Leipzig, 1910). There, one finds in the fourth section the following radical thesis: "The fact of science must be understood as a 'fieri' (becoming). The 'fieri' alone is the fact: all being which science attempts to fix, has to be resolved in the great current of becoming. Only this becoming can finally be said to be." This paper is, of course, not the appropriate place to deal with the Neo-Kantian philosophy of science and mathematics in any greater depth. I would like to call attention to the fact that the Neo-Kantian approach emphasized the conceptual unity between science and mathematics, which has gone largely unnoticed in the traditional positvist philosophy of science.
22. Cf. footnote 15.
23. Lakatos took this fact into consideration later in his "Methodology of Scientific Research Programmes."

REFERENCES

Bloor, D., 1991 (1976) *Knowledge and Social Imagery*, Chicago, The University of Chicago Press.

Brandon, R.N. (1996) *Concepts and Methods in Evolutionary Biology*, Cambridge, Cambridge University Press.

Cohen, H. 1902 (1977) *Die Logik der reinen Erkenntnis* (Hermann Cohen Werke 6), Hildesheim and New York, Olms.

Corfield, D. (1997) Assaying Lakatos's Philosophy of Mathematics, *Studies in History and Philosophy of Science* 28, 99–121.

Edalat, A. (1997) Domains for Computation in Mathematics, Physics and Exact Real Arithmetics, *The Bulletin of Symbolic Logic* 3, 401–452.

Elstrodt, J. (1996) *Maß- und Integrationstheorie*, Berlin, Springer.

Ernest, P. (1997) The Legacy of Lakatos: Reconceptualizing the Philosophy of Mathematics, *Philosophia Mathematica* 5, 116–134.

Fine, A. (1981) Conceptual Change in Mathematics and Science: Lakatos' Stretching Refined, Proceedings of the 1978 Biennial Meeting of the Philosophy of Science Association (PSA), vol. 2, 328–341.

Hamilton, W.R. (1967) *The Mathematical Papers of Sir William Rowan Hamilton*, Vol. 3, Algebra, Cambridge, Cambridge University Press.

Hahn, Hans, 1997 (1914) Über Annäherung an Lebesguesche Integrale durch Riemannsche Summen, in *Gesammelte Abhandlungen* 3, ed. by L. Schmetterer and K. Sigmund, Wien, Springer, 17–48.

Hahn, Hans, 1997 (1915) Über eine Verallgemeinerung der Riemannschen Integralfunktion, in *Gesammelte Abhandlungen* 3, ed. by L. Schmetterer and K. Sigmund, Wien, Springer, 49–64.

Hawkins, Th. (1970) *Lebesgue's Theory of Integration, Its Origins and Developments*, Madison, The University of Wisconsin Press.

Henstock, R. (1991) *The General Theory of Integration*, Oxford, Clarendon Press.

Koetsier, T. (1991) *Lakatos' Philosophy of Mathematics. A Historical Approach*, Amsterdam, North-Holland.

Lakatos, I. (1976) *Proofs and Refutations: The Logic of Mathematical Discovery*, Cambridge, Cambridge University Press.

Lakatos, I. (1978) *Philosophical Papers*, Volumes 1 and 2, Cambridge, Cambridge University Press.

Larvor, B.P. (1997) Lakatos as Historian of Mathematics, *Philosophia Mathematica* (3) 5, 42–64.

Natorp, P. (1910) *Die logischen Grundlagen der exakten Wissenschaften*, Leipzig und Berlin, Teubner.

Pickering, A. (1995) *The Mangle of Practice, Time, Agency, and Science*, Chicago, The University of Chicago Press.

Rota, G.C. (1997) I*ndiscrete Thoughts*, ed. by F. Palombi, Boston, Birkhäuser.

Toulmin, St. (1972) *Human Understanding* I. General Introduction and Part I: The Collective Use and Evolution of Concepts, Princeton, Princeton University Press.

University of the Basque Country (EHU/UPV)
Department of Logic and Philosophy of Science
PO Box 1249
E-20080 Donostia-San Sebastián
Spain

MICHAEL STÖLTZNER

WHAT LAKATOS COULD TEACH THE MATHEMATICAL PHYSICIST*

Since the outset of modern physical science, its close relationship to mathematics has distinguished it from other natural sciences. Once a science has reached a certain stage of formalisation, it always applies mathematics, but no science finds itself more frequently in the vanguard of genuinely mathematical developments than physics. The cases of general relativity and quantum mechanics are paradigmatic for what Eugene P. Wigner once called "the unreasonable effectiveness of mathematics in the natural sciences."[1] In both cases, the basic equations of the theory made major use of mathematical structures discovered shortly before: Non-Euclidean geometry and matrix-valued differential equations. Since the early 1980s, however, the rapidly growing and extremely fertile interaction between string theory and differential geometry has inverted the direction of inspiration. Theoretical physicists constantly reveal genuinely mathematical results and propound beautiful mathematical structures undreamed of before. In this sense, Arthur Jaffe, a renowned mathematical physicist and editor of the discipline's leading journal, has recently proposed to modify Wigner's famous dictum into the "unreasonable effectiveness of theoretical physics in mathematics."[2]

In 1990, Edward Witten was awarded the Fields Medal for his contributions to geometry, which were largely stimulated by string theory. This immediately prompted many private debates among mathematicians and mathematical physicists as to how Witten's mostly conjectural and intuitive results ought to be appraised. No mathematician doubted that Witten's representation of the Jones invariants of knots using Chern-Simons field theory was a major breakthrough that connected two whole hitherto unrelated subjects. Moreover, considerable parts of the results were quickly proven by leading pure geometers. At this point, two closely related questions arose: Should mathematical claims from string theory be accepted, without further ado, as genuine mathematics? Who is to be credited for the mathematical result, if it is eventually proven?

In their 1993 article "'Theoretical Mathematics': Toward a Cultural Synthesis of Mathematics and Theoretical Physics" published in the *Bulletin of the American Mathematical Society*, Arthur Jaffe and Frank Quinn proposed a set of prescriptions for the interaction between mathematicians and theoretical physicists that should foster mathematicians' receptivity of

G. Kampis, L. Kvasz and M. Stöltzner (eds.),
Appraising Lakatos: Mathematics, Methodology and the Man, 157–187.

ideas from physics by safeguarding mathematical rigour against *uncontrolled* speculation. "[M]athematical authors [including theoretical physicists writing mathematical papers] should make a choice: either they provide complete proofs, or they should agree that their work is incomplete [conjectural] and the essential credit will be shared. Referees and editors should enforce this distinction, and it should be included in the education of students."[3] Evidently, the Jaffe-Quinn prescriptions presuppose that, even in a rapidly growing field, it can be decided, in each case, whether a proof is rigorous or merely conjectural, heuristic. But, as I will argue, this decidability thesis does not hold without qualifications. If my claim holds true, then prescriptions for a fruitful 'cultural synthesis' between mathematics and theoretical physics are less obvious, as well.

The Jaffe-Quinn theses provoked a rather broad controversy that is documented in no less than 16 responses by leading mathematicians and the authors' summary rejoinder in the next volume of the *Bulletin*.[4] The matter was taken up in the May 1997 issue of *Synthese*. I will discuss the positions taken on both occasions together, because they share three main tenets of the debate: First, descriptive and normative, internalist and externalist (societal) perspectives are neatly interwoven. Second, the debate is almost exclusively led by mathematicians who implicitly or explicitly (in the *Synthese* papers) employ philosophical arguments. Third, throughout the papers no foundational problem is profoundly at stake – except for Hintikka's which, on the other hand, does not mention the Jaffe-Quinn debate at all.

Among all contributors, only Gian-Carlo Rota approaches the debate from a genuinely philosophical viewpoint. Most contributors, however, begin with mathematical, historical and sociological considerations that also include the prospective effects of new ways of communication and publishing, which might threaten mathematical rigour.[5] Yet a philosophical analysis of the debate and a classification of the various opinions is still lacking.

For this purpose, I will rely upon Imre Lakatos' philosophy of mathematics, inasmuch as his intention of devising a rational methodology of mathematical progress is precisely the issue of the Jaffe-Quinn debate.[6] To render Lakatos' methodology – which was shaped mainly by 17th and 18th century mathematics – applicable to "one of the most refreshing events in the mathematics of the 20th century,"[7] modifications are called for that are quite in accordance with the recent literature on Lakatos' philosophy of mathematics. First, one has to reconstruct the application of the methodology of scientific research programmes (MSRP) to mathematics – an application which Lakatos had in mind at the time of his untimely death. (At any rate, one issue, which is essential to mathematical physics, is missing in Lakatos' account: how to appraise the contact between two different research programmes – a mathematical and a physical one – that do not mutually compete on an identical factual basis?) Second, it should be recognised "that the appropriate use of rigorous definition and axiomatization has not acted as a hobble on the creativity of mathematicians, but rather an invaluable tool in the forging of new mathematical theories and the extension of old ones."[8] Despite his repeated criticisms of Euclideanism and his insistence on the principal fallibility of all mathematics, Lakatos, later in his life, expressed more sympathy toward logic than he had before. Third, although he already criticised "Popper's overkill of inductivism,"[9] Lakatos' apology of induction is limited to the very general level and cannot be used to justify the effectiveness of theoretical physics in mathematics. But to my mind, conceding a certain dose of inductive reasoning from physical models to mathematical concepts fits very well to the quasi-empirical character of mathematics Lakatos has argued for.

This paper is organised in three parts: After portraying the various stances taken in the Jaffe-Quinn debate, I outline the lessons to be drawn for it from Lakatos' published works. Sections 1–1.4 and 2.1–2.4 respectively are thematically connected. The third section examines which modifications of Lakatos' account are in place to enable a fruitful assessment of present mathematical physics.

1. THE JAFFE-QUINN THESIS: APPRAISING 'THEORETICAL MATHEMATICS'

Jaffe and Quinn inquire as to whether speculative mathematics, as it occurs in current interactions between physics and mathematics, is at all dangerous, and as to how those unpleasant side effects known from the history of informal mathematics may be avoided. Hence, their analysis contains a descriptive and a normative part. The latter consists in the quoted prescription to always openly distinguish between speculative mathematics (conjectures) and rigorously proven theorems. Can this norm be complied to without exception? The answer depends on whether there exists an unambiguous and universal criterion to neatly separate rigorous proof from informal and conjectural arguments.

1.1. The ontology of 'theoretical mathematics' and the role of proof

Here the Jaffe-Quinn thesis starts: "Modern mathematics is nearly characterized by the use of rigorous proofs. This practice, the result of literally thousands of years of refinement, has brought to mathematics a clarity and reliability unmatched by any other science."[10] The authors distinguish two stages of the mathematical research process: "First, intuitive insights are developed, conjectures are made, and speculative outlines of justifications are suggested. Then the conjectures and speculations are corrected; they are made reliable by proving them. We use the term *theoretical* mathematics for the speculative and intuitive work; we refer to the proof-oriented phase as *rigorous* mathematics."[11] This terminology expresses a functional analogy between rigorous proof and experimental physics. Both correct, refine and validate the claims of their theoretical counterparts.

Proofs serve two main purposes. First, they "provide a way to ensure the reliability of mathematical claims"[12] – justification, in Lakatosian terms. "Second, the act of finding a proof often yields, *as a byproduct*, new insights and unexpected new data."[13] Hence, what Lakatos calls heuristics is, in Jaffe and Quinn's view, only subordinated to the justificatory role of proof. 'Theoretical Mathematics' is built on an asymmetry of proof and conjecture. Posing the latter does not necessarily involve proof.

Despite this asymmetry, 'Theoretical Mathematics' tacitly requires a 'quasi-empirical' ontology – quite in accordance with Lakatos' views. "For if we don't assume that mathematical speculations are about 'reality' then the analogy with physics is greatly weakened – and there is no reason to suggest that a speculative mathematical argument is a theory of anything, any more than a poem or novel is 'theoretical'"[14] writes Morris W. Hirsch in his Response. Saunders MacLane calls for a structuralist austerity doctrine: "If a result has not yet been given valid proof, it isn't yet mathematics: we should strive to make it such."[15] To his mind, all other assertions root in the misconception of "set theory as THE foundation of mathematics, and so sometimes [philosophers of mathematics] eagerly spread the gospel that mathematics is the study of an ideal realm of sets – set theoretic platonism."[16]

Notwithstanding his Incompleteness Theorems, Kurt Gödel openly subscribed to the latter view and preserved Hilbert's optimism that any well-defined mathematical problem is decidable by finding an appropriate axiom system. Hence, William Thurston cannot rely on Gödel when he claims that "Gödel's incompleteness theorem implies that there can be no formal system that is consistent, yet powerful enough to serve as a basis for all of the mathematics that we do."[17] MacLane is right to stress that "what we do" concerns suitably restricted systems. Hence, Gödel's theorems do not entail, as René Thom erroneously claims, "that rigor can be no more than a local and sociological criterion."[18]

Jaffe and Quinn are aware of the point where the analogy terminologically breaks down: "we are not suggesting that proofs should be called 'experimental' mathematics. There is already a well-established and appropriate use of that term, namely to refer to numerical simulations as tests of mathematical concepts."[19] Still, many working mathematicians dispute the reliability of computer proofs although, as Thurston remarks, much more correctness and completeness is needed to make a computer programme run a proof than to have a corresponding theorem accepted by experts. Armand Borel criticises *where* the functional distinction is situated.

> [Mathematics,] in analogy with physics, has an experimental and a theoretical side, but operates in an intellectual world of objects, concepts and tools. Roughly, the experimental side is the investigation of special cases ... and the theoretical side is the search of general theorems. In both, I expect proofs of course, and I categorically reject a division into two parts, one with proof, the other without.[20]

This Bourbakian position is rather suitable for unificatory programmes where relatively stable axiomatisations and standards exist for the subdisciplines, but it seems too austere for the developing field which Jaffe and Quinn had in mind.

Morris W. Hirsch, on the other hand, emphasises that "the nonrigorous use of mathematics by scientists, engineers, applied mathematicians and others, out of which rigorous mathematics sometimes develops, is in fact more complex than simple speculation"[21] and involves the use of mathematical language for 'narrative purposes.' Ultimately, validation comes from experiment – possibly on a computer. Karen Uhlenbeck criticises that the broad role of applied mathematics in physics is narrowed down. " '[T]heoretical mathematics' already exists. It is called 'applied mathematics,' a much bigger field than pure mathematics ... Only the combined elitism of very pure mathematics and high-energy fundamental physics would claim that its own brand of speculative and applicable mathematics should have a special name."[22] What about non-linear dynamics or mathematical biology?[23] A representative of this alleged elite, the string theorist Albert Schwarz, also considers the terminology inappropriate as a common name for heuristic mathematics and theoretical physics. But he does stress the peculiarity of string theory within applied mathematics: Today, theorists "are not able to extract reliable predictions from string theory because this is connected with enormous mathematical difficulties. The physicists have chosen the only possible way: to analyze carefully the mathematical structure of string theory."[24] This is exactly Jaffe and Quinn's point: Theoretical physicists "have found a new 'experimental community': mathematicians ... who provide them with reliable new information about the structure they study."[25]

But Witten cherishes higher aspirations: "when a mathematical result is really relevant to a physics problem it often happens that, turning things around, the result can be *deduced* from the behavior of the physics problem."[26] From such claims, to the effect that the putative truth of string theory be more than heuristics, ontological problems arise. First of all, any quasi-empirical ontology that renders these conjectures meaningful rests on toy-models and highly idealised cases, because string theory "concern[s to a large extent] experimentally inaccessible events: particles of incredible energy, movement on the scale of the universe, or creation of new universes."[27] This empirical inaccessibility marks a great difference to applied mathematics. If the applied mathematician has good reasons to believe that a precisely measured physical system is described by a complicated equation, he or she might take the existence of a solution for granted, although the existence proof does not obtain, and start studying other properties of the equation and its solutions.

It is important to note, already at this stage, that the quasi-empirical ontology of mathematics needed to make 'Theoretical Mathematics' meaningful is neither tied to Platonism, nor does it render mathematics an empirical science. As Lakatos stressed, merely the flow of truth is at stake:

"a theory which is quasi-empirical in my sense may be either empirical or non-empirical in the usual sense."[28] While *Euclidean* theories are built on indubitable axioms from which truth flows down through valid inferences, in quasi-empirical theories truth is injected at the bottom by virtue of a set of accepted basic statements. In the latter case, truth does not flow downward from the axioms, but falsity is retransmitted upward. "[I]n a quasi-empirical theory the (true) basic statements are *explained* by the rest of the system."[29] Theoretical physics is, of course, quasi-empirical in this sense; it is also empirical in the usual sense. Among the possible axioms of a theory which 'explain' physical facts, genuinely mathematical ones can be found nevertheless, especially in mature theories. Thus, these axioms will figure in physicists' non-rigorous reasoning.

But 'Theoretical Mathematics' marks out a domain for conjectural results *within mathematics*. To the Euclideanist, and even to the non-Euclidean structuralist, such as MacLane, there is no such territory except in an ephemeral sense: "Mathematics rests on proof – and proof is eternal."[30]

To Jaffe and Quinn, only 'Theoretical Mathematics,' as opposed to rigorous mathematics, is quasi-empirical in Lakatos' sense, and hence speculative and fallible. They do not address foundational questions, but their thesis presupposes that, if no errors are present, rigorous proofs are final in an appropriate sense. But, the boundary between both fields marked by the method of verification is not absolute. "Scientific verification and mathematical proof differ as much as ice and water can coexist although they are qualitatively different states of matter."[31] Still, phase transitions are hard to pin down. Opinions of experts diverge – as is documented, for instance, in the debate over Thurston's geometrisation programme. Jaffe and Quinn provide two arguments as to why a tight border should nonetheless be maintained. First, rigour is, historically, the core of mathematics. Second, finality of proof is a norm for mathematicians' success, rather than logical necessity. Rigour guarantees the reliability of the scientific literature, which is a precondition for mathematical progress. Otherwise, further work is confused and students are misled.

Also, in Lakatos' view, the borderline, if any, is drawn by the mode of verification: "If mathematics and science are both quasi-empirical, the crucial difference between them, if any, must be in the nature of their 'basic statements' or 'potential falsifiers'."[32] Singular spatio-temporal statements, of course, do not falsify a mathematical theory. Contradictions are possible *logical* falsifiers.

> But if we insist that a formal theory should be the formalization of some informal theory, then a formal theory may be said to be 'refuted' if one of its theorems is negated by the corresponding theorem of the informal theory. One could call such an informal theorem a *heuristic falsifier* of the formal theory. Not all formal theories are in equal danger of heuristic refutation in a given period. For instance, *elementary group theory* is scarcely in any danger: in this case the original informal theories have

been so radically replaced by the axiomatic theory that heuristic refutations seem to be inconceivable.[33]

On the other hand, after the destruction of naive set theory by logical falsifiers, one cannot speak any more of set theoretical facts. Nevertheless, one might still continue to consider it to be the unifying basis of mathematics. Hence, the question of mathematical facts rests upon a subtle interaction between the informal and the formal level. For the Jaffe-Quinn debate, this entails that those objects of informal 'Theoretical Mathematics,' which are blatantly inconsistent, can hardly count as quasi-empirical mathematical facts in Lakatos' sense.

If one considers informal mathematics as a developmental stage in its own right, there arises the question of when to axiomatise an informal theory. Axiomatised theories, if consistent, are exempted from logical falsification, because they no longer have any counterexamples formalisable within the system. "[B]ut we have no guarantee that our formal system contains the full empirical or quasi-empirical stuff in which we are really interested and with which we dealt in the informal theory. There is no formal criterion as to the correctness of formalization."[34] Premature formalisation has its dangers. Probability theory without the Lebesque integral, or algebra without complex numbers would be much poorer theories and lack key theorems. But, the mitigated Euclideanist might respond to Lakatos' claim that axiomatisation could have made clear the need for these concepts quite early on (See Section 3 below). Formalisation is, on the other hand, not the ultimate end of mathematics. Post-formal mathematics studies the consistency of the axiomatic system and provides a suitable metatheory. The main problem is the classification of possible representations. "[A]xioms in the most important mathematical theories implicitly not just define one, but quite a family of structures."[35] The Peano axioms, for instance, are not fulfilled only by the familiar natural numbers, but also by the so-called Skølem functions. Gödel's Incompleteness Theorems show that the problem of unintended structures is quite generic in mathematics. It is almost omnipresent in mathematical physics, where many representations might be mathematically reasonable, but unphysical – a condition which is not always easily formalisable. Note that post-formal metamathematical arguments are usually informal and fallible, but can nevertheless propel a revision of axiomatisation.

1.2. The linear model of mathematical progress, or: what we learn from history

Jaffe and Quinn, instead, adhere to a linear growth model, as far as rigour is concerned. A considerable part of their paper and some responses deal with the controversial lessons certain historical examples teach. Aside from undisputed success stories, there are also some 'cautionary tales' that demonstrate that relaxing standards and relying on intuitions occasionally was a hindrance – or even disastrous – for a budding research programme.

The ideal attitude in the contact between mathematics and physics was assumed by mathematical physicists, such as "D. Hilbert, F. Klein, H. Poincaré, M. Born, and later H. Weyl, J. von Neumann, E.P. Wigner, M. Kac, A.S. Wightman, R. Jost, and R. Haag ... These people often worked on questions motivated by physics, but they retained the traditions and values of mathematics,"[36] to wit: rigour, scholarship, and knowledge of the literature. Their speculations, on the other hand, were addressed to physicists. The main rule of conduct in such contacts is to treat the counterpart in the same manner as theoretical and experimental physicists respect each other.[37] As Jaffe sketches in his *Synthese* paper, it is mainly physicists' attitudes[38] that have changed during this century. When the successful rules of quantum field theory became hopelessly difficult for mathematicians to tackle, a general antipathy towards allegedly unnecessary rigour emerged, one that lasted until the 1960s. Many mathematicians, on the other hand, preferred the 'better-safe-than-sorry' strategy.

Jaffe and Quinn discuss essentially three types of success stories. (i) Brilliant conjectures have inspired the development of whole fields. "The Hilbert problem list, of amazing breadth and depth, has been very influential in the development of mathematics in this century."[39] The Wightman axioms pioneered research in axiomatic quantum field theory. Now they have been supplanted by Haag's nets of local algebras. (ii) Conjectures that were accompanied by technical details or even an outline of a proof, such as the Weil conjectures, have initiated entire research programmes. (iii) Famous conjectures can be highly motivating if they turn out to be the corollary of a general theorem. In the case of the programme to prove Fermat's 'Last Theorem,' this conjecture was reduced to another one, that was eventually proven in 1994.

"Most of the experiences with theoretical mathematics have been less positive ... Straightforward mistakes are less harmful ... Weak standards of proof cause more difficulty. [(i)] In the eighteenth century, casual reasoning led to a plague of problems in analysis concerning issues like convergence of series and uniform convergence of functions."[40] (ii) At the beginning of the 20th century, the 'Italian school' of algebraic geometry "collapsed after a generation of brilliant speculation ... In 1946 the subject was still regarded with such suspicion that Weil felt he had to defend his interest in it."[41] (iii) In the *Analysis Situs,* "Poincaré claimed too much, proved too little, and his 'reckless' methods could not be imitated. The result was a dead area which had to be sorted out before it could take off."[42] I will discuss the first example, one of Lakatos' case studies, below. It roughly boils down to what Jeremy J. Gray, a historian of mathematics, emphasises for the Italian school: "[B]y modern standards it seems to lack rigour – but this perception is modern, and due to Zariski, who also brought new questions to bear (such as arbitrary fields). What it seemed to contain at the time was a rich mixture of results and problems."[43] Hirsch responds that Poincaré ... proved quite a lot by the standards of the day – but there was little use for it because the mathematics which could use it was not sufficiently developed."[44]

In some cases, issues of publication and society are involved. René Thom proposed catastrophe theory to explain forms of physical phenomena. "The application was mathematically theoretical, and its popularisation, particularly by E.C. Zeeman, turned out to be physically controversial."[45] In his Response, Zeeman classifies catastrophe theory as rigorous mathematics, and sustains that the controversy was mainly provoked by "a few journalists and mathematicians who were ignorant of the science and did not fully understand thc mathematics."[46] It is noteworthy, indeed, that Fermat's Last Theorem, which represented probably the best publicly known open problem of mathematics, was proven in the old-fashioned style "featur[ing] one or two lone persons sharing their partial results only with the closest circle of friends over an eight year period and publishing the complete, final work but not intermediate ideas."[47] Thurston's 'geometrisation theorem' was a "grand insight delivered with beautiful but insufficient hints, the proof was never fully published."[48] Hence, the literature did not contain a reliable basis, upon which to build. Theoretical mathematicians, on the other hand, collaborate world-wide by email and constantly race for priority in unrefereed electronic bulletin boards.

The way how Jaffe and Quinn classify successes and deficiencies in mathematics suggests characterising their picture of history as follows: (i) They prefer a linear growth that is initiated by a conjecture or a theoretical research programme which, though slowly and piecemeal, immediately takes off and yields rigorous results. (ii) Ruptures should best be avoided, because they hamper progress in the long run, although momentary growth might be considerable. (iii) Lasting historical progress requires establishing scholarship and a reliable literature. Moreover, Jaffe and Quinn expect scientific journals to provide the same reliability as textbooks. To sum up: although the authors do not write a piece of Whig historiography, they seem to wish that the growth of mathematics be guided to conform to a Whiggish account.

In his Response, Michael Atiyah lodges a protest against the linear growth model:

> [Jaffe and Quinn] present a sanitized view of mathematics which condemns the subject to an arthritic old age. They see an inexorable increase in standards and are embarrassed by earlier periods of sloppy reasoning. But if mathematics is to rejuvenate itself and break new ground it will have to allow for the exploration of new ideas and techniques which, in their creative phase, are likely to be dubious as in some of the great eras of the past. Perhaps we now have high standards of proof to aim at but, in the early stages of new developments, we must be prepared to act in more buccaneering style.[49]

"However, a buccaneer is a pirate, and a pirate is often engaged in stealing," writes MacLane in answer to Atiyah's thunder, and furthermore, "[b]uccaneers have no place in mathematics."[50] MacLane's argument quoted in Section 1.1 suggests that these pirates steel single cases and claim possession of genuine mathematical truth, although such truth is only attributable to general structures that are independent of the cases picked. To a buccaneer, Euclideanism seems to have a smack of maritime law. In Section 2.2, I will argue with Lakatos that mathematical buccaneers, as well, obey certain progressive rules of conduct.

1.3. Conjectures and research programmes

The historical part of the Jaffe-Quinn debate does not concern single theorems, but typical mathematical research programmes, initiated with a conjecture or a problem, and often supplemented with a proof proposal. These are a substantial, yet non-exhaustive (See 2.3) part of the hard core. The various proof techniques and concepts used, which often come from other branches of mathematics, represent the positive heuristics. According to Barry Mazur, conjectures "*provide the skeletal architecture of a Theory*."[51] Such architectural conjectures allow the formation of a unified research programme that is conditional on their truth. Hence, a research programme "is an attempt of the present to lay down architectural plans for the construction of future theories ... The program for the classification of simple groups is a perfect example."[52] Jaffe and Quinn hold similar views and stress the organisational aspect: "The importance of large-scale, goal-formulating work (necessarily theoretical) is growing. We are in the age of big science, and mathematics is not an exception. The classification of finite simple groups, for instance, is estimated to occupy 15,000 journal pages!"[53] But, asks Hirsch, "[h]as the classification been rigorously proved? ... Is there an expert who claims to have read it all and verified it?"[54] Still worse, "some of the parts that were farmed out by the organizers of the project have never in fact been completed."[55]

MacLane takes this case as an indication that the sequence for the understanding of mathematics is: intuition, trial, error, speculation, conjecture, proof. "The mixture and the sequence of these events differ widely in different domains, but there is general agreement that the end product is rigorous proof – which we know and can recognize, without the formal advice of the logicians."[56] Recalling MacLane's ontology cited above, it is doubtful whether he can really appraise the not yet completed classification programme as a genuine part of mathematics, just because of unifying inspiration and organisation. That the importance of informal methods depends on the branch of mathematics, is quite in accordance with Lakatos' remarks about heuristic refutation (See 1.1).

Jaffe and Quinn hold – as most participants do – that any mathematical methodology must be universal for the whole discipline. In this respect, mathematics is more unified that physics. But, "mathematics is much *more finely subdivided into subdiciplines than physics.*"[57] Moreover, some mathematicians change subject frequently or work simultaneously in more than one field. Indeed mathematics is acquitted of one important element of physical theory, to wit, the approximation of physical facts by a mathematical model. This concentrates a substantial number of theoretical physicists around some important problems or theoretical obstructions, such as the unification of elementary forces, the solar neutrino problem, or string theory.

Can 'Theoretical Mathematics' itself be understood as a research programme? Taken in the narrow sense of the intersection of string theory and geometry, the answer seems to be affirmative. Its hard core would contain the implicit conviction that the basic concepts of the physical world, the strings, are also fundamental in a purely mathematical sense. If, on the other hand, 'Theoretical Mathematics' should be a separate domain within mathematics, it can hardly be considered a research programme because speculation alone (and across the finer subdisciplines) does not make up a hard core.

How broad are mathematical research programmes in general? Lakatos and Mazur agree that they evolve around an initial conjecture which, in the course of the programme, becomes modified by the hidden lemmas that are unearthed when attempting a proof. But how to weigh the respective import of conjecture and proof? When Jaffe and Quinn remark: "Conjectures range from brilliant to boring, from impossible to obvious,"[58] the same holds for proofs. Proving deep conjectures, such as the Weil conjecture, is of great value because they entail many other theorems. Fermat's Last Theorem represents the opposite extreme. As Rota notes, "Wiles' proof appeals to an astonishing variety of distinct pieces of mathematics ... . But the magnitude of this triumph brings out in stark contrast the insignificance of the conjecture that was proved."[59]

1.4. Prescriptions, conventions and crediting

Jaffe and Quinn issue three sets of prescriptions:

1. "Theoretical work should be explicitly acknowledged as theoretical and incomplete; in particular a major share of credit for the final results must be reserved for the rigorous work that validates it."[60] There might even be Theoretical Mathematics journals or electronic bulletin boards.

2. The reliability of the literature should be secured by a standard nomenclature that unambiguously flags 'theoretical' results as 'conjectures' (instead of 'theorems') that 'predict' (instead of 'show'), etc.

3. "Research announcements should not be published, except as summaries of full versions that have been accepted for publications."[61] The quick communication facilities on the Internet invite research announcements' being exaggerated in order to secure priority. After a

proposal by Seiberg and Witten, "[b]asically ten years of Donaldson theory were re-established, revised, and extended during the last three weeks of October 1994."[62] This break-neck speed was mainly due to rapid communication on the Web.

As heuristic proofs, unlike their rigorous counterparts, are a matter of degree, Albert Schwarz suggests introducing coloured flags and a much more refined terminology that embraces the words pretheorem (if supported by details), fact (if reliable), statement (if heuristic proof), conjecture (if mere analogy), etc. His elaborate proposal on flagging[63] is so complicated, that it would require an immense bureaucracy to classify work or approve classification of the authors. René Thom waxes satirical at the thought:

1) a crib ... denoting 'live mathematics,' allowing change, clarification, completing of proofs, objection, refutation. 2) the tombstone cross. Authors pretending to full rigor, claiming eternal validity, may use this symbol as freely as they wish. This kind of work would constitute 'graveyard mathematics.' 3) the Temple. This would be a label delivered by an external authority, the 'body of high priests' ...[64]

Any such flagging requires an arbiter and a metatheory of rigour. Even if this could be carried through, and flags unanimously set, who is allowed to conjecture? Historically, conjectures were a privilege of those working in the field over a long time, not a matter for freshmen or students. If the conjecture is set, only a few experts will be likely to prove the conjecture. Hence 'Theoretical Mathematics' cannot be a real branch to specialise in or lecture about, nor is the promise of a corresponding journal very great.

Jaffe and Quinn reserve a major part of the credit for the person who finally proves the conjecture. But some conjectures are very detailed, and the decisive concepts and techniques were already present in them. Benoit Mandelbrot regards this crediting rule as being appallingly one-sided.

Anyone who will give rigorous solutions to this enormously difficult problem [of percolation] will deservedly be hailed in mathematics, even if the work only confirms the physicists' intuition. I hope that proper credit will be given for the individual physicists for their insights, but fear it will not. Will this mathematics be also noticed by physicists? Only if they prove more than was already known, or if the rigorous proofs are shorter and/or more perspicuous than the heuristics.[65]

If we assume that the analogy between rigorous mathematics and experimental physics is true, then Mandelbrot is making a very Lakatosian point: in research programmes, theory is quite autonomous with respect to experimental testing.

1.5. Demarcationism versus elitism – the role of the individual

Jaffe and Quinn's prescriptions also raised some general protests: Atiyah "rebel[s] against their general tone and attitude which appears too authoritarian."[66] Armand Borel disagrees with their general thrust: "what mathematics needs least are pundits who issue prescriptions or guidelines for presumably less enlightened mortals."[67] In Lakatosian terminology, Jaffe and Quinn are outspoken *demarcationists,* who enact norms for mathematics' being regarded as scientific, in order to guarantee – among other things – a democratic progress of science by securing the reliability of the literature, such that people outside great research centres or privileged discussion groups can also participate.

Lakatos juxtaposes demarcationism and elitism. According to this latter view, "science can only be judged by case law, and the only judges are the scientists themselves."[68] A large part of scientific knowledge is inarticulable, at least outside the élite, such that the producer, not the product, is assessed in first place. The standards of a scientific elite are only descriptively accessible to psychology and sociology. But, according to Lakatos, "[e]veryone, whether élitist or not, is bound to use normative third-world criteria, whether explicit or hidden, in establishing criteria for a scientific community."[69] For, what if experts disagree? – as, for instance about many examples raised in the Jaffe-Quinn debate. Lakatos does not accept Kuhn's thesis that consensus is always quickly reached, even after scientific revolutions. Instead, the elitist has to resort to a Darwinian struggle of ideas, or to Hegel's 'cunning of reason.' "It is only this thin, *ad hoc* authoritarian/historicist doctrine which separates élitists of this [Kuhnian] kind from the sceptic"[70] – Feyerabend, for instance.

Some disputants in the Jaffe-Quinn debate are elitists to various degrees. Mandelbrot writes that his "reading of history is that mankind continually produces some individuals with the highest mathematical gifts who will not (or cannot) bend to pressures like those proposed by JQ."[71] Or, yet to a lower degree, Thurston, who outlines a rather different conception of the mathematical enterprise and the individuals advancing it. Rota holds similar views. Both papers contain views strikingly similar to those of Lakatos, although at first glance their focusing on understanding or enlightenment contrasts with Lakatos' orientation on statements.

Thurston's picture of mathematics is based on individual understanding – a doctrine closely related to elitism, on Lakatos' account.[72] "We are not trying to meet some abstract production quota of definitions, theorems and proofs. The measure of our success is whether what we do enables *people* to understand ... mathematics."[73] Accordingly, the communication of ideas is an essential part of mathematics that is by no means exhausted by written sources and formal proofs, which Jaffe and Quinn regard as primary.[74] "By informal contact, people learn to understand and copy each other's way of thinking, so that ideas can be explained clearly and easily."[75] Mathematicians (students foremost) continually learn new formal languages and build mental models of mathematical facts. "When the idea is clear, the formal setup is usually unnecessary and redundant."[76] Thus, "the flow of ideas and the social standard is much more reliable than formal documents. People are not very good in checking *formal correctness* of proofs, but they are quite good at detecting potential weaknesses or flaws in proofs."[77]

On this basis, Thurston criticises "our strong communal emphasis on theorem-credits [which, contrary to Jaffe and Quinn's rules] has a negative effect on mathematical progress. If what we are accomplishing is advancing human understanding of mathematics, then we would be much better off recognizing a far broader range of activity ... Soccer can serve as a metaphor"[78] for the various roles single mathematicians play in a successful team. Some 'point people' in this 'network of ideas' will be more likely to score than others. But, "there are theorems in the path of these advances that will almost inevitably be proven by one person or another."[79] This rather Duhemian account of scientific progress makes clear that the soccer metaphor also refers to a third world of ideas. The 'points' are the steps in an advancing research programme.

Thurston's personal experiences illustrate two different models of mathematical progress. He rapidly proved some dramatic theorems about foliations and "documented [them] in a conventional, formidable mathematician's style,"[80] without providing the background which permitted people to follow his thoughts and gain their own credits by proving new results. This eventually led to an evacuation of the field. When posing the geometrisation conjecture, Thurston, instead, focused on providing the intellectual infrastructure and the context for this result.

> What mathematicians most wanted and needed from me was to learn my ways of thinking, and not in fact to learn my proof of the geometrization conjecture for Haken manifolds. It is unlikely that the proof of the general geometrization conjecture will consist of pushing the same proof further. ... Not all proofs have an identical role in the logical scaffolding we are building for mathematics. This particular proof has only temporary logical value, although it has a high motivational value in helping support a certain vision for the structure of 3-manifolds.[81]

When the general proof of the geometrisation conjecture is found, "proofs of special cases are likely to become obsolete."[82] At this point, Thurston's conception of proof appears rather Lakatosian. Proofs generally make theorems reliable, but their heuristic value may become even more important. Even if they always remain formally true, theorems and proofs have a limited lifetime within the mathematical research programme. Thus, progress cannot be measured by the number of theorems deduced, but by their – usually informal – contextual significance within a research programme.

Rota demands that "[a]ll normative assumptions shall be weeded out."[83] Even more typical for an elitist position – in Lakatosian terms – is his explicit adherence to the phenomenological *Verstehen*-doctrine. "[M]athematicians are not satisfied with proving conjectures. They want the reason"[84] which corresponds to the definiteness of proof. Mathematicians were, for instance, not satisfied with the Cartan classification of simple Lie groups until they discovered the reason for the existence of five exceptional groups in a very unusual property of the group SO_8. Such proofs convey enlightenment which, accordingly, becomes an *objective* property of mathematical statements that keeps the discipline alive. Mathematicians, so Rota, typically eschew talk about enlightenment because it admits degrees. Instead, they resort to degreeless beauty and commit the 'light-bulb mistake'; to wit, they disguise extensive conceptual buildups and long-winded arguments afterwards as instantaneous realisations. "Mathematical beauty and mathematical truth share in the fundamental property of all objectivity, that of being inescapably context-dependent."[85] In order to prevent arbitrariness, elitist Rota has to assume substantial agreement among mathematicians of a given historical period and context as to which theorems are beautiful or rigorously proven, thus presupposing what Lakatos called the 'authoritarian/historicist doctrine.'

Husserl's phenomenology emphasises mathematical progress – as Lakatos' research programmes do. Rota mentions two aspects. *Ugliness* is typically a major motivation to refine theories and arguments. *Possibility* is yet more central. "[T]he reasons for a theorem are found only after digging deep and focusing upon the possibilities of the theorem. Once such reasons are found, the choice of particular formal statements that express them is secondary."[86] Rota even proposes "that a rigorous version of the notion of [objective] possibility be added to the formal baggage of metamathematics."[87] The main import of Wiles' proof of Fermat's Last Theorem were the possibilities it opened up. Although Rota does not mention the Jaffe-Quinn debate, the possibilities revealed by Witten's 'theoretical' results enjoy the same property. Rota's notion of possibility seems rather close to Lakatos' excess content and excess corroboration that make a research programme progress. Rota's view of the relation between proof and conjecture is again Lakatosian, though with a somewhat different justification. "The error [of the standard 'Euclidean' conception] lies in assuming that a mathematical proof has been devised for the explicit purpose of proving what it purports to prove."[88] Instead, lemmas may be more important than the conjecture, and the proof may be reorganised to make this manifest. The initial conjecture then becomes a simple corollary. In contrast to Lakatos, Rota holds that the dialectics between conjectures and proofs is supported by formalisation because "exchangeability of theorem and proof"[89] is a main virtue of the axiomatic method.

2. WHAT LAKATOS CAN CONTRIBUTE TO THE DEBATE

In my discussion of the various positions taken in the Jaffe-Quinn debate, I have repeatedly alluded to Lakatos' philosophy. It is high time now for a concise and coherent account of the method of proofs and refutations and the MSRP. In this, I will focus on issues important for mathematical physics which – in line with Uhlenbeck and others – is understood in a wider sense than one which embraces only string theory and geometry.

2.1. The various stages of the interrelation between conjectures and proofs

Jaffe-Quinn and Lakatos share a common starting point: 'Proof' stands "for a *thought-experiment – or 'quasi-experiment' – which suggests a decomposition of the original conjecture into subconjectures or lemmas,* thus *embedding it* in a possibly quite distant body of knowledge."[90] Thought-experiments follow an initial naive trial and error phase. In the famous example of the Euler-Poincaré conjecture, around which *Proofs and Refutations* is built, they correspond to a stretching and a triangulation of the polyhedron. The lemmas, or subconjectures, ensure that the 'thought-experimental' steps of the proof are permissible, such that the proof can validate the conjecture. But, quite similar to the development of experimental techniques in science, the decomposition is informative even if validation is not obtained.

Lakatos modifies Popper's critical fallibilism with regard to two core aspects. First, fallibilism extends to mathematics because the latter is quasi-empirical. Second, refutation does not entail immediate rejection – as it was the case in Popper's Darwinist account. Just as most scientific theories are born refuted, mathematical theorems are heuristically defective because they typically constrain the informal stuff admitted by the conditions. Refutations are suggested by counterexamples that either concern the conjecture (*global* counterexamples) or the lemmas (*local* counterexamples). Hence, there are three possible types. (i) Global, but not local counterexamples logically refute the conjecture. They are what most mathematicians would call a counterexample. (ii) If a global counterexample is also local, it does not refute the theorem, but confirms it. (iii) Local, but not global counterexamples show a weakness of the theorem, such that one has to search for modified lemmas. Cases (ii) and (iii) are not genuinely logical, but *heuristic* counterexamples.

The imaginary class of *Proofs and Refutations* discusses various strategies if counterexamples emerge. The first strategy, *monster-barring,* rejects the global counterexample as "a *monster*, a pathological case"[91] of a polyhedron by modifying the latter's definition in a suitable way. It decreases the domain of validity of the conjecture – only. But counterexamples abound despite such linguistic ad hoc remedies. Theoretical physicists often apply this strategy by providing physical reasons for a 'natural definition' of the mathematical concept in question or, more operationally, by rejecting 'pathologies' as beyond the scope of the studied model. This method conveys a justification that cannot be obtained in mathematics proper, because there is no factual basis for specifying some property as a concept's essence. The history of theoretical physics teaches us that excluding pathologies is never final, because almost all mathematical structures eventually turn out to be of use to physicists.

The next strategy, *exception-barring,* restricts the domains of both the conjecture and of the guilty lemma. But the withdrawal could have been too radical, and the method still does not exploit the proof. *Monster-adjustment*, the third strategy, also concerns the domain of validity of the basic concepts.

"Monsters don't exist, only monstrous interpretations,"[92] says pupil Rho. This therapeutic method strives to see an example in the alleged counterexample by finding a suitable interpretation. In a footnote, Lakatos rejects monster-adjustment: "Nothing is more characteristic of dogmatist epistemology than its theory of error."[93] Later, however, he noted that monster-adjustment could be empirically progressive in science.[94] Indeed, empirical science might ease the main problem of the monster-barrer. If a variety of interpretations of a mathematical structure are more or less on a par, physical facts might single out the proper one for mathematical physics. In quantum field theory, for instance, the choice of a representation of the symmetry group is literally unavoidable, because there are infinitely many inequivalent representations. Still, one might consider this (quasi-)empirical aspect to be an unwanted bias for mathematics. MacLane holds "that a mathematical structure is a scientific structure but one which has many different empirical realizations ... A careful formulation must rest on ideas independent of any chosen examples."[95] Nonetheless, the recent developments in geometry have shown that physics, at least, provides heuristic motivation.

In Cauchy's time, mathematics matured through an appropriate estimation of proof-analysis. The method of *lemma-incorporation* "upholds the proof but reduces the domain of the main conjecture to the very domain of the guilty lemma."[96] In this way, the lemma refuted by the counterexample is built into the conjecture. Hence, proofs improve a conjecture, even if they do not prove it. This "displays the fundamental dialectical unity of proof and refutations," that is ultimately rechristened as the method of proof and refutations.[97] But on principle, one has to incorporate expectable, but not yet known, counterexamples. This reveals that lemma incorporation proceeds through constant overstatements, by attempting to keep as much as possible from the initial thought-experiment and its heuristics. While this method emphasises the heuristic role of proof, exception-barring focuses on validity and advances through a series of understatements. Accordingly, it corresponds to the 'better-safe-than-sorry' strategy prescribed by Jaffe and Quinn. Lakatos stresses that a careful proof-analysis, which constantly suspends unnecessary restrictions, makes the method of proof and refutations "a limiting case of the exception-barring method."[98] It is important to note that exception-barring and lemma-incorporation already take place on the level of research programmes – a term coined only after *Proofs and Refutations.* In fact, hidden lemmas made explicit by incorporation often enter a programme's hard core.

As a good demarcationist, Lakatos turns the historical lessons into five methodological rules:[99]

1. If you have a conjecture, set out to prove it and to refute it. Inspect the proof carefully to prepare a list of non-trivial lemmas (proof-analysis); find counterexamples both to the conjecture (global counterexamples) and to the suspect lemmas (local counterexamples).

2. If you have a global counterexample discard your conjecture, add to your proof-analysis a suitable lemma that will be refuted by the counterexample, and replace the discarded conjecture by an improved one that incorporates that lemma as a condition. Do not allow a refutation to be dismissed as a monster. Try to make all 'hidden lemmas' explicit.
3. If you have a local counterexample, check to see whether it is not also a global counterexample. If it is, you can easily apply Rule 2.
4. If you have a counterexample which is local but not global, try to improve your proof analysis by replacing the refuted lemma by an unfalsified one.
5. If you have a counterexample of any type, try to find, by deductive guessing, a deeper theorem to which they are counterexample no longer.

In a later footnote, Lakatos recalls his "deliberate *mixed* usage of the justificationist term 'proof' and of the heuristic term 'proof'."[100] More than being a mere by-product (as Jaffe-Quinn hold), heuristics stands on a par with rigorous justification. Rigour decreases the heuristically gained content, such that rule 4 becomes a necessary counterweight. As with the various levels of heuristic insight gained during the development of modern mathematics, there also occured different levels of rigour. These "*differ only about where they draw the line between the rigour of proof-analysis and the rigour of proof*, i.e. *about where criticism should stop and justification should start*."[101] At this point, the editors have added a footnote stating that Lakatos underplayed the achievements of rigorous proofs. By accepting a rather modest logical metatheory, all the doubt is cast on the justification of the axioms.[102] To my mind, this important qualification cannot easily be turned into a clear distinction between both levels because, according to Lakatos' anti-inductivist and anti-intuitionist methodology, it is mainly proofs that contribute to concept-formation, on which the axioms are based.

Exploiting the heuristic possibilities of the proof might overthrow the initial conjecture and supplant it by newly formulated theorems. "Descartes' and Euler's 'naive conjecture' ... was irrelevant and superfluous for the later development."[103] Although of considerable motivational value, it ceased, in rational reconstruction, to belong to the core of the research programme that led to algebraic topology. Moreover, "different proofs [better: 'improofs'] of the same naive conjecture lead to quite different theorems."[104] Understanding rule 4 in this way leads to the method of proof*s* and refutations. The important point for the Jaffe-Quinn debate is that Lakatos' methodology of mathematical growth does not play down the role of proofs. Emphasising their heuristic role, on the contrary, puts proof thought-experiments into the core of historical development. The 'speculations' that make up 'theoretical mathematics' only concern single conjectures – be they supplemented with a proof-technique or not.

According to Lakatos, not even naive conjectures start from facts – be it inductively or intuitively. "[T]he proof is already there!,"[105] pupil Zeta remarks. "The naive logic of conjectures and refutations has no starting point – but the logic of proofs and refutations has: it starts with the first naive conjecture to be followed by a thought-experiment."[106] By keeping the latter despite already existing refutations by counterexamples, deductive guessing replaces naive guessing. Later, Lakatos considered it to be a leitmotiv of *Proofs and Refutations* that "*one may bravely – and profitably – go on to 'explain' a hypothesis known to be false*."[107] One may not even need a conjecture to start proving or testing by means of thought-experiments. Being ready to give up the naive conjecture, a more general theorem might be easier to prove. An extended version of the method of analysis-synthesis driven by heuristic and validating thought-experiments contains possible *occult hypotheses*. For the fundamental theorem of algebra, or for the evaluation of certain real integrals, the introduction of complex numbers is a necessary prerequisite. For geometry, string theory has recently provided brilliant occult hypotheses and new concepts that were generated by Witten's heuristic proofs. Hence, from a Lakatosian perspective, 'Theoretical Mathematics' is not an intermediate step in the sequence from conjecture to proof, but rather an indispensable element of the extended analysis-synthesis circuit.

Over the course of history, concepts grow – sometimes even by rather wild extensions, such as the Peano curve. Lakatos argues that typically the monster barrers do not contract concepts, but that the refutationists expand them to cover objects unintended by the naive conjecture. "*Often, as soon as concept-stretching refutes a proposition, the refuted proposition seems such an elementary mistake that one cannot imagine that great mathematicians could have made it*."[108] But such an accusation, which also characterises Jaffe and Quinn's 'cautionary tales,' neglects precisely that sort of concept growth that reaches beyond a mere change in rigour. In this respect, most negative Responses to 'Theoretical Mathematics' still have too narrow a focus.

2.2. How to historically appraise inconsistent mathematics

Lakatos severely criticises the widespread view that an 'informal' proof is a formal proof with gaps. Proofs containing logical gaps – or, more precisely: if experts agree that the argument will fail if logically formalised – are better called 'quasi-formal.' "[T]o suggest that an informal proof is just an incomplete formal proof seems to me to be to make the same mistake as early educationalists did, when, assuming that a child was merely a miniature grown-up, they neglected the direct study of child-behaviour."[109] History of mathematics can be appraised only by a philosophy that sufficiently acknowledges informal growth. Justificationists unduly maximise historical continuity by separating the hard formal kernel that holds true still today from the erroneous 'metaphysical' interpretation. In the same manner, Jaffe and Quinn separated the two groups of informal reasoning as to whether the conjecture in question could subsequently be proven, and whether it fostered continuous growth. This leads to the problem "of how to appraise inconsistent theories like ... Dirac's delta function. Are invalid theories beneath contempt? ... Can they be appraised only if a posthumous reconstruction has saved them and proved them to be, if not respectable, at least excusable ancestors of respectable theories which look consistent and rigorous by the standards of the day?"[110] So Lakatos' criticism of an exclusive focus on rigour extends to the past. Hence, Poincaré's results are still not fully assessed if one insists (against Jaffe-Quinn) that they matched the standards of the day – as some respondents did. Historical appraisal cannot be guided by the justificationism of logical positivists.[111] Instead, heuristic power – or possibilities, in Rota's sense – decides the faith of a research programme that was ventured from an initial conjecture. Formal arguments alone, however, cannot make up the core of a research programme without taking into account 'metaphysical' heuristics. The latter might even be linked to the heuristic support derived from physics – as could have been the case with Dirac's delta function, which had become part and parcel of field theory long before it was made mathematically rigorous by contriving the notion of generalised functions.

2.3. On the methodology of scientific research programmes

It is well-known that, at the time of his death, Lakatos had planned to apply MSRP to the history of mathematics. A footnote in the seminal paper "Falsification and the methodology of scientific research programmes" reads as follows: "Unfortunately in 1963–4 I had not yet made a clear terminological distinction between theories and research programmes, and this impaired my exposition of a research programme in informal, quasi-empirical mathematics."[112] Recent investigations[113] show that qualifications of MSRP are called for in order to apply it to mathematics. Before dwelling upon this point, I shall give a brief outline of the original MSRP in empirical science because the present paper, after all, deals with the interactions between theoretical physics and mathematics.

A research programme is defined by its hard core, which is tenaciously protected by negative heuristics. It is surrounded by a protective belt of quite flexible positive heuristics, which constantly put forward auxiliary hypotheses against anomalies. The programme supplies a conceptual framework and contains a powerful problem solving machinery. As MSRP accepts the conventionalist position that there is no sharp distinction between theory and experiment, these do not encounter each other one-by-one; rather, sequences of theories within a research programme compete with rivals in the face of a larger body of empirical evidence. This makes it possible to establish internal criteria of progress. A programme is progressive if each theory has excess empirical content over its predecessors, and if some of the predicted novel facts are corroborated. A programme is degenerating if its theories are only fabricated to accommodate known facts by way of a (content-decreasing) linguistic reinterpretation.

So far, all this fits quite neatly to Lakatos' philosophy of mathematics, which emphasised the inseparability of conjecture and proof. Mathematical progress was fostered by content-increasing heuristics, not by linguistic monster-barring. Lakatos' concept of growth elucidates why string theory has to seek the union of mathematics to get its excess content corroborated – at least as a mathematical quasi-fact. MSRP seems to be a reasonable approach, if theories abound. If, in contrast, theories grow more slowly than empirical facts are provided, Lakatos could call hardly any programme progressive. Being aware of this difficulty does not cure it: "if exactly the same theories and the same evidence is rationally reconstructed in different time orders, they may constitute either a progressive or a degenerative shift."[114] To Lakatos, unifications or classifications that are usually held in high regard by mathematicians cannot count as progressive problemshifts, unless they lead to concepts or results unknown so far, such as the five exceptional Lie groups.

The ambiguity of the notion of progress is mitigated by Lakatos' patience in letting a theory develop. "One must treat budding programmes leniently: programmes may take decades before they get off the ground and become empirically progressive. Criticism is not a Popperian quick kill by refutation. Criticism is always constructive: there is no refutation without a better theory."[115] Content-decreasing stratagems can be temporarily employed, if anomalies abound and technical difficulties slow down possible predictions. Then one does not accept anomalies as genuine counterexamples, and one allows for a certain autonomy of theory. "*Mature science consists of research programmes in which not only novel facts but, in an important sense, also novel auxiliary theories, are anticipated; mature science – unlike pedestrian trial-and-error – has 'heuristic power'* ... [which] generates *the autonomy of theoretical science.*"[116] Hence, MSRP justifies the autonomy of 'theoretical mathematics' – even if it is not empirically progressive due to failed proofs – but rejects its neat separability from rigorous mathematics.

By virtue of this autonomy, no instant rationality of science obtains, and there exists only a regulative principle. "This *requirement of continuous growth* is my rational reconstruction of the widely acknowledged requirement of 'unity' or 'beauty' of science."[117] Here, Lakatos' and Jaffe-Quinn's intentions converge. Although the former allows vastly more ruptures in the life and competition of theories than the latter, on a very large scale, he too enacts a norm to avoid evolutionary decay. To my mind however, it is not clear whether this regulative principle is strong enough to guarantee the balance between theory and experiment, or between conjecture and proof, as MSRP tacitly presumes.

After all, it seems that the rationality scheme of progress versus degeneration can be quite easily translated from science to mathematics. A programme is theoretically progressive if it proposes fruitful concepts and techniques; it progresses empirically if it solves interesting problems (particularly ones that were posed in another field). Increasing the stock of theorems by number is a bad measure, because something follows from any consistent concept. This would correspond to switching on a measurement apparatus and waiting for anything whatsoever to be recorded.

The issue of what stuff, exactly, a methodology of mathematical research programmes could be all about, seems less clear. Here, Lakatos' quasi-empirical ontology needs further specification. In a recent paper, David Corfield has argued that "rivalry between research programmes concerns high level issues."[118] These levels come about because, in comparison to physical science, "[m]athematics appears to have an extra degree of freedom at this [basic] level [where battles are usually fought out] which makes it improbable that programmes will be in direct competition for precisely the same territory."[119] Hard cores do not simply boil down to axioms, and there are no universally agreed-upon facts – as it was the case in the paradigmatic competition between ondulatory and emission theory in 17th and 18th century optics. High level questions, beliefs or general aims, might enter the hard core, shifting emphasis away from conjectures as the sole driving force of research programmes. In fact, conjectures might "be decided one way or the other by an uninformative proof or an uninstructive counterexample."[120]

Corfield's proposal goes beyond Lakatos' above-quoted observation that proofs and refutations might even start without a conjecture, and it brings understanding and enlightenment into the game. This "would bring the hard core and positive heuristic closer, thereby threatening to collapse the whole construction."[121] But this move seems necessary, if one wants to assess the phenomenon – quite common in mathematics – of two theories' emerging from a single common problem, or converging to form one, although they do not dispute the same area of quasi-empirical facts. One might, from this point of view, answer Jaffe-Quinn's remark that "mathematics is much *more finely subdivided into subdiciplines than physics*, because the methods have permitted a deeper penetration into the subject matter,"[122] by saying that the subject matter is more finely subdivided, as well. One might adopt Koetsier's (1991) terminology here, distinguishing global research traditions, research programmes and particular research projects. On all three levels, nonetheless, mathematics is tied together by several links, be they the foundation of most disciplines on set theory, rather common standards of rigour, or many theorems that link entire fields. Thus, David Sherry (1997) goes a bit far in stressing that the plurality of theories in mathematics, which is due to the larger degree of freedom in creating new concepts, supersedes any brand of fallibilism.

2.4. The role of time scales and the social effects of research programmes

As discussed in Section 1.5, Lakatos favoured demarcationism without, however, designating any authority that enacts and controls the norms – a role Jaffe-Quinn attributed to the discipline's leading journals. The main rule in MSRP is that all research programmes openly admit their current scores and that their representatives precisely specify the conditions under which they are willing to consider a claim refuted. This closely corresponds to Jaffe's "plea for 'Truth in Advertising'."[123]

There is no absolute and final truth, because "the hallmark of scientific behaviour is a certain scepticism even toward one's most cherished theories."[124] Thus, it is even advisable to

> issue only temporary certificates of acceptance: if a theory is accepted$_1$ [because of its excess empirical content] but does not become accepted$_2$ [corroborated] within *n* years, it is eliminated; if a theory is accepted$_2$ but has had no lethal duels for a period of *m* years, it is also eliminated.[125]

Similarly, Morris W. Hirsch proposes to date proofs:

> If after ten years no errors have been found, the theorem will be generally accepted ... If any thirty year period elapses without publication of an independent proof, belief in the theorem's correctness will be accordingly diminished. This would allow for the reality that concepts of proof and rigor change.[126]

Datedness of acceptance should be viewed as a corrective to autonomy of theory. This is particularly pressing in mathematics, where the larger degree of freedom allows the proof of many quite idiosyncratic and uninteresting theorems. But the time scale depends on the subdicipline; in string theory, even one month is quite long. Lakatos emphasises that priority disputes are perfectly functional if two research programmes compete. But one might wonder if this extends down to the level of research projects, research groups or individuals, because elitism could re-enter through the back door by allowing only some research centres to compete. On the other hand, overly bold claims of key scientists might lead astray a whole industry. For this reason, Jaffe and Quinn rightly insist on maintaining independent control in the age of the Web.

3. What Lakatos Could Have Contributed to the Debate

In this final section, I will discuss three topics where modifications of Lakatos' methodology be required in order to make it applicable to mathematical physics.

3.1. *Axiomatisation and the benefits of Euclideanism*

It has not escaped most readers' attention that there is a rupture between the first and the second part of *Proofs and Refutations*. With Poincaré, a conjecture about polyhedra changed into a theorem about the dimensions of certain vector spaces by "translating non-Euclidean informal theories with vague, obscure terms and uncertain inferences ... into these already Euclidean theories,"[127] such as arithmetic, geometry, logic, or set theory. The teacher's attitude is a mixed one: "I do not like your philosophy, but I still may like your proof."[128] It seems highly arguable whether such an attitude is consistent. Already, the editors of Lakatos' work noted his later appreciation of formal logic and ended the dialogue in this spirit. David Corfield collects further textual evidence "that in many respects he [Lakatos] inclined further towards the logicist pole than has been thought."[129]

Within the Popperian framework, another problem emerged: the irrefutability of 'rubber-Euclideanism.' "[O]ne can always stick to one's hopes of deriving them from some deeper layer of self-evident foundations."[13] Lakatos was also unaware of the impacts of the axiomatic method on natural science. So Gamma comments that the translation programme "certainly will not cover physics. You will never translate wave-mechanics into geometry."[131] As a matter of fact, mathematical physics has, to date, succeeded in achieving considerable part of this goal.

The main theme within Lakatos' philosophy of mathematics is the relation between formal and informal theories. "Up to now no informal mathematical theory could escape being axiomatized."[132] Lakatos warns against premature formalisation, which might not exploit all the possibilities of the informal theory. Jaffe-Quinn, on the contrary, caution not to wait too long. Is, at the bottom line, *all* formal mathematics contingent on some quasi-empirical stuff? When Lakatos claims that "[t]here is indeed no respectable formal theory which does not have in some way or another a respectable informal ancestor,"[133] Corfield "challenge[s] anyone who holds this view to try to find an informal ancestor of an Eilenberg-MacLane space or a spectral sequence dating from the pre-axiomatic stage of algebraic topology."[134] Indeed, axiomatisation seems to be a precondition of mathematical growth in the 20th century that cannot be reduced to the structuralist austerity doctrine. Many conjectures (and their heuristics) are only expressible after an axiomatic framework has been set up. Mazur even considers both to be twins, because "[p]erhaps the growth of conjecture just mirrors the growth of modern axiomatic method"[135] since Hilbert. Yet, although this parallel might hold for conjectures that "provide the skeletal architecture of a Theory,"[136] it does not hold for the humbler Lakatosian ones, which dialectically interact with single proofs. Another virtue of axiomatisation is that it "facilitates interaction between apparently different areas of mathematics [on the linguistic level], an essential feature of this century, and has helped to overcome the compartmentalization of earlier centuries."[137] But – one might ask – what if two research projects (or programmes) in different stages of formalisation meet? To this I shall now turn.

3.2. Coexistence of research programmes

Although MSRP does not consider mere unification a progressive problem-shift, it is beyond doubt that the linking of two entire areas of mathematics represents major progress. This arouses suspicion that MSRP eventually cannot accommodate any relationship between theories other than that of competition – at least after a due grace period for the younger one. Not even theoretical and empirical identity prevent this, in the case that two such theories belong to different research programmes. Admittedly, if competition leads to the elimination of pointless formal results that just multiply language, then this amounts to a wholesome purge. On the other hand, independent evidence is perhaps the best support any scientific fact could obtain. The fact that the Loschmidt-Avogardro number could be measured by several methods that belonged to rather distinct fields of physics provided strong support for atomism. The coexistence of theoretical frameworks, such as Schrödinger's and Heisenberg's equivalent formulations of quantum theory, allows one to choose the easier method for computation, thereby increasing the heuristic power. Why halve your tool kit?

Although mathematical theories hardly dispute an identical territory of facts, independent proofs play a major role – not only in justification, but also in simplification. First proofs are usually very clumsy. Considerable credit is still given for subsequent simplifications that unearth new quasi-facts hidden in the theorem. Even if a certain level of definiteness is reached, two or more equally simple and beautiful proofs may, in some cases, remain, connecting the theorem to different fields of mathematics. In *Proofs and Refutations*, Lakatos was able to play down this possibility, because all alternatives proposed were weaker in content than those deriving from the main heuristic. But already his insistence on the quasi-empirical character of mathematics suggests that independent proofs increase verisimilitude, just as independent experiments do.

To apply the present considerations to the Jaffe-Quinn debate, one has to understand what happens if two proofs belong to research programmes that are at a different stage of axiomatisation – such as geometry and string theory. Should the stricter of the two standards prevail consistently throughout the argument? This, however, might considerably hamper overall progress by restricting the heuristic power of the budding programme to which the less rigorous proof belongs. So it seems closer to the spirit of Lakatos' philosophy of mathematics to equilibrate reliability and heuristic power, and set standards of rigour in accordance with the specific aims of the intersection of the two programmes, even before this field has itself become a programme. For example, where 'Schrödinger Operators,' a functional-analytic technique in studying atomic physics, are at issue, heuristics does not have much say, because the field has a clear-cut conceptual framework, and successful proofs usually prove what they intend to prove. String theory, on the contrary, has still to discover its basic physical and mathematical concepts. What distinguishes it from other encounters between mathematics and physics, is the lack of foreseeable empirical corroboration. Lakatos' quasi-empirical ontology would, in principle, allow for the compensation of lacking mathematical rigour by way of empirical corroboration. For this to happen, however, a core Popperian dogma has to go.

3.3. Lakatos' rejection of induction

Proofs and Refutations contains a sequence in which induction is rejected as a way to find conjectures. After pupil Beta's attempt to list all possible polyhedra is turned down, he claims that at some point induction has to appear:

Beta [to Zeta]: ... You only pushed back the inductive starting point further: now to the statement that $V–E=1$ for any edge whatsoever. Did you prove or did you *observe* that?

Zeta: I proved it. I knew of course that for a single vertex $V=1$ (Fig. 19). My problem was to construct an analogous relation ...

Fig 19.

Beta (furious): Didn't you observe that for a point $V = 1$?

Zeta: Did *you*? (*Aside, to Pi*): Should I tell him that my 'inductive starting point' was empty space? That I began by 'observing' *nothing*?[138]

Is this really a *reductio ad absurdum* of mathematical induction? The teacher believes so: "Facts do not suggest conjectures and do not support them."[139] Even if one concedes to Lakatos that deductive guessing and thought-experiments reach back to the earliest stage of mathematics, and that the initial conjectures are quickly forgotten, it seems implausible – if such an inductive term is permitted here – that trial and error would be entirely blind. In his later years, Lakatos himself was more liberal:

There is then a pattern by which one gets from naive Popperian guessing to the method of *proofs* and refutations (*not* conjectures and refutations), and then one step further, to mathematical research programmes. This patters refutes the philosophical claim that the heuristic source of research programmes is always some big metaphysical vision. A research programme may be of humbler origin: it may originate in low-level generalisations. My case-study in a sense rehabilitates inductive heuristic: it is frequently the study of facts and the practice of low-level generalisations which serve as the launching pad for programmes. Mathematics and science are importantly inspired by facts, factual generalizations and then by this imaginative deductive analysis.[140]

He also explicitly keeps his distance from Popper in this respect: "The infallibilist heuristic of deducing theories from facts certainly failed – but to replace it by the Popperian heuristic of speculations and refutations was to pour the baby along with the bathwater."[141]

On the general level of reliability of science as a whole, Lakatos gave induction a role that was even justificatory. In "Changes in the problem of inductive logic," he introduces the notion of acceptance$_3$: the estimation of a theory's future performance.[142] Its reliability requires a general inductive principle, which links corroboration and verisimilitude, which represents a bit of fallible speculative metaphysics. Applied to mathematics, this is still a far cry from Platonism.

NOTES

* I thank Thomas Hoffmann-Ostenhof, for pointing me towards this debate already at the time when it was emerging, as well as David Corfield, for detailed and illuminating criticism.

1. Quoted according to (Jaffe, 1997), p. 136.
2. (Jaffe, 1997), p. 138.
3. (Jaffe and Quinn, 1993), p. 10.
4. (Atiyah et al., 1994), (Thurston, 1994); (Jaffe and Quinn, 1994). Occasionally I will also use the minutes of a roundtable on 'Physics and Mathematics' that was organised at the XIth congress of the International Association of Mathematical Physicists, published in (Iagolnitzer, 1995).
5. (Jaffe, 1997), p. 140.
6. Lakatos is mentioned by Mazur, but his concept of conjecture is misunderstood there. Recently, David Corfield (1997), p. 118, has also cited the Jaffe-Quinn debate.
7. Michael Atiyah in (Atiyah et al., 1994), p. 179.
8. (Corfield, 1997), p. 100.
9. "Anomalies versus 'Crucial Experiments' (A Rejoinder to Professor Grünbaum)" in (Lakatos, 1978b), p. 213, fn. 3.
10. (Jaffe and Quinn, 1993), p. 1.
11. Ibid.
12. Ibid., p. 2.
13. Ibid., my emphasis.
14. (Atiyah et al, 1994), p. 186.
15. (MacLane, 1997), p. 151.
16. Ibid.
17. (Thurston, 1994), p. 171.
18. (Atiyah et al., 1994), p. 203.
19. Ibid., p. 2.
20. (Atiyah et al., 1994), p. 180.
21. Ibid., p. 186.
22. Ibid., p. 202.
23. See also Thom, ibid., p. 204, in this respect.
24. Ibid., p. 197.
25. (Jaffe and Quinn, 1993), p. 3.
26. (Iagolnitzer, 1995), p. 704, my emphasis.
27. Witten in (Atiyah et al., 1994), p. 206.
28. Lakatos: "A Renaissance of Empiricism in the Recent Philosophy of Mathematics", in: (Lakatos, 1978b), p. 29.
29. Ibid., p. 28f.

30. (Atiyah et al., 1994), p. 193.
31. (Jaffe, 1997), p. 135.
32. Lakatos: "A Renaissance of Empiricism ...", p. 35.
33. Ibid., p. 36.
34. Lakatos: "What does a Mathematical Proof Prove?", in (Lakatos, 1978b), p. 67.
35. Ibid., p. 69.
36. (Jaffe and Quinn, 1993), p. 4.
37. I have some doubts as to whether Jaffe and Quinn's account of the relationship between theoreticians and experimentalists really tallies with the facts.
38. Perhaps, it should be stressed that this concerns, above all, frontier physicists (cf. Uhlenbeck's point, cited above).
39. (Jaffe and Quinn, 1993), p. 6.
40. Ibid., p. 7.
41. Ibid.
42. Ibid.
43. (Atiyah et al., 1994) p. 185.
44. Ibid., p. 187. Gray's remarks on p. 185 support this point.
45. (Jaffe and Quinn, 1993), p. 7f.
46. (Atiyah et al., 1994), p. 207.
47. (Jaffe, 1997), p. 141.
48. (Quinn and Jaffe, 1993), p. 8.
49. (Atiyah et al., 1994), p. 178.
50. (MacLane, 1997), p. 150.
51. (Mazur, 1997), p. 198.
52. Ibid., p. 202.
53. (Jaffe and Quinn, 1993), p. 6.
54. (Atiyah et al., 1994), p. 188.
55. Ibid.
56. Ibid., p. 191.
57. (Jaffe and Quinn, 1993), p. 2.
58. Ibid., p. 6.
59. (Rota, 1997b), p. 188.
60. (Jaffe and Quinn, 1993), p. 10.
61. Ibid., p. 11.
62. (Jaffe, 1997), p. 143.
63. (Atiyah et al., 1994), p. 200.
64. Ibid., p. 204.
65. Ibid., p. 194.
66. Ibid., p. 178.
67. Ibid., p. 180.
68. Lakatos: "The Problem of Appraising Scientific Theories: Three Approaches", in (Lakatos, 1978b), p. 111.
69. Ibid., p. 114.
70. Ibid., p. 117.
71. (Atiyah et al., 1994), p. 195.
72. Lakatos: "The problem of appraising ...", p. 111, fn. 4, writes: "Elitism is closely related to the doctrine of *Verstehen*."
73. (Thurston, 1994), p. 163.
74. See their "Reply to Thurston" in (Atiyah et al., 1994), p. 210.
75. (Thurston, 1994), p. 166.
76. Ibid., p. 167.
77. Ibid., p. 169.
78. Ibid., p. 172.
79. Ibid.
80. Ibid., p. 173.
81. Ibid., p. 176.
82. Ibid.
83. (Rota, 1997b), p. 184.

84. Ibid., p. 187.
85. (Rota, 1997a), p. 176.
86. (Rota, 1997b), p. 191.
87. Ibid.
88. Ibid., p. 190.
89. Ibid.
90. (Lakatos, 1976), p. 9. Lakatos gives two historical sources for this terminological analogy to science: *Thought-experiments* ('deiknymi') prevailed, according to Árpád Szabó, in pre-Euclidean Greek mathematics. The term *quasi-experiment* stems from an editorial summary preceding Euler's "Specimen de usu Observationum in Mathesi Pura".
91. Ibid., p. 14.
92. Ibid., p. 31.
93. Ibid., fn. 3.
94. See fn. 3 on p. 63 of "Falsification and the Methodology of Scientific Research Programmes" in (Lakatos, 1978a)
95. (MacLane, 1997), p. 150.
96. (Lakatos, 1976), p. 34.
97. Ibid., p. 37.
98. Ibid.
99. Rules 1–3 on p. 50, rule 4 on p. 58, rule 5 on p. 76 of (Lakatos, 1976).
100. Fn. 3 on p. 135 of "Changes in the Problem of Inductive Logic" in (Lakatos, 1978b).
101. (Lakatos, 1976), p. 56.
102. See Worrall and Zahar's editors' note on p. 138 of (Lakatos, 1976).
103. (Lakatos, 1978a), p. 62 ("Falsification and ...").
104. (Lakatos, 1976), p. 65.
105. Ibid., p. 73.
106. Ibid., p. 74.
107. Fn. 3 on p. 176 of "Changes in the Problem of Inductive Logic" in (Lakatos, 1978b).
108. (Lakatos, 1976), p. 87, fn.1.
109. "What does a Mathematical Proof Prove?", in (Lakatos, 1978b), p. 63.
110. "Cauchy and the Continuum", in (Lakatos, 1978b), p. 59.
111. Lakatos' picture of Logical Empiricism and the Vienna Circle was shaped by Popper's deliberately selective reading of these movements – the only exception being Victor Kraft, the first philosopher he met after leaving Hungary in 1956.
112. (Lakatos, 1978a), p. 52, fn. 1.
113. See (Koetsier, 1991) and (Corfield, 1997).
114. "Falsification ...", in (Lakatos, 1978a), p. 66, fn.2.
115. "Introduction: Science and Pseudoscience", in (Lakatos, 1978a), p. 6.
116. Ibid., p. 88.
117. Ibid.
118. (Corfield, 1998), p. 276.
119. Ibid., p. 295.
120. Ibid., p. 280.
121. Ibid., p. 281.
122. (Jaffe and Quinn, 1993), p. 2.
123. (Iagolnitzer, 1995), p. 700.
124. Lakatos: "Introduction: Science and Pseudoscience", in (Lakatos, 1998a), p. 1.
125. "Changes in the Problem of Inductive Logic", in (Lakatos, 1978b), p. 175.
126. (Atiyah et al., 1994), p. 189.
127. (Lakatos, 1976), p. 123.
128. Ibid., p. 109.
129. (Corfield, 1997), p. 108.
130. Lakatos: "A Renaissance of Empiricism ...", p. 31.
131. (Lakatos, 1976), p. 123.
132. "What Does A Mathematical Proof Prove", in (Lakatos, 1978b), p. 66.
133. Ibid., p. 62.
134. (Corfield, 1997), p. 112.
135. (Mazur, 1997), p. 200.

136. Ibid., p. 198; italics of the original removed.
137. Ibid., p. 114.
138. (Lakatos, 1976), p. 72. Interestingly, in some editions of *Proofs and Refutations* figure 19 consists of a blot of decent size •. The 1994 reprint with corrections (as of 1981) – which I am using – agrees with the original figure in *The British Journal for the Philosophy of Science* 14 (1964), p. 302. This non-figure of empty space is also much closer to the anti-inductivist thrust of Lakatos' argument. But with a dialectical smile, one might grant some figurative representation for Beta's view. (I thank Kluwer's British Production Unit for pressing me with this figure.)
139. Ibid., p. 73.
140. Lakatos: "The Method of Analysis-Synthesis", in (Lakatos, 1978b), p. 97.
141. Ibid., p. 89. An interesting study of Popper's position is (Schurz, 1994).

REFERENCES

Michael Atiyah et al. (1994) "Responses to 'Theoretical Mathematics: Toward a Cultural Synthesis of Mathematics and Theoretical Physics'", *Bulletin of the American Mathematical Society* 30, pp. 178–211.

David Corfield (1997) "Assaying Lakatos's Philosophy of Mathematics", *Studies in the History and Philosophy of Science* 28, pp. 99–121.

David Corfield (1998) "Beyond the Methodology of Mathematical Research Programmes", *Philosophia Mathematica* III, 6, pp. 272–301.

Jaakko Hintikka (1997) "A Revolution in the Foundations of Mathematics", *Synthese* 111, pp. 155–170.

Daniel Iagolnitzer (ed.) (1995) *Proceedings of the XIth International Congress of Mathematical Physics.* Cambridge, MA: International Press.

Arthur Jaffe and Frank Quinn (1993) "'Theoretical Mathematics': Toward a Cultural Synthesis of Mathematics and Theoretical Physics", *Bulletin of the American Mathematical Society* 29, pp. 1–13.

Arthur Jaffe and Frank Quinn (1994) "Response to Comments on 'Theoretical Mathematics'", *Bulletin of the American Mathematical Society* 30, pp. 208–211.

Arthur Jaffe (1997) "Proof and the Evolution of Mathematics", *Synthese* 111, pp. 133–146.

Teun Koetsier (1991) *Lakatos' Philosophy of Mathematics: A Historical Approach,* Amsterdam: North-Holland.

Imre Lakatos (1976) *Proofs and Refutations. The Logic of Mathematical Discovery,* edited by John Worrall and Elie Zahar, Cambridge: Cambridge University Press.

Imre Lakatos (1978a) *The Methodology of Scientific Research Programmes* (Philosophical Papers Volume 1), edited by John Worrall and Gregory Currie, Cambridge: Cambridge University Press.

Imre Lakatos (1978b) *Mathematics, Science and Epistemology* (Philosophical Papers Volume 2), edited by John Worrall and Gregory Currie, Cambridge: Cambridge University Press.

Saunders MacLane (1997) "Despite Physicists, Proof is Essential in Mathematics", *Synthese* 111, pp. 147–154.

Barry Mazur (1997) "Conjecture", *Synthese* 111, pp. 197–210.

Gian-Carlo Rota (1997a) "The Phenomenology of Mathematical Beauty", *Synthese* 111, pp. 171–182.

Gian-Carlo Rota (1997b) "The Phenomenology of Mathematical Proof", *Synthese* 111, pp. 183–196.

Gerhard Schurz (1994) "Karl Popper und das Induktionsproblem", in: Martin Seiler and Friedrich Stadler (eds.), *Heinrich Gomperz, Karl Popper und die 'Österreichische Philosophie',* Amsterdam-Atlanta, GA: Rodopi, 1994, pp. 147–161.

David Sherry (1997) "On Mathematical Error", *Studies in the History and Philosophy of Science* 28, pp. 393–416.

William Thurston (1994) "'Theoretical Mathematics': Toward a Cultural Synthesis of Mathematics and Theoretical Physics", *Bulletin of the American Mathematical Society* 30, pp. 161–177.

Institut Wiener Kreis
Museumstr. 5/2/19
A-1070 Wien
Austria

Teun Koetsier

Lakatos' Mitigated Scepticism in the Philosophy of Mathematics

1. Introduction

Lakatos liked to view his work in the philosophy of mathematics against the background of the traditional epistemological battle between dogmatists and sceptics. Dogmatists are those who hold A) that we can attain truth and B) that we can know that we have attained truth. Sceptics are those who hold A) that we cannot attain truth, or at least B) that we cannot know that we have attained truth. Lakatos himself represented a form of mitigated scepticism (often called critical fallibilism). Like the sceptics, he held A) that we cannot attain truth, or at least B) that we cannot know that we have attained truth, but he held in addition – and in this respect he distinguished himself from extreme sceptics – C) that we can improve our knowledge and know that we have improved it.

In this paper I will consider the position of Lakatos against the background of this battle between sceptics and dogmatists. I will distinguish Lakatos$_1$, the early Lakatos, who mainly wrote about the philosophy of mathematics, from Lakatos$_2$, the later Lakatos, who developed the methodology of scientific research programmes and wrote almost exclusively about the philosophy of science. I will argue that in this battle, Lakatos$_1$ sided with the sceptics and attacked the dogmatists. A major goal of his reconstruction of developments in mathematics was to show the fallibility of mathematical knowledge. The reconstructions reveal a rich logic of mathematical discovery, the methodology of proofs and refutations (MP&R), but for Lakatos$_1$, at heart, the rationality of the development of mathematics needed no defence. The position of Lakatos$_2$ against the background of the battle between sceptics and dogmatists is entirely different. Lakatos$_2$ no longer took the rationality of the development of science and mathematics for granted. On the contrary, his primary goal was now to defend that rationality against extreme scepticism. It was no longer the fallibility of knowledge but the mechanism of the improvement of knowledge that he concentrated on. The methodology of scientific research programmes (MSRP) was Lakatos$_2$'s answer to the extreme sceptics.

Rational reconstructions are Lakatos' main tool to defend both the fallibility of mathematical knowledge and the rationality of its development.

G. Kampis, L. Kvasz and M. Stöltzner (eds.),
Appraising Lakatos: Mathematics, Methodology and the Man, 189–210.

In the present paper I will concentrate on the reconstruction that $Lakatos_1$ gave of the developments in the foundations of mathematics at the beginning of the 20th century in a 1962 paper called "Infinite Regress and the Foundations of Mathematics."[1] I will also look at these developments from the point of view of $Lakatos_2$ and conclude that they can be reconstructed much better in terms of the MSRP. Finally I will draw the conclusion that $Lakatos_2$ represents a form of mitigated scepticism that does not contradict, but rather supplements the mitigated dogmatism that many mathematicians are inclined to adhere to after failed attempts to give all of mathematics once and for all a crystal clear foundation.

2. FROM PYRRHO TO LAKATOS

I will first put Lakatos' scepticism in a historical perspective. The fullest early presentation of antique scepticism can be found in the work of Sextus Empiricus (second or third century), of the Pyrrhonist school in Alexandria. The Pyrrhonists attributed their origins to Pyrrho of Ellis (c. 360–c. 270 BC). Sextus suspends judgement about all dogmatic claims to knowledge extending beyond immediate experience. One of his arguments casting doubt runs as follows. Ask for the criterion on the basis of which a claim is made. The dogmatist will find the criterion, for example, logical inference, evident. However, one can then ask for the criterion on the basis of which he finds it evident. Inevitably the dogmatist will be faced with begging the question or with infinite regress. The Pyrrhonist advice to man was to live peacefully and follow natural inclinations without subscribing to definite views.

Sceptical criticism always hurts those who believe in indubitable knowledge. It is a fundamental problem of all dogmatist positions. Yet the sceptical position is not unproblematic either. The fundamental problem in a consistent sceptical position concerns the conflict between the sceptical philosophical view and the natural inclinations of the philosopher. A serious sceptic must solve this problem in some way.[2] The fundamental problems of both the dogmatist and sceptical positions were clearly described by Pascal. In the *Pensées* Pascal writes that he cannot imagine a consistent Pyrrhonist: "What will a man in this state do? Will he doubt everything? Will he doubt whether he is awake, or being pinched or being burned? Will he doubt whether he is doubting? Or whether he exists? One cannot go that far and I state that there never existed a complete factual Pyrrhonist. Nature supports the powerless reason and prevents it from extravagating that far."[3] Pascal uses the argument to show that reason, whether in its sceptical or in its dogmatist form, is at heart powerless: nature confuses the sceptics and doubt confuses the dogmatists. Pascal's advice is to listen to God.

As Richard Popkin has shown,[4] religious controversy and the rediscovery of the classical sceptical texts led to a revival of scepticism in the sixteenth and seventeenth centuries. It was also a period of great scientific successes, which no one could ignore. The sceptics Gassendi and Mersenne, in the first half of the seventeenth century, granted that the problems raised by Sextus could not be solved in a fundamental sense, but they asserted that the theories of natural science – data about appearances and deductions from hypotheses – could be pragmatically used as a guide for living in the world. Popkin describes the views of Mersenne as follows: "The dogmatist and the destructive sceptic were both wrong, the former insisting that we can and must have knowledge of reality, the latter insisting that everything is in doubt. Between the two views lies a new outlook, constructive scepticism, doubting our abilities to find grounds for our knowledge, while accepting and increasing the knowledge itself."[5] The major eighteenth century sceptical philosopher, David Hume,[6] also struggled with the fundamental problem of the sceptical position. His answer was a theoretical scepticism combined with a pragmatic dogmatism. On the one hand, Hume could not help believing what we all believe concerning reality, but, on the other hand, he denied that those instinctive beliefs can be rationally justified. Our allegedly rational conclusions concerning reality are in Hume's view based upon various psychological and biological factors.

The remarkable thing about the twentieth century is that there are many philosophers who represent sceptical points of view. Often, however, they do not call themselves sceptics. I will restrict myself here to Popper's critical fallibilism, which had a great impact on Lakatos. As we saw above, a critical fallibilist accepts the sceptic's view that we cannot attain truth or at least cannot know whether we have attained it, but the critical fallibilist also argues that we can improve our knowledge and that we can know that we have improved it. The critical fallibilist is a sceptic of certainty; for him or her knowledge consists – at its heart – of guesses. However, the critical fallibilist at the same time argues that guesses can be improved through some "logic of discovery." Popper himself wrote that his critical fallibilism is not radically different from the positions of some of the ancient sceptics.[7] Moreover, one of the starting points of Popper's philosophy is the work of Hume himself. Popper agreed with Hume's sceptical criticism of the method of induction. He disagreed, however, with Hume's conviction that because of that, our "knowledge" of the world is a rationally indefensible belief.[8] Seen in terms of the battle between dogmatists and sceptics, both Popper's and Lakatos' critical fallibilism are attempts to provide a new solution to the fundamental problem of scepticism.

3. LAKATOS$_1$

3.1. Proofs and refutations

Lakatos$_1$ challenged dogmatism in the philosophy of mathematics by means of two rational reconstructions of developments in mathematics. Mathematics has always been a stronghold of dogmatism, and in order to shake the received view Lakatos selected developments that reveal the fallible character of mathematics: he concentrated on informal mathematics. In "What does a mathematical proof prove?,"[9] written probably between 1959 and 1961, Lakatos emphasises that informal proofs or rather pre-formal proofs should be distinguished from "quasi-formal" proofs. A quasi-formal proof or "a formal proof with gaps"[10] is "a proof in an axiomatized mathematical theory which has already taken the shape of a hypothetico-deductive system, but which leaves its underlying logic unspecified."[11] It is a fact that most of twentieth century mathematics takes place in axiomatized theories that leave the underlying logic unspecified. So it is not amazing that Lakatos concentrated on pre-twentieth-century developments and on developments in the foundations of mathematics.[12] First there is the reconstruction of the 19th century history of Euler's formula for polyhedra in "Proofs and Refutation."[13] In the introduction of "Proofs and Refutations," Lakatos placed his work in the context of the argument between dogmatists and sceptics. "The dogmatists hold that – by the power of our human intellect and/or senses – we can attain truth and know that we have attained it. The sceptics, on the other hand, hold either that we cannot attain truth at all (unless with the help of mystical experience), or that we cannot know if we can attain it or that we have attained it."[14] By elaborating the point that informal mathematics grows through the incessant improvement of guesses by the logic of proofs and refutations, he challenged directly mathematical formalism, which identifies mathematics with its twentieth century formal axiomatic abstraction. However, because a similar pattern of guesses and improvement of guesses occurred in the foundations of mathematics, he – indirectly – challenged contemporary mathematical dogmatism, as well. The latter challenge was elaborated upon in "Infinite Regress" in which Lakatos reconstructed the developments in metamathematics in the first half of the century.

"Proofs and Refutations" is an important contribution to the methodology of mathematics. Lakatos discusses several of the mechanisms of growth in the development of mathematics, without really doubting that the development of mathematics constitutes progress. Hegelian remarks on the "autonomous dialectic of mathematics"[15] clearly indicate that Lakatos$_1$ took the overall rationality of the development of mathematics pretty much for granted. His major concern, of course, was to demonstrate the fallibility of mathematics. The reconstruction shows how proved theorems are repeatedly refuted and replaced by other proved theorems. The question as to whether these replacements lead to progress, to the improvement of mathematical knowledge, is posed but not conclusively answered. "Proofs and Refutations" is very rich in content, but there is nothing an extreme sceptic would worry about. The character Kappa, who represents such a cynical sceptic, is given a lot of space to make his points in the last seven pages of the dialogue; the subsequent exclamation of Beta, which concludes the conversation,[16] "But I had no problems at the beginning! And now I have nothing *but* problems!" is very much to the point. It is easy to read "Proofs and Refutations" as an illustration of the fact that in informal mathematics A) we cannot attain truth, or at least B) we cannot know that we have attained truth.

Elsewhere, I have compared Lakatos' rational reconstruction of the history of Euler's formula with the real history, and I concluded that the real history is considerably less dramatic than its reconstruction.[17] Lakatos makes his point that mathematical knowledge is fallible, but even in 19th century mathematics this fallibility is remarkably weak. I argued that real history shows that older theories are repeatedly refuted in the sense that, from a new perspective, old definitions appear to be one-sided; originally unintended interpretations of the old definitions yield counter-examples to the older theories. The resulting changes can usually be characterised as a "refinement"; the new theory is more refined but usually it does not really refute the old theory. The fact that old informal proofs like, for example, Cauchy's proofs of Euler's formula, can still be used in a pedagogical context without any sense of contradiction also illustrates this. Lakatos' point is that informal mathematics is fallible. He is right. However, the fallibility of informal mathematics is very weak.

3.2. "Infinite regress and the foundations of mathematics"

In "Infinite Regress," $Lakatos_1$ also challenged the dogmatists in the philosophy of mathematics. In the first section of that paper, Lakatos briefly discusses the general situation of the epistemology of science by means of a rational reconstruction of its history as a battle between dogmatists and sceptics. The major dogmatist enterprise of saving science from the sceptics and giving it a firm foundation is what Lakatos calls Euclideanism. Euclideanism is the attempt to develop science as a "Euclidean theory," i.e. a deductive system with injections of infallible truth that inundate the whole system from the top.[18] In Lakatos' opinion, Euclideanism in science has been on the retreat since the seventeenth century, and in the end, all attempts to give science once and for all a certain foundation collapsed because of sceptical criticism. For $Lakatos_1$, the way out is Popperian critical fallibilism, in which the search for an ultimate foundation is given up and the falsification of theories – crucial injections of the truth-value 'False' from the bottom – plays a central role. The question *How do we know*?, which causes only pseudo-problems, is replaced by the question *How do we improve our theories?* In the second and third sections of the paper, Lakatos discusses the logicist, intuitionist and formalist foundational attempts in order to show that even in mathematics, a traditional dogmatist bulwark, Euclideanism is on the way out. By means of what he calls logico-trivialization, the logicists, that is in Lakatos' description in particular Russell, tried to create an infallibly true foundation and to stop in that way the infinite regress generated by sceptical criticism. They did not succeed. Lakatos' description of the subsequent changes in Russell's views over the course of time is fascinating. First, according to Lakatos, Russell was convinced that he could give mathematics a firm Euclidean foundation. Soon, however, the Euclidean "honeymoon" gave place to "intellectual sorrow."[19] Lakatos quotes Russell writing in 1924 that, "if we are to believe in the truth of pure mathematics, it cannot be solely because we believe in the truth of the set of premises. Some of the premises are much less obvious than some of their consequences, and are believed chiefly because of their consequences."[20] In the end, according to Lakatos, Russell gave up the philosophy of mathematics. Lakatos writes here: "Let us now draw some of the conclusions which Russell refused to draw. The infinite regress in proofs and definitions in mathematics cannot be stopped by a Euclidean logic. Logic may *explain* mathematics but cannot prove it. It leads to sophisticated speculation which is anything but trivially true. The domain of triviality is limited to the uninteresting decidable kernel of arithmetic and of logic – but even this trivial kernel might some time be overthrown by some detrivializing sceptic criticism."[21] In Lakatos' reconstruction, Russell's attempts to give mathematics a firm foundation are followed by a brief discussion of the intuitionist programme. Intuitionism offered no solution either, according to Lakatos. Lakatos writes: "Pure-grained sceptics are rare: we find, however, that pessimistic dogmatists are virtually sceptics. These pessimistic dogmatists demand that we should abandon speculation and restrict our attention to some narrow field which they gracefully – but without any real justification – acknowledge to be safe. In modern mathematical philosophy Intuitionism represents this brand of destructive, sceptical dogmatism, 'a treason to our science,' as Hilbert put it [...]."[22] In Lakatos' rational reconstruction, Hilbert's formalist programme is the next attempt to turn mathematics in a Euclidean theory. The idea was to prove the consistency of mathematics by means of meta-considerations "so obviously true that not even sceptics can doubt them."[23] However, according to Lakatos, Gödel's discoveries in particular made Hilbert's programme collapse.

The reconstruction in "Infinite Regress" shows us the history of the foundations of mathematics as consisting of theories that are framed, refuted, and replaced by other theories that are in their turn replaced by others. The story is illustrated with well chosen quotations that show the doubts of the mathematicians involved or their contradicting views on, for example, the status of axioms. Lakatos succeeds in making his point, namely, that the absolute certainty of mathematical theories is a myth: the dogmatists do not succeed in giving mathematics its final and firm foundation. On the other hand, the fallibility that the reconstruction shows is again remarkably weak. Lakatos concentrates on the failures and does not mention the successes that were achieved. The question is whether the developments in the foundations of mathematics really show that Euclideanism is on the way out. I will argue below that this interpretation is much too simple.

4. LAKATOS$_2$

4.1. Lakatos opposes the sceptics

Lakatos$_1$ concludes the first section of "Infinite Regress" with the following remark: "The indefatigable sceptic will ask again: "How do you know that you improve your guesses? But the answer is easy: 'I guess.' There is nothing wrong with an infinite regress of guesses."[24] This makes some sense. If there is no claim to indubitable knowledge the sceptic's question "How do you know?" misses its target. Yet in the argument the fallibilist stops claiming anything at all. Saying that there is nothing wrong with "I guess that I improved my guess" is in fact not much different from joining the sceptics. But Lakatos$_1$ was not primarily interested in such matters; it was his goal to challenge the dogmatists, not to defend fallibilism from the extreme sceptics. However, a mitigated scepticism like fallibilism must provide an answer to the question: How can mathematicians be sure that their science is improving? Part of the answer was given by Lakatos$_2$, who no longer took the overall rationality of the development of science and mathematics for granted, but felt a strong need to defend that rationality. For example, in his "Popper on Demarcation and Induction,"[25] Lakatos expresses his concern that before 1960, Popper never said anything against scepticism, and that he did not distinguish scepticism from fallibilism.[26] According to Lakatos, Popper's "Logik der Forschung" contains nothing with which the most radical sceptic need disagree.[27] In that book, scientific progress is the increased awareness of ignorance, rather than the growth of knowledge. "It is '*learning*' without ever knowing."[28] The shift of 1960 in Popper's work was, according to Lakatos, the result of Tarski's rehabilitation of the correspondence theory of truth. When Popper became aware of Tarski's theory, he realised that it was possible to define progress in terms of the degree of correspondence to the truth of theories, where the verisimilitude of a theory equals truth content minus falsity content. However, Lakatos is not too happy with Popper's final position. These critical remarks of Lakatos with respect to Popper and scepticism are related to Lakatos' other work at that time. Lakatos$_2$ was engaged in a vehement battle with what he considered to be extreme sceptics like Kuhn and Feyerabend. Kuhn later denied that he had intended to defend an extreme scepticism, but it is very understandable that Lakatos interpreted him in that way. If a scientific revolution is an irrational change, comparable to a religious conversion, then we cannot distinguish science from pseudo-science – and the extreme sceptics are right. Lakatos$_2$ felt challenged to give a precise answer to the question of how scientists and mathematicians improve their science. His answer was the Methodology of Scientific Research Programmes.

4.2. The MSRP applied to mathematics

4.2.1. The MSRP

The Methodology of Scientific Research Programmes boils down to the following: A scientific research programme manifests itself in the form of a series of successively developed theories that have two elements in common: a hard core and a positive heuristic. The hard core consists of conjectures that are irrefutable within the programme. The positive heuristic consists of rules that generate new and better theories within the programme. If new predictions are generated within the programme, that is called theoretical progress; if the predictions are confirmed, that is empirical progress. A rational scientific community tends to join the most progressive research programmes, and progressive research programmes replace stagnating ones.[29] The MSRP represents – with respect to mathematics – a step forward. In order to show this, I will rationally reconstruct, in terms of the MSRP, the history of the foundations of mathematics as described by Lakatos in "Infinite Regress." The view that the development of the foundations of mathematics in the first half of this century consists of three failed attempts to give a Euclidean foundation to mathematics is too much a caricature of the real developments; an application in terms of the MSRP makes this clear. Such an application of the MSRP to mathematics obviously requires some modifications. For example, the notions of theoretical and empirical progress are not immediately applicable to mathematics. However, in this paper I will use the notions of hard core, positive heuristic, progress and stagnation rather loosely.[30] I will, for example, measure the progress of research programmes in mathematics in terms of the new results that are being generated by a programme, without attempting to give a precise definition of a result. I will distinguish three major foundational programmes in mathematics at the beginning of this century: Russell's logicist programme, Brouwer's intuitionist programme and Hilbert's formalist programme. The three programmes appeared at approximately the same time. Russell defined his programme in 1900 after reading Peano's work, and in his "The Principles of Mathematics"[31] of 1903 the first version of the logicist solution to the problems of the foundation of mathematics was given. Brouwer defined the intuitionist programme in his doctoral dissertation "Over de Grondslagen der Wiskunde" of 1907 and Hilbert gave the first outline of the formalist programme in his paper "Über die Grundlagen der Logik und Arithmetik" of 1904. However, while the logicist and the intuitionist programmes were developing independently of each other and of Hilbert's programme, Hilbert first turned his attention to other matters. One reason is that Russell and Brouwer were primarily interested in the foundational problems, while Hilbert was primarily interested in mathematics as a whole. I will argue below that all three programmes failed in the sense that they did not quite yield what their intellectual fathers expected of them. Yet it would be totally wrong to describe them as three failures rooted in a wrong view on the nature of mathematics, Euclideanism. I will argue that all three programmes contributed considerably to mathematics as a whole. The competition of the three programmes and their successes and failures must be seen against the background of the development of mathematics altogether. It is tempting to view Hilbert's mathematical work as representing a huge overall "structuralist" mathematical research programme that, after a period of competition, came to dominate twentieth-century mathematics. The position of Hilbert's foundational programme is special in this respect because it represented the mathematics of the future, while Russell's and Brouwer's programmes had very close ties to nineteenth century non-structuralist mathematics and, to a certain extent, represented the past. I will finally argue that the behaviour of the mathematical community with respect to the three programmes was quite rational from the point of view of Lakatos' MSRP.

4.2.2. The logicist foundational programme

Russell once wrote: "What I myself have had to say whether about mathematics or about physics [...] has proceeded always by a certain method. [...] what are the minimum hypotheses from which this broad measure of truth will result?"[32] This statement was not merely incidental; it refers to a fundamental element of Russell's way of working. Paul Hager argues, in an interesting book about continuity and change in the development of Russell's philosophy, that the method of analysis is one of Russell's fundamental tools.[33] This is, for example, clear in the 'Preface' of "Principia Mathematica." There, Russell states that the task of mathematical philosophy is "to analyse existing mathematics, with a view to discovering what premises are employed, whether these premises are mutually consistent, and whether they are capable of reduction to more fundamental premises." Moreover, writes Russell, once "we have decided upon our premises, we have to build up again as much as may seem necessary of the data previously analysed, and as many other consequences of our premises as are of sufficient general interest to deserve statement."[34] Hager stresses that for Russell an analysis is unlikely to be final – there is no reason to suppose that it is impossible to find simpler ideas and axioms – and the premises established by analysis are far from being self-evident.[35] There are many quotations that support Hager's view. I consider the method of analysis, in this sense, in combination with the conviction that mathematics is logic in disguise as the central element of the positive heuristic of Russell's foundational programme. I will now briefly follow the development of that programme, starting from the moment when Russell, after reading Peano's work in 1900, discovered independently (Frege had already discovered it earlier) that the natural numbers could be defined as equivalence classes by means of the equivalence relation: set x is equivalent to set y iff there exists a one-to-one correspondence between x and y. Peano's analysis had led to three primitive notions (0, number and successor) and 5 postulates. Russell's further analysis could reduce the natural numbers to "such purely logical terms as *or*, *not*, *all* and *some*."[36] In that same year – 1900 – Russell wrote, in a euphoric mood, a first draft of "The Principles of Mathematics."[37] The overall positive heuristic in combination with a specific idea of how to reduce arithmetic, and in the end all of mathematics to logic, generated results. He wrote later: "Every day I found myself understanding something that I had not understood on the previous day."[38] This was the "intellectual honeymoon" that Lakatos mentioned, and it was indeed followed by "intellectual sorrow" in the next year.[39] Cantor's proof that there is no greatest cardinal number led Russell to the paradox that is named after him. While he searched for a solution, three prerequisites were on his mind: i) the contradictions should disappear, ii) the solution should leave intact as much of mathematics as possible, iii) "the solution should, on reflection, appeal to what may be called 'logical common sense' – i.e. that it should seem, in the end, just what one ought to have expected all along."[40] Russell drew the conclusion that the paradoxes were caused by the assumption that: for every propositional function, there exists the class of objects satisfying it.[41] In a 1907 paper, he discussed two possibilities of defining a criterion to restrict the existence of classes or concepts in general, which he called the zig-zag theory and the theory of limitation of size. Gödel wrote in 1944 that the axiomatic set theory developed later by Zermelo can be considered as an elaboration of the second idea. Quine's 1937 system shares some essential characteristics with the zig-zag theory. However, Russell's logicist convictions led him in another direction and he based himself on a more radical idea, the no-class theory: classes and concepts never exist as real things, they are just a manner of speaking about things. In other words "a notion is a symbol together with a rule for translating sequences containing the symbol into such sentences as do not contain it, so that a separate object denoted by the symbol appears as mere fiction."[42] According to Gödel, the no-class theory led Russell to the vicious circle principle and the type theory, as exposed in detail in what was originally meant to be the second volume of "The Principles of Mathematics," but became the first edition of "Principia Mathematica" (1910–13), which Russell wrote together with Whitehead. "Principia Mathematica" contains valuable contributions to logic. However, there was a major problem, which already became clear in the first edition. Many fundamental theorems in the theory of real numbers seemed essentially to require the so-called axiom of reducibility. In the extra introduction, which distinguishes the second edition of the "Principia Mathematica" (1925–27) from the first edition, Whitehead and Russell described the problem of the axiom of reducibility as follows: "This axiom has a purely pragmatic justification: it leads to the desired results, and no others."[43] Their remarks make it clear that development of the programme was stagnating. The programme ran into a problem for which the logicist positive heuristic had no solution. Whitehead and Russell refer to the work of L. Chwistek, who attempted to dispense with the axiom of reducibility, but was forced to sacrifice a great deal of ordinary mathematics.[44] Other possible solutions also led to the collapse of important parts of classical mathematics. Reacting to the second edition of "Principia Mathematica," Ramsey wrote two papers, published in the 1926, in which the problem of the axiom of reducibility was taken up and another way out was suggested. In a sense, the logicist programme stopped developing further after those papers. Logicism did not succeed in solving the problem of the axiom of reducibility. Although the axiom did not lead to contradictions, it appeared in no way possible to view it as appealing "to what may be called 'logical common sense' – i.e. that it should seem, in the end, just what one ought to have expected all along." Because classes and concepts introduced on the basis of the no-class theory do not have all the properties required for their use in ordinary mathematics, the axiom of reducibility is needed to guarantee the existence of certain concepts. Ramsey's work avoids the axiom of reducibility, but it is based on a different assumption contrary to the spirit of the no-class theory: that one can form propositions of infinite and even non-denumerable length. Moreover, the axiom of reducibility was not the only problem that the logicist programme ran into. Russell was also forced to accept an axiom guaranteeing the existence of an infinite set and the axiom of choice. In the course of time, the idea that mathematics as a whole can be reduced in a simple way to logic seemed to be wrong, and was no longer seen as a viable option. The logicist programme has exerted great influence, in particular on the development of logic. However, with respect to the ultimate goal to reduce all of mathematics to logic, the progress of the programme stagnated and it was shelved.

4.2.3. The intuitionist foundational programme[45]

In Brouwer's life (1881–1966) as an active mathematician, three periods can be distinguished: i) Early intuitionist work (1905–1908), ii) Topological work (1908–1913) and iii) Later intuitionist work (1917–1919). All three periods are dominated by Brouwer's intuitionist views. Brouwer's starting point is totally different from that of Russell. In Russell's view of mathematics, logic and linguistic constructions play a major role. In Brouwer's view of mathematics, the role of logic and language is absolutely secondary. In Brouwer's dissertation,[46] the intuitively given continuum is the starting point. For Brouwer, the basic intuition of mathematics is the continuous flow of time, in which we can distinguish different moments. The continuum is like a "matrix" that can be filled with points, not arbitrarily, but in accordance with the nature of the continuum. In order to deal with the continuum mathematically, Brouwer applies a potentially infinite system of points (cuts) on the continuum that he calls the "dual scale." By means of this dual scale, he turns the intuitively given continuum into a "measurable continuum" as follows. In the intuitive continuum, he constructs "a sequence of points having the order type of the positive and negative whole numbers." This sequence of points is potentially infinite, and so is the sequence of intervals that it yields. Each of the intervals is now cut up by means of repeated dichotomies. Brouwer writes: "then if we add a point in every interval, then again in each of the intervals so obtained, and so on, we obtain the order type η on the continuum, which in this way comes to correspond with the system of the finite dual fractions."[47] Those dual fractions are the elements of the set of rational numbers, written in the binary system. At the time, Brouwer believed that he could handle the continuum by means of this potentially infinite scale. He remarks that such a dual scale need not be dense everywhere, which means by definition that it does not necessarily penetrate into every segment of the continuum. However, he says, we can contract every segment not penetrated by the scale into one point. The result is, says Brouwer, that by definition the scale will be dense everywhere. In this way, the potentially infinite "dual scale" turns the intuitive continuum into the "measurable continuum," which is in fact the standard-continuum and can be identified with the system of real numbers. The example illustrates Brouwer's positive heuristic: mathematics consists of mental constructions on the basis of a fundamental mathematical intuition.

In 1907, it was not at all Brouwer's intention to do away with the results of classical mathematics. On the contrary: he considered most of classical mathematics to be sound. He still assumed, in the words of Van Stigt, "that the classical treatment of 'real number' can be justified and form a sound basis for analysis and function theory."[48] It is interesting to see how Brouwer introduces the arithmetic operations on the measurable continuum, the construction of which we described above, and how he introduces the various geometries. He did it, namely, by means of groups of one-to-one continuous transformations and well-chosen topological restrictions. Moreover, he felt this to be the only correct way. He wrote: "The arithmetical operations on the measurable continuum ought to be defined by means of group theory."[49] Brouwer extended this approach to the different geometries. With respect to geometry the idea – that a continuum can be given 'structure' by group theory and topological restrictions – already occurs in a 1902 paper by Hilbert[50] concerning Hilbert's fifth problem, to which Brouwer refers. For Hilbert, it was only one possible way to study geometry. For Brouwer the situation was totally different. After the discovery of non-Euclidean geometries, an introduction of geometry on the basis of some Euclidean intuition was for Brouwer out of the question. The axiomatic approach, which Hilbert applied in his famous "Grundlagen der Geometrie" (axioms are essentially arbitrary statements that implicitly define a geometry) was also radically opposed to Brouwer's view of mathematics. For Brouwer, the idea of looking at groups of one-to-one continuous mappings and by means of the right topological restrictions getting the geometry that one wants, was completely in accordance with his positive heuristic.

In the period of 1908–1913, Brouwer did not suspect that a consistent application of his constructivist ideas would imply that the classical continuum and the classical notion of function would have to be given up. He assumed that classical mathematics only needed a better foundation. And, on the other hand, he also assumed that classical logic could still be applied in mathematics, provided the argument referred directly to mentally constructed objects. The situation that Brouwer faced in 1907, after the defence of his dissertation, was as follows. The dissertation was unfinished in several respects. On the one hand, the sketch that Brouwer gave in the first chapter of an intuitionist construction of mathematics was very incomplete and, for example, Hilbert's fifth problem remained for Brouwer's intuitionist construction a major unsolved problem. On the other hand, the intuitionist philosophy and the related general methodological considerations were also unfinished. Obviously, both aspects of Brouwer's research programme required further elaboration. However, Brouwer decided to work on the former aspect first. That is how Brouwer was very naturally drawn into the problem of the Lie-groups and, eventually, into topology. I will not discuss Brouwer's topological work here, but it is constructivist in the sense of the first part of the dissertation. Brouwer's style in topology is such that there is first of all a strong visual aspect, as the mental images corresponding to the topological transformations are obviously related to the intuitive continuum. Brouwer's topology is mental construction, although not always in the strict sense of Brouwer's later constructionism. The proofs possess a great conceptual clarity based on clear images. Moreover, manifolds are constructed out of simplices and manifolds and continuous mappings are handled by means of potentially infinite systems of approximations similar to the way in which, in the dissertation, the continuum is handled by means of the dual scale. Brouwer's topology is not abstract; his notions always refer to mathematical systems that can be considered to be mentally constructed.

In 1908, a year after the publication of his dissertation, Brouwer returned to the question of the logical principles in "De onbetrouwbaarheid van de logische principes" ("The unreliability of the logical principles").[51] Contrary to what he had assumed in the dissertation, Brouwer argues here that, with respect to finite systems, all principles can be applied, but with respect to infinite systems the principle of the excluded third is not reliable. Yet, he writes, "we shall never, by an unjustified application of the principle, come up against a contradiction and thereby discover, that our reasoning was badly founded."[52] The reason is that Brouwer accepted the principle of contradiction which says that p and non-p cannot be true at the same time. The unjustified conclusion p from a proof of non-non-p is unjustified intuitionistically, but it can never lead to a contradiction because non-non-p excludes non-p as a possibility. Most of Brouwer's topological work, which was all done before 1917, in the period from 1909 to 1913, is perfectly in harmony with these ideas. In that topological work, he repeatedly applies the principle of the excluded third. At the time, however, he had no reasons to suspect that it would not be possible to eventually justify all of his topological results in an intuitionistically satisfactory way.

In 1912, Brouwer became a professor at the University of Amsterdam. It meant the fulfilment of a great ambition. It also coincided with a shift of interest. After several years of intensive activity in topology, Brouwer's interest shifted back to the foundations of mathematics; in 1917, he started to develop intuitionist mathematics. Brouwer had become aware of the full implications of intuitionism, and he had realised that mathematics had to be rebuilt completely. That is what he set out to do. At that moment, his mathematics started to diverge seriously from the mainstream. The intuitive continuum disappears from the scene. The generating processes of the order types of ω and η are explicitly extended to include free choices and a new notion of set, the "Brouwer species," is introduced. Real numbers are introduced by means of choice sequences, and as a result, classical real analysis has to be sacrificed to a considerable extent. Illustrative is the fact that in intuitionist real analysis, every full function is uniformly continuous. A full function is a function defined on a closed unit interval. During and after 1917, Brouwer's programme still progressed, but in the course of time the programme stagnated: it became clear that many classical results could not, or only with great difficulty, be obtained intuitionistically. The intuitionist programme contributed considerably to the development of mathematics through the topological results it yielded in the period 1909–13. And the mature intuitionism from after 1917 contributed to mathematics as a whole: intuitionist logic played an important role in mathematical logic and the intuitionist results are a valuable contribution to constructivist mathematics. However, as an attempt to give a solid foundation to mathematics as a whole, intuitionism ultimately failed.[53]

4.2.4. The formalist foundational programme

It is not difficult to view Hilbert's mathematical work as the result of an extremely successful research programme with the axiomatic method as its central element. In his obituary of Hilbert,[54] Hermann Weyl distinguishes six main periods in Hilbert's life (1862–1943) as an active mathematician: i) Theory of invariants (1885–1893), ii) Theory of algebraic number fields (1893–1898), iii) Foundations of geometry (1898–1902), iv) Integral equations (1902–1912), v) Physics (1910–1922), and vi) Foundations of mathematics in general (1922–1930). The climax of the first period consists in two papers on the theory of invariants: "Über die Theorie der algebraischen Formen" of 1890, and "Über die vollen Invariantensysteme" of 1893. The first period made Hilbert famous. Even Gordan, who objected to a non-constructive proof in the second paper with the famous words, "This is not mathematics, it is theology!,"[55] never denied the value of Hilbert's contributions. In a way, the papers represented the end of developments in the theory of invariants, at least for the time being. Van der Waerden wrote later about the two papers: "Their essential core, which Hilbert consciously formulates in his second paper, consists in the application of arithmetic methods to algebraic problems. By considering the field of invariants as a special case of a field of functions in his deliberations, he stands at the turning point of a historical development: before him, the interest of algebrists was focused mainly on the tabulation – in the most explicit possible way – of all invariants of given fundamental forms, while those algebrists who came after him concentrated more on the *general* arithmetic and algebraic characteristics of systems of rational and algebraic functions. It was out of this range of ideas that the general theory of abstract fields, rings and modules later emerged."[56] The first period shows Hilbert's inclination to approach a problem from a general, abstract point of view. The same inclination is present in the second period. Hasse wrote about Hilbert's number-theoretic works: "On the one hand, they take a clear, generalized view in their examination of the problems upon which the work of the number theorists of the former century was based, and deal with these problems on this generalized level using, for the most part, new methods far superior to earlier ones, thereby pointing the way, on the other hand, to developments that would appear in the new century ... ."[57] I would like to combine these remarks about the first two periods with the following well-known anecdote. In 1891, when Hilbert was still working on the theory of invariants, he attended a lecture in Halle by Hermann Wiener on the foundations and structure of geometry. Hilbert was impressed by Wiener's abstract point of view in dealing with geometric entities. In the railway station in Berlin, on his way back to Königsberg, he is reported to have remarked to his companions: "One must be able to say at all times – instead of points, straight lines and planes – tables, chairs, and beer mugs."[58] This statement contains the axiomatic point of view in a nutshell. In "Grundlagen der Geometrie" of 1899, he would apply this point of view to geometry,[59] and in 1900 Hilbert applied the axiomatic method to arithmetic, i.e. the system of the real numbers, which he characterised as a real archimedean field that cannot be extended to a larger field of the same kind.[60] Hilbert had become aware of the power of the axiomatic method. In Hilbert's work in natural science, the axiomatic method plays a central role: a major objective was the axiomatization of physics.

The formalist foundational programme dates from 1904. In that year, at the third International Congress of Mathematicians in Heidelberg, Hilbert delivered an address in which he sketched a formalist consistency proof of arithmetic. The paper introduces the basic elements that would return in his further work on foundations: the assumption of the extra-logical existence of basic objects, like "1," and the reduction of arithmetic to a collection of formulae. Also, the development of a proof theory in logic (reduction of logic to the manipulation of formulae) and, moreover, the basic idea of the consistency proof of the formal system: prove that all formulae of a collection possess a certain property by showing that initial formulae have it and that the rule transmits it.[61] The positive heuristic of Hilbert's programme is, in that way, explicitly given. The consistency of a theory should be proved in two steps: i) define a formal system that captures the theory, ii) prove the formal consistency of the system.

In the 1920s, Brouwer's papers on intuitionist mathematics started to appear in international journals. As a result, the relationship between Hilbert and Brouwer, which had been cordial for many years – in 1919, Hilbert had even offered Brouwer a chair in Göttingen – rapidly deteriorated. Hilbert, who greatly respected Brouwer as a mathematician, realised that the intuitionist attack on the principle of the excluded third was an attack on a substantial part of classical mathematics. He experienced Brouwer's work as a serious threat and decided to act.[62] The fact that Hermann Weyl joined Brouwer, made the threat even more serious. In a series of papers in 1922, 1923, 1925 and 1927, Hilbert expanded his 1904 ideas and made them more precise. In the first 1922 paper, the clear separation of meta-mathematics from formalized actual mathematics is present. What meta-mathematics consists of is not precisely defined, but illustrated: it would be based on an intuition ("Anschauung") more or less in the sense of Kant. A nice survey of the development of Hilbert's foundational research until the beginning of the thirties is given by Bernays.[63] Hilbert's programme led in that period to a great number of interesting results by Hilbert and Bernays, but also by, for example, Ackermann, Von Neumann, Herbrand, Gentzen and others. Undoubtedly Gödel's incompleteness theorems made it clear that Hilbert's original aim had been too ambitious. Even after 1935, however, the programme continued to yield fruit. In 1988, Simpson wrote an interesting paper on partial realisations of Hilbert's programme.[64] Simpson defines Hilbert's goal as follows: i) to isolate the unproblematic "finitistic" portion of mathematics (Simpson identifies finitistic mathematics with the formal system of primitive recursive arithmetic [PRA]), ii) to reconstitute infinitistic mathematics as a big formal system (Simpson identifies the big system as Z_2, i.e. second order arithmetic), and iii) to give a finitistically correct consistency proof for Z_2. Gödel showed that the wholesale reduction which Hilbert envisioned is impossible. However, Simpson argues that the remaining question is how much can be reduced, and he states that recent investigations have shown that it is quite a large part. Many theorems of classical infinitistic mathematics can be reduced: Heine-Borel for closed bounded subsets of Euclidean n-space, the separable Hahn-Banach theorem, the existence and uniqueness of prime ideals in countable commutative rings, etc.

Simpson also shows that several other reductionist programmes have been defined. He mentions Gödel's higher type functionals, Feferman's predicativism, Myhill's intuitionistic set theory and Gentzen's work on the consistency of the Peano arithmetic. Those programmes can be seen as having succeeded in validating certain infinitistic results on the basis of precisely defined infinitistic extensions of finitistic mathematics.

4.2.5. The rationality of the development of mathematics

What logicism and intuitionism have in common is the goal to found mathematics in something like intuitive certainty. Russell expressed it as follows: "the solution should, on reflection, appeal to what may be called 'logical common sense' – i.e. that it should seem, in the end, just what one ought to have expected all along."[65] I take this to mean that if one is confronted with the solution, one knows intuitively that it is correct. This kind of intuition is not precisely defined, and it need not be of a "visual," Kantian kind. For Brouwer, the certainty of mathematics is based upon a more precise notion of intuition: the clarity of mental construction. Hilbert's mathematical structuralist research programme, of which his formalist foundational programme is a part, is different in this respect. In principle, every system of axioms defines a mathematical structure, provided the system is consistent. In principle, this means that the mathematician is extremely free. Mathematical theories need not at all refer to some intuitively given subject matter. The only limitation left is that an axiom system should be consistent. The structuralist programme was extremely progressive; Hilbert's own mathematical work already shows this clearly. The programme also absorbed successfully the results of other programmes. For example, Brouwer's topological results were immediately incorporated. Understandably, many mathematicians were attracted by the structuralist research programme, and the fact that it dominated 20th century mathematics for quite some time only shows that the mathematical community behaves rationally in the sense of Lakatos' MSRP.[66] None of the three foundational programmes completely solved the foundational problems of mathematics, but Hilbert's foundational programme was part of a much bigger, highly progressive research programme. That fact may also be a reason why Hilbert's foundational programme received considerable attention. Yet, even if one considers the progress of the three foundational programmes without taking the broader background into account, Hilbert's foundational programme progressed more than the two other programmes.

4.2.6. Fallibility

The example of the developments in the foundations of mathematics in the first decades of the 20th century shows that developments in mathematics can be reconstructed along the lines of Lakatos' MSRP. This implies that the MSRP incorporates important aspects of the global rationality of the development of mathematics. Moreover, our reconstruction of the developments in the foundations of mathematics once more shows the weak fallibility of mathematics. The three foundational programmes were not logically refuted. Moreover, they do not logically contradict each other either. In fact, one could easily argue that they supplement and support each other. For example, the title of Gödel's famous 1931 paper, "On formally undecidable propositions of Principia Mathematica and related systems," illustrates the role that the formal system of Principia Mathematica played within the formalist programme. The three foundational programmes were all Euclidean in the sense of Lakatos' "Infinite Regress." They represent Euclideanism in mathematics. Lakatos emphasised the fact that none of the three programmes succeeded in completely realising in mathematics the Euclidean ideal. On the other hand, it needs to be emphasised that none of the three programmes can be viewed as a total failure. On the contrary, all three yielded many new results, and the formalist programme, in particular, was extremely fertile. The great successes of Euclideanism in mathematics in combination with the very weak fallibility of mathematics require an explanation, which Lakatos' developmental rationality of the MSRP does not give.

5. MITIGATED DOGMATISM

It is interesting to confront our conclusions about Lakatos' philosophy of mathematics with non-sceptical views thereof. In his paper on partial realisations of Hilbert's programme that we mentioned above, Simpson explicitly takes a dogmatist position and attacks the sceptics in the philosophy of mathematics. In Simpson's opinion, mathematics is true and therefore useful – and not the other way around. He argues as follows. In the last analysis the only way to show the truth of mathematics is to demonstrate that it refers to reality. The formulas of finitistic mathematics refer in a relatively unproblematic way to various real-world processes. Moreover, finitistic mathematics is as firmly grounded as a science can be. The reduction of infinitistic proofs to finitistic ones shows that the finitistically meaningful end-formulas of infinitistic proofs are true in the real world. Moreover, the formulas that occur in such proofs become more reliable when they are seen to correspond to reality. In fact, Simpson distinguishes two levels of certainty: the level of finitistic mathematics and the level of infinitistic mathematics (insofar as it has not yet been reduced to finitistic mathematics). His goal is to reduce as much infinitistic mathematics as possible to finitistic mathematics, thereby extending as much as possible the certainty of finitistic mathematics to infinitistic results. I shall call Simpson's position a mitigated dogmatism. He believes A) that we can attain truth and B) that we can know that we have attained truth. This makes him a dogmatist. However, Simpson also believes C) that there are degrees of certainty with respect to B).

In "A Renaissance of Empiricism,"[67] Lakatos_1 attacked some dogmatist reactions to the uncertainty in the foundations of mathematics. He writes: "*How can one take fallibilism seriously without taking the possibility of refutations seriously?* One should not pay lip-service to fallibilism: 'To a philosopher there can be nothing which is absolutely self-evident' and then to go to state: 'But in practice there are, of course, many things which can be called self-evident [...] each method of research presupposes certain results as self-evident'."[68] Here Lakatos_1 quoted Bernays, and also mentioned Kreisel and Gödel, as proponents of a wrong view of mathematics. According to Lakatos_1, they represented a soft fallibilism, which divorces fallibilism from criticism. Bernays, Kreisel and Gödel all represent forms of mitigated dogmatism. I doubt whether Lakatos_2 would still have attacked them in this way. Lakatos_1 was still very much under the influence of Popper's falsificationism, in which logical refutations play a central role. In the MSRP, Lakatos_2's own brand of falsificationism, the emphasis is no longer on theories and refutations, but on research programmes and their competition. The MSRP is an overall developmental rationality that does not exclude the Euclidean rationality of particular mathematical research programmes. The immense successes of this Euclidean methodology in several subsequent research programmes, in combination with weak fallibility, suggest that the view that results from Lakatos' mitigated sceptical work is no longer radically opposed to forms of mitigated dogmatism.

On the other hand, if mitigated dogmatists would accept the important role of criticism in the development of mathematics, if they would accept that concepts, conjectures, theorems, proofs, axioms, definitions, whole theories, research programmes etc. are always subjected to a critical evaluation and after some time replaced by others, they would be very close to the mitigated sceptic, who acknowledges the successes of the Euclidean methodology and is aware of the very weak fallibility of mathematics.

The way in which I compare Lakatos' mitigated scepticism with, for example, Simpson's mitigated dogmatism, is restricted to the methodological level. I abstract from possible differences of opinion concerning the nature of mathematical truth. The reason is that neither Simpson nor Lakatos possesses a very clear view in this respect. Simpson relates the truth in finitistic mathematics to empirical reality, without further elaboration. Lakatos only left us a fragmentary philosophy of mathematics and simply never gave a final answer to questions like: what is mathematics about? In "A Renaissance of Empiricism"[69] he pointed out that platonistic intuition, construction, convention and empirical reality could all play a role.

On the methodological level, however, the situation reminds one of the following remarkable quotation from David Hume's "Dialogues Concerning Natural Religion":

> No philosophical dogmatist denies that there are difficulties both with regard to the senses and to all science; and that these difficulties are in a regular, logical method, absolutely insolvable. No sceptic denies, that we lie under an absolute necessity, notwithstanding these difficulties, of thinking, and believing, and reasoning with regard to all kinds of subjects, and even of frequently assenting with confidence and security. The only difference, then, between these sects, if they merit that name, is, that the sceptic, from habit, caprice, or inclination, insists most on the difficulties; the dogmatists, for like reasons, on the necessity.[70]

NOTES

1. Imre Lakatos, "Infinite Regress and the Foundations of Mathematics", in: John Worrall and Gregory Currie (eds.) *Imre Lakatos, Philosophical Papers*, Vol. II, Cambridge [Etc.]: Cambridge University Press 1978, pp. 3–23. I will denote the paper as "Infinite Regress."
2. By ignoring it, the sceptic runs the risk of becoming slightly ridiculous. There is a story about Pyrrho, which recounts that his friend once saw how he was chased by a dog and could only save himself by climbing a tree. When his friends pointed out that he had obviously taken the dog to be very real, he is said to have remarked that the philosopher can never totally get rid of the human being.
3. Blaise Pascal, *Penseés*, Texte établi par Léon Brunschvicg, Paris: Garnier-Flammarion 1976, pp. 172–173.
4. Richard H. Popkin, *The History of Scepticism from Erasmus to Spinoza*, Berkeley: University of California Press 1979.
5. Ibid., p. 140.
6. Unlike many other sceptics, Hume also explicitly held sceptical views about mathematics. He held, in Popkin's words, "that we can never have adequate grounds for maintaining that any mathematical proposition is true", Richard A. Watson, James E. Force (eds.), *The High Road to Pyrrhonism*, papers by Richard Popkin, San Diego: Austin Hill Press, Inc. 1980, p. 143.
7. Karl R. Popper, *Objective Knowledge, An Evolutionary Approach*, revised edition, Oxford: Clarendon Press 1979, p. 99.
8. Ibid., Chapter I.
9. In J. Worrall and G. Currie (eds.) *Imre Lakatos, Philosophical Papers*, Vol. II, Cambridge University Press 1978, pp. 61–69.
10. Op. cit., p. 63.
11. Op. cit., p. 62–63.
12. Corfield rightly criticises Lakatos for underestimating the fertility and the importance of quasi-formal mathematics in David Corfield, "Assaying Lakatos' Philosophy of Mathematics", *Studies in the History and Philosophy of Science* 28, 1997, pp. 99–121.
13. Imre Lakatos, "Proofs and Refutations", *British Journal for the Philosophy of Science* 14, 1963–64, pp. 1–25, 120–139, 221–245, 296–342. References are to the reprint in John Worrall and Elie Zahar (eds.), *Imre Lakatos, Proofs and Refutations, The Logic of Mathematical Discovery,* Cambridge [Etc.]: Cambridge University 1976.
14. Op. cit., pp. 4–5.
15. Op. cit., Appendix 2, p. 146. There are some interesting remarks on the influence of Hegel on Lakatos in John Kadvany, "The Mathematical Present as History," *The Philosophical Forum* 26, pp. 263–287. John Kadvany, *Imre Lakatos and the Guises of Reason*, Duke University Press, 2001, appeared after this paper was written. See also the paper by M. Motterlini in this volume.
16. I am referring here to the end of the original 1963–64 paper.
17. Teun Koetsier, *Lakatos' Philosophy of Mathematics, A Historical Approach*, Amsterdam: North-Holland 1991.
18. "Infinite Regress," p. 4.
19. Op. cit., p. 14.
20. Op. cit., p. 17.
21. Op. cit., p. 19, italic is Lakatos'.
22. Op. cit., p. 19.
23. Op. cit., p. 20.
24. Op. cit., p. 10.
25. Written in 1970–71 and published in 1974; references are to the reprint in J. Worrall and G. Currie (eds.) *Imre Lakatos, Philosophical Papers*, Vol. II, Cambridge University Press 1978, pp. 139–167.
26. Op. cit., p. 157.
27. Op. cit., p. 156.
28. Op. cit., p. 155; italics is Lakatos'.

29. I will not go into the question of whether the MSRP is a satisfactory answer to an extreme sceptic. It is very difficult to satisfy an extreme sceptic. For a more general defence of critical fallibilism against the sceptics, I refer to Alan Musgrave, *Common Sense, Science and Scepticism, A Historical Introduction to the Theory of Knowledge*, Cambridge: Cambridge University Press 1993
30. Elsewhere, I modified the MSRP in order to be able to use it in a more precise way to describe some developments in mathematics. I called the result the Methodology of Mathematics Research Traditions. Cf. Koetsier, op. cit.. For a different view and a detailed discussion of the application of the MSRP to mathematics I refer, in particular, to Ladislav Kvasz' contribution in this volume. For the general problem of the applicability of the MSRP to mathematics see also E. Glas, "Testing the Philosophy of Mathematics in the History of Mathematics", *Studies in the History and Philosophy of Science* 20, 1989, pp. 115–131 and pp. 157–174, and D.D. Spalt, *Vom Mythos der Mathematischen Vernunft*, Darmstadt: Wissenschaftliche Buchgesellschaft 1981.
31. Bertrand Russell, *The Principles of Mathematics*, Cambridge University Press, 1903 (Second edition: London, 1937).
32. Bertrand Russell, *My Philosophical Development*, London: Unwin Books 1975, p. 162.
33. Paul J. Hager, *Continuity and Change in the Development of Russell's Philosophy*, Dordrecht [Etc.]: Kluwer Academic Publishers 1994.
34. Ibid., p. 7.
35. Ibid., pp. 6–7.
36. Russell 1975, op. cit., pp. 54–55.
37. Ibid., p. 56.
38. Ibid., p. 56.
39. Ibid., p. 56.
40. Ibid., p. 61.
41. Here I follow Gödel's 1944 paper "Russell's Mathematical Logic", reprinted in P. Benacerraf and H. Putnam, *Philosophy of Mathematics, Selected Readings*, Second Edition, Cambridge [Etc.]: Cambridge University Press 1983, pp. 447–485.
42. Gödel, op. cit., p. 456.
43. Alfred North Whitehead and Bertrand Russell, *Principia Mathematica*, Vol. I, Cambridge: University Press 1950 (reprint of the second 1925 edition), p. xiv.
44. Ibid.
45. My description of Brouwer's programme is based upon Teun Koetsier and Jan van Mill, "General Topology, in Particular Dimension Theory, in the Netherlands: the Decisive Influence of Intuitionism", in C.E. Aull and R. Lowen (eds.), *Handbook of the History of General Topology*, Vol. I, Dordrecht: Kluwer Academic Publishers 1997, pp. 135–180.
46. Luitzen Egbertus Jan Brouwer, *Over de Grondslagen der Wiskunde*, Amsterdam, 1907. The references will be to the English translation: L.E.J. Brouwer, "On the Foundation of Mathematics", in: Arend Heyting (ed.), *L.E.J. Brouwer Collected Works, Vol. 1, Philosophy and Foundations of Mathematics*, Amsterdam [Etc.]: North-Holland 1976, pp. 15–101.
47. Brouwer, op. cit., pp. 17–18.
48. Walter P. van Stigt, *Brouwer's Intuitionism*, Amsterdam: North Holland 1990, pp. 320–321.
49. Brouwer, op. cit., p. 98.
50. D. Hilbert, "Über die Grundlagen der Geometrie", *Mathematische Annalen* 56, 1902. Reprinted as an appendix in D. Hilbert, *Grundlagen der Geometrie*, 4th edition, Leipzig, 1913, pp. 163–218.
51. L.E.J. Brouwer, "De onbetrouwbaarheid der logische principes", *Tijdschrift voor Wijsbegeerte* 2, 1908, pp. 152–158. References are to the English translation: L.E.J. Brouwer, "The unreliability of the logical principles", in: *L.E.J. Brouwer Collected Works, Vol. 1, Philosophy and Foundations of Mathematics*, Amsterdam [Etc.]: North-Holland 1976, pp. 107–111.
52. Brouwer, op. cit., p. 110.
53, Actually, work in intuitionism never completely stopped. However, it is presently an extremely small research area.
54. An abbreviated version of the obituary is included as an appendix of Constance Reid, *Hilbert*, Berlin [Etc.]: Springer-Verlag 1970, pp. 245–283.
55. Reid, op. cit., p. 251.

56. David Hilbert, *Gesammelte Abhandlungen*, Zweiter Band, [Repr.] New York: Chelsea Publ. Corp. 1965, p. 401 (italics is mine – T. K.).
57. David Hilbert, *Gesammelte Abhandlungen*, Erster Band, p. 528.
58. Reid, op. cit., p. 57. In his obituary of Hilbert, Weyl writes that, according to Blumenthal, it must have been 1891 and, moreover, that Wiener's paper was on the role of Desargues' and Pappus' theorems. (Reid, op. cit., p. 264).
59. Weyl wrote: "There could not have been a more complete break than the one dividing Hilbert's last paper on the theory of number fields from his classical book, *Grundlagen der Geometrie*, published in 1899" (Reid, op. cit., p. 264). This statement does not take into account that in his number theoretic papers Hilbert does not deal with foundational matters. *Grundlagen der Geometrie* and the number theoretic papers have in common that the problem that Hilbert intends to solve is solved by considerations that are structural, and more abstract than many of his contemporaries were capable of exploiting. It is only natural that concentration on the foundations leads to a more explicit formulation of that structural point of view.
60. See p. 197 of Paul Bernays, "Hilberts Untersuchungen über die Grundlagen der Arithmetik", in David Hilbert, *Gesammelte Abhandlungen*, Dritter Band, [Repr.] New York: Chelsea Publ. Corp. 1965, pp. 196–216.
61. J. Van Heijenoort (ed.), *From Frege to Gödel: A Source Book in Mathematical Logic, 1879–1931*, Cambridge: Harvard University Press 1967, pp. 129–138.
62. Hilbert acted in different ways. The confrontation of the two mathematicians reached a climax in 1928 when Hilbert was ill and worried that after his death his fellow-editor of the Mathematische Annalen, Brouwer, would gain too much influence. Hilbert dismissed Brouwer from the board of the Annalen. For an extensive treatment of the 'Crisis of the Mathematische Annalen' we refer to D. van Dalen, "The War of the Frogs and the Mice, or the Crisis of the Mathematische Annalen", *The Mathematical Intelligencer* 12, Number 4, 1990, pp.17–31. A Lakatosian approach requires that one concentrates on the internal history. That is what I am doing in this paper. Brouwer's dismissal as editor of the Annalen is part of the external history.
63. Bernays, op. cit.
64. Stephen G. Simpson, "Partial Realisations of Hilbert's Program", *The Journal of Symbolic Logic* 53, 1988, pp. 349–363.
65. Russell, 1975, op. cit., p. 61.
66. An important book about the competition between pre-twentieth century mathematics and modern structuralist mathematics is Herbert Mehrtens, *Moderne Sprache Mathematik*, Frankfurt am Main: Suhrkamp 1990. Mehrtens describes the transition as a very complex process. From my point of view, he insufficiently emphasises the fact that, at heart, the structuralist programme came to dominate because of the large number of results it produced and the ease with which it incorporated other results.
67. Imre Lakatos, "A Renaissance of Empiricism", in: J. Worrall and G. Currie (eds.) *Imre Lakatos, Philosophical Papers*, Vol. II, Cambridge University Press 1978, pp. 24–42.
68. Op. cit., p. 42; italics is Lakatos'.
69. Op. cit., p. 40.
70. David Hume, *Dialogues Concerning Natural Religion* (edited with commentary by Nelson Pike), Indianapolis and New York, 1970, p. 113 footnote.

Division of Mathematics and Computer Science
Faculty of Science, Vrije Universiteit
De Boelelaan 1081
1081 HV Amsterdam
The Netherlands

Ladislav Kvasz

Lakatos' Methodology Between Logic and Dialectic

The aim of this paper is a critical analysis of the methodology of Imre Lakatos. We will try to show that the full potential of Lakatos' methodological ideas were not allowed manifest itself, because of their being confused with dialectic. By separating the *hard core* of Lakatosian methodology from the dialectical heritage of his marxist past, we believe there can be created a better-functioning – and more effective – methodology in the same spirit. Thus, even if we start with a criticism of Lakatos' ideas, we will not stop there, but will try to turn our criticism into some positive emendations.

It is generally accepted that Lakatos' method of rational reconstruction of the history of mathematics has, aside from its many merits, some weaknesses, as well. We would like to show that these weaknesses were caused by his confusion of dialectic with logic. In this way, Lakatos developed an appealing and interesting theory, which, at least at first glance, has the advantages of both – the liveliness of dialectic and the soundness of logic. Unfortunately, this attempt to combine dialectic with logic also has one disadvantage. The focus on logic restricts severely the scope of the changes, to which this method can be applied. That is why Lakatos was forced to neglect in his rational reconstruction many episodes in the history of mathematics, which simply do not fit into his scheme. But on the other hand, dialectic gives his theory the illusion of universality, for which reason, perhaps, he seemed to be unaware of his omissions. Thus, we see dialectic as being the main hobble on the development of Lakatos' methodology.

We interpret dialectic very broadly as a current of philosophical thought, which tries to interpret the growth of knowledge using a prescribed pattern of stages, methods, or laws of development of knowledge. Most often, there are three such stages (Hegel's *thesis*, *antithesis*, and *synthesis*; Kuhn's *normal science*, *crisis*, and *revolution*; or Lakatos' stages of *naive trial and error*, *proof-procedures*, and *research programme* – see Lakatos, 1978, pp. 93–103). Usually, the dialecticians believe that the pattern of the development of knowledge is of a logical nature (Hegel's idea of dialectical logic, Popper's logic of scientific discovery, or Lakatos' logic of mathematical discovery), what creates tension between the development of knowledge and formal logic. Finally, dialecticians believe that the pattern of development of knowledge is universal, and thus applicable to nearly every field of human knowledge. We believe that in Lakatos' methodology we can find all these characteristic features of dialectic and so Lakatos is, at least in our broad

G. Kampis, L. Kvasz and M. Stöltzner (eds.), Appraising Lakatos: Mathematics, Methodology and the Man, 211–241.

sense of the word, a dialectician. Nevertheless, it seems that it was precisely this dialectical heritage, which hindered him in the development of his theory and prevented him from using all of its potential.

The aim of our paper is to separate the creative core of Lakatos' ideas from their dialectical cover and in this way to make a positive problemshift in Lakatosian methodology. Our strategy is to offer for each of the three abovementioned "dialectical faults" (i.e. the existence of prescribed patterns, their logical nature and universal character) an alternative reconstruction. For each case, then, we present the classical Lakatosian reconstruction of the development of some theory, and confront it with a different reconstruction of the same material. This approach is usually applied in order to assess the Lakatosian reconstruction of history. We will not proceed in this manner. Our aim is not to assess Lakatos. We, rather, use the confrontation with alternative ways of reconstruction to separate Lakatos' methodology from its dialectical features, in order to find some positive possibilities for its further development.

Our paper has three parts, which correspond to the three features of dialectic described above. In the first part of the paper, we will try to show that the basic patterns of the methodology of proofs and refutations (*monster-barring*, *exception-barring*, and *lemma-incorporation*) can be supplemented by further ones. Thus we can raise the question of describing and classifying all such patterns, instead of presenting only some of them, as Lakatos did.

In the second part of the paper, we will analyse the development of mathematics from geometry to topology, the reconstruction of which, we believe, is a central weakness of Lakatos' *Proofs and Refutations*. We will present an alternative reconstruction of the development of this field. Our reconstruction is based on Wittgenstein's concept of the form of language from the *Tractatus;* thus, it is neither dialectical nor logical. It is based on the picture theory of meaning and is, therefore, semantic in nature. On the basis of this comparison, it will become obvious that Lakatos' method of proofs and refutations applies only to a very restricted variety of changes in the development of mathematics. Perhaps that is the reason why, even if Lakatos proclaimed it to be a universal approach to understanding changes in mathematics, it did not find many followers and slowly petered out.

In the third part of the paper, we turn to Lakatos' methodology of scientific research programmes (MSRP). We will point to an omission, which is in many respects similar to the omission of the reconstruction of the transition from geometry to topology in *Proofs and Refutations*. We will again confront the Lakatosian MSRP with another means of reconstruction, based on a classification of epistemic ruptures in the scientific language. We suggest that there are four different types of changes in science. Lakatos deals in his methodology only with one of them, but presents his results as being universally valid. We believe that if we supplement Lakatos' theory with the three other kinds of change, the fundamental problems with the identification of research programmes in mathematics can be solved.

1. Lakatos' Methodology of Proofs and Refutations

Lakatos' *Proofs and Refutations* (1976) are written in the form of a dialogue, which takes place in a classroom. But it is not an ordinary classroom. Lakatos has brought together the greatest mathematicians of the past, who contributed to the theory of polyhedra – from Cauchy, Lhuilier and Hessel to Abel and Poincaré. It is exciting to imagine what would happen if all the participants in a scientific debate, which lasted over two centuries, could meet and discuss the problems together. How would Cauchy react to the counterexamples to his theorem? Would he accept Poincaré's topological proof? So, the very idea of such an imaginary dialogue is interesting. But Lakatos achieved even more. He succeeded in distilling from this history some basic patterns of thought, which can be found in many other areas of mathematics. These are his famous monster barring, exception barring, and lemma incorporation. We will present them briefly.

1.1. Monster barring, exception barring, and lemma incorporation

The discussion in the classroom is of Euler's theorem, which says that for all polyhedra the number of vertices V, number of edges E, and number of faces F fulfil a simple relation: $V - E + F = 2$. After the teacher has presented a proof – actually, the classical proof stemming from Cauchy – some counterexamples appear. We will not present all the counterexamples discussed in the book; we select just a few of them here, and present the basic ideas.

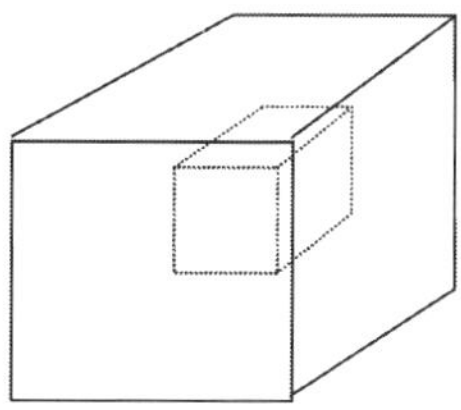

Example 1. A cube, which has inside of it an empty hole in the form of a smaller cube. It is easy to see that, in this case, $V - E + F = 4$, and not 2, as it should, according to the theorem.

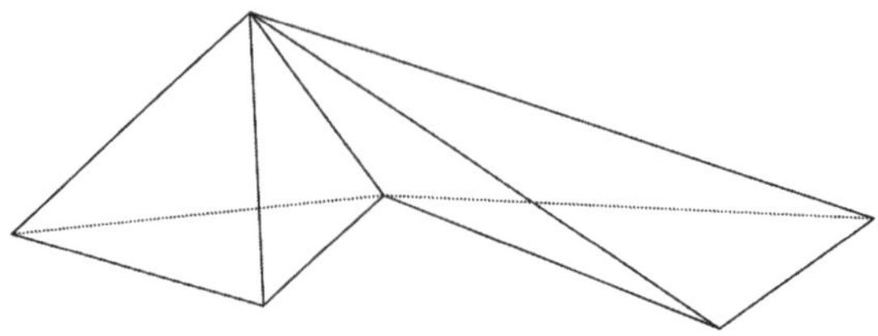

Example 2. Two tetrahedra which have one edge in common. Here we have $V - E + F = 3$

Surely, these examples represent, so to speak, principles of how to construct some fairly strange objects! In this sense, it is not difficult to imagine an object having many holes, or to build a whole chain of different polyhedra in which any two neighbours have a common edge. Now the question is how to deal with these objects, which contradict Euler's theorem.

The first strategy described by Lakatos is that of *monster-barring*. These strange objects are surely not what we have in mind when we speak of polyhedra. They are some monsters, and we should not allow them to enter into our considerations. They are of no theoretical interest, and no normal mathematician would ever think of them as polyhedra.

The second strategy Lakatos calls *exception-barring*. According to it, we admit that these objects are genuine polyhedra, and therefore real counter-examples to the theorem. The theorem, as it was stated originally, does not hold. It is not so general as we have thought. We have to restrict our theorem in such a way that all of these exceptions would fall outside its domain. It is obvious that all the examples mentioned above are not convex. Thus, if we restrict ourselves to convex polyhedra, the theorem is rescued.

The third strategy described by Lakatos is that of *lemma incorporation*. In both previously described cases, we have not learned much new from our new objects. In monster-barring, we just ignore them and state the theorem more generally than it was originally. On the other side, in exception-barring, we restricted the theorem too much. We should not restrict ourselves to the original theorem, but should, rather, try to find a more general one, which would also include the strange objects. Only in this way, can we learn something really new. Thus, we should try to understand, in which way a common edge or hole changes the resulting statement of the theorem, and we should find a way of incorporating them into the theorem. Our task is not to find safe ground, upon which the theorem holds (as in exception-barring). We have first to understand what new things we can learn from these objects.

1.2. Lemma exclusion as a further candidate

Teun Koetsier, in his book, *Lakatos' Philosophy of Mathematics* (Koetsier, 1991) compared Lakatos' reconstruction of the development of the theory of polyhedra from *Proofs and Refutations* with the actual history. He gives the following account:

> There is some resemblance between Lakatos's reconstruction concerning the formula of Euler for polyhedra and the real history. ... Yet the rational construction deviates considerably from the chronological order in which things actually happened. ... As far as chronology is concerned, there is not much in common between dialogue and real history. There is no doubt that; *Proofs and Refutations* contains a highly counterfactual rational reconstruction. (Koetsier, 1991, p. 42).

We agree with this assessment as well as with Koetsier's statement that "*Proofs and Refutations* is mainly convincing because it shows recognisable mathematical behaviour" (p. 44).

Nevertheless, there is another question, as well. Even if we admit that Lakatos' rational reconstruction deviates from actual history, we must ask: does it present a true picture of mathematical behaviour itself? Doubts in this regard were expressed recently by David Corfield (Corfield, 1997), who argued that the counterexamples do not play as important a role in mathematics as Lakatos gives them in his rational reconstruction. He cites the example of Poincaré's *Analysis situs*, the proofs of which were changed to a large extent; there was, nevertheless, only one counterexample to a single lemma in all 300 pages of the text (Corfield, 1997, p. 108). Thus, the reformulations of Poincaré's proofs and theorems were not the result of the discovery of counterexamples. That means that there are other patterns of mathematical behaviour, which were omitted in Lakatos' reconstruction.

One such clearly recognisable pattern can be found in Koetsier's book. We have in mind the proof of the interchangeability theorem for partial differentiation by H.A. Schwarz. Schwarz first stated the theorem of interchangeability with six conditions, proved it, and then attempted to drop as many of the conditions as possible. He succeeded in dropping three of the six conditions, and ended with a much stronger theorem than the one which he proved at the beginning (Koetsier, 1991, p. 268–271). We suggest calling this method *lemma-exclusion*, and to consider it a counterpart to Lakatos' lemma incorporation. If we incorporate this fourth method into Lakatos' theory, we get the schema overleaf.

According to this schema, we have two possible reactions to the appearance of a counterexample. The one is to ignore the counterexamples as monsters, the other is to consider them as exceptions and restrict the theorem to safe ground. Nevertheless, the monster-barrer states the theorem more generally than it really holds. On the other hand, the exception-barrer often restricts the theorem too strongly (for instance, Beta in *Proofs and Refutations* p. 28, who restricted Euler's theorem to convex polyhedra, or H.A. Schwarz, who in his first theorem restricted the interchangeability theorem

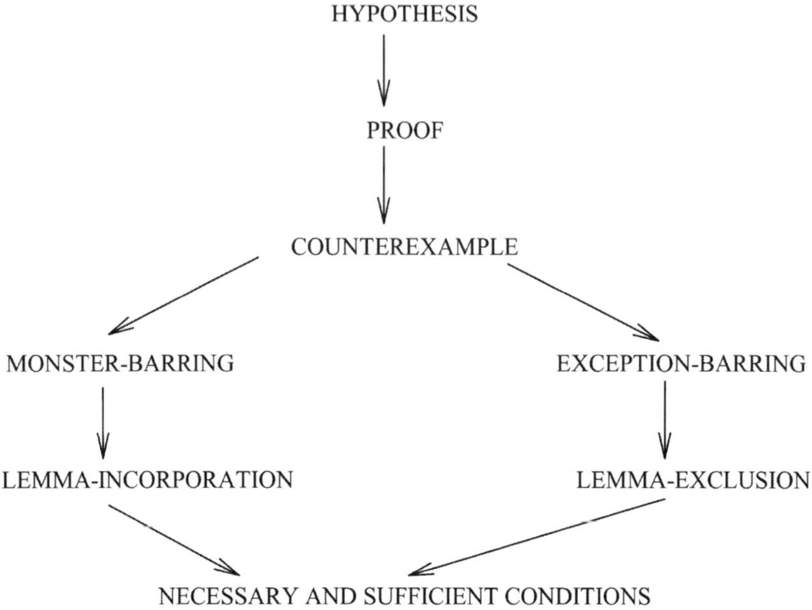

only to functions fulfilling all the six conditions). After some time, these first reactions are overcome. The *monster-barring* method is followed by *lemma-incorporation*, where the aim of having the theorem be as general as possible is still preserved, but the counterexamples are no longer ignored. On the other hand, *exception-barring* is followed by *lemma-exclusion* (as the case of H.A. Schwarz suggests), where the aim of being constantly on safe ground is preserved, but the too-strong restrictions of the domain of the theorem are weakened step by step. In the ideal case, *lemma-incorporation* and *lemma-exclusion* meet each other when the necessary and sufficient conditions of the theorem are found.

1.3. Dialectic versus history in the methodology of proofs and refutations

One possible objection against including the method of *lemma-exclusion* into the methodology of proofs and refutations is that in this method, the counterexamples do not have such an important role as they have in the other three methods. Nevertheless, it is important to realise that the stress placed on counterexamples stems from the dialectical background of Lakatos' thought, rather than from the analysis of the development of mathematics itself. Even if we do not deny the heuristic value of dialectic for Lakatos' methodology, we are not compelled to also accept its limiting consequences.

We believe that in order to answer the questions about Lakatos' methodology, some further case studies are needed. Some research has been already done (see Koetsier, 1991), but it was mainly in the framework of the methodology of scientific research programmes. An interesting topic for further development of the methodology of proofs and refutations would be a thorough study of the various proofs of Euclid's fifth postulate, from Proclus up to Saccheri. There, the situation is analogous to that of Euler's theorem, as analysed by Lakatos – a long series of proofs and disproofs of a rather simple statement. Nevertheless, as the fifth postulate is not provable, the role of counterexamples might be quite different than it was in the proofs of Euler's theorem. The mistakes in the proofs of the fifth postulate were most often found through the discovery of a circularity in the proof, rather than by a counterexample.

To what extent, then, the method of *lemma-exclusion* fits into the Lakatosian methodology of proofs and refutations is an open question. We presented it only in order to show that the possibilities of this methodology are far from having been exhausted. It is natural to ask what other methods exist, which belong to the methodology of proofs and refutations. Is our reconstruction of mathematical practice complete, or we are still occupied only with a fragment of it? What are the relationships between the various methods?

2. The Conflict Between Dialectic and Logic in Proofs and Refutations

The methods described in the previous section are by no means bound to the theory of polyhedra. Rather, they describe universal strategies, which mathematicians used in many areas. The term "monster" was used to describe the new functions of real variable, discovered in the nineteenth century, which have many strange properties, and which are now called fractals. The first reaction of many leading mathematicians was just to ignore them as monstrosities. The first reaction to complex numbers, or to the discovery of incommensurability in ancient Greece, was very much the same. That is why any mathematician feels true enjoyment when reading the first part of the *Proofs and Refutations*. Nevertheless, as one reaches the second part (p. 106), some problems appear.

2.1. Lakatos' Proofs and Refutations *– some problems*

The first problem is actually a technical one. On page 113, Lakatos claims that the heptahedron (i.e. the projective plane) bounds.

> The next term to be elucidated is bound. I shall say that a k-circuit bounds, if it is the boundary of a $(k+1)$-chain. ... Now it is absolutely clear that for instance the heptahedron bounds.

Here, Lakatos contradicts his own statement from page 110, where he (correctly) asserted, that the heptahedron is a one-sided surface, for which reason there exists no geometrical body, which the heptahedron could bound. It is surprising that the editors John Worrall and Elie Zahar, who completed the book with detailed commentary, let this problem go without any notice. Also Mark Steiner, who in his paper "The philosophy of mathematics of Imre Lakatos" (Steiner, 1983) presented a nice exposition of Poincaré's proof, and filled in the gaps of Lakatos' original presentation, did not mention this inconsistency of Lakatos. This inconsistency in the second part of the *Proofs and Refutations* is probably a result of the fact that Lakatos did not want to complicate his text with details concerning the orientation of surfaces. But if we define the concept of boundary without orientation, the whole theory becomes obscure, and it is easy to make mistakes. Steiner introduced the concept of orientation, at least implicitly, when speaking of the "sum" and "difference" of schemes (Steiner, 1983, p. 515).

The second problem is a deeper one. Aside from the appendices, the book consists of two major parts. In the first – geometrical – part, Lakatos presents a detailed analysis of *monster-barring, exception-barring,* and *lemma-incorporation.* The second major part of the book, containing Poincaré's proof of Euler's theorem, is not so thoroughly worked out. Nevertheless, we can assume that if Lakatos had lived longer, he would have applied his arsenal of heuristic strategies to this material, as well. What is striking is the lack of any attempt to connect the two parts of the book. Lakatos, who always stressed the necessity to reconstruct the circumstances in which new concepts emerged, and who harshly criticised mathematicians – like Hilbert or Rudin – who presented formal definitions without any historical background (see Lakatos, 1976, pp. 15 and 145), suddenly pulls out of his top-hat the basic concepts of algebraic topology without the slightest comment, and pretends that everything is all right.

2.2. An alternative reconstruction of the development of geometry

The aim of this section is to introduce Wittgenstein's concept of the form of a language into geometry, and to show how it can be used to achieve a better understanding of the historical development of geometry. The basic idea is to use the Picture Theory of Meaning to understand the pictures of geometry. According to Wittgenstein's *Tractatus,* the form of a language consists of those signs and expressions which have no real denotation. They do not denote things, but their function is to make denotation possible. I would like to examine the language of the pictures of geometry, and try to find the development of its form. A more complete exposition of this method of reconstruction can be found in "History of geometry and the development of the form of its language" (Kvasz, 1998). Here we restrict ourselves to the analysis of Renaissance paintings and projective geometry, which are closely related to Wittgenstein's *Tractatus*. Then we turn to the works of Riemann and Poincaré, which are relevant to our discussion of Lakatos. The leap from Desargues to Riemann is, perhaps, a big one; we hope that the basic idea will, nonetheless, be clear enough. This idea is that the development of geometry consisted in the incorporation of the form of its language into the language itself.

2.2.1. The language of perspectivist paintings

The painters of the Renaissance approached painting with a new goal. They began to paint the world as they saw it, to paint it from a particular point of view, to paint it *in perspective*. They wanted to paint objects in such a way, that the picture would evoke in the spectator the same impression as if he were looking at the real object. It had, therefore, to produce the illusion of depth. To reach this goal, the painter had to follow three principles of perspective:

Perspective of size – remote objects are to be painted smaller

Perspective of colours – remote objects are to be painted with dimmer colours

Perspective of outlines – remote objects are to be painted with softer outlines

By following these principles, a special line appears in the painting – the horizon. In fact, the painter is not allowed to create it by a stroke of his brush. He is not permitted to paint the horizon, which shows itself only when the picture is completed. According to proposition 2.172 of the *Tractatus* ("A picture cannot, however, depict its pictorial form: it displays it."), the horizon belongs to the form of the language. It corresponds to the boundary of the world pictured by the painting, and therefore, according to proposition 5.632 ("The subject does not belong to the world: rather, it is a limit of the world"), the horizon belongs to the subject. So aside from the signs of the iconic language, which express definite objects, there are expressions on the painting connected not with the objects, but with the subject, which is the bearer of the language.

Albrecht Dürer (1471–1528) showed us in one of his drawings a method, by which it is possible to create a perspectivist painting. It was by a similar procedure that the Renaissance painters discovered the principles of perspective. Among other things, they discovered that in order to evoke the illusion of two parallel lines, for instance two opposite sides of a ceiling, they had to draw two convergent lines. They discovered this, but did not know why it was so. The answer to this question was given by projective geometry.

2.2.2. The language of projective geometry

Gérard Desargues (1593–1662), the founder of projective geometry, came up with an excellent idea. He replaced the object with its picture. So while the painters formulated the problem of perspective as a relation between the picture and reality, Desargues formulated it as a problem of the relation between two pictures.

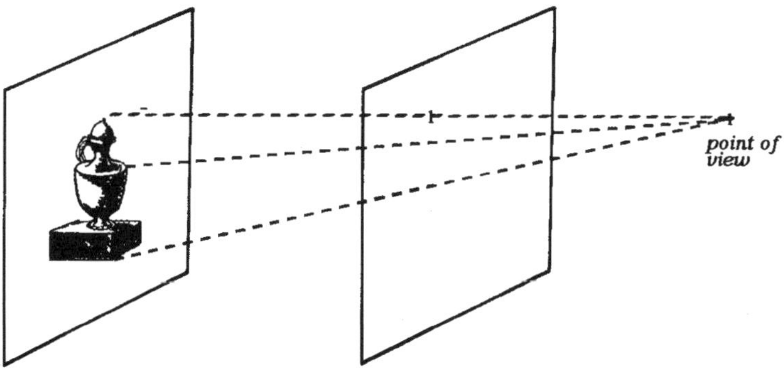

The advantage brought by Desargues' idea is that, instead of the relation between a three-dimensional object and its two-dimensional picture we have only to deal with a relation between two two-dimensional pictures. After this replacement of the object by its picture, it is easy to see that Dürer's dotting procedure becomes a central projection of one picture onto the other, with its centre in our eye. We have mentioned all this only to make clear that the centre of projection represents the point of view from which the two pictures make the same impression.

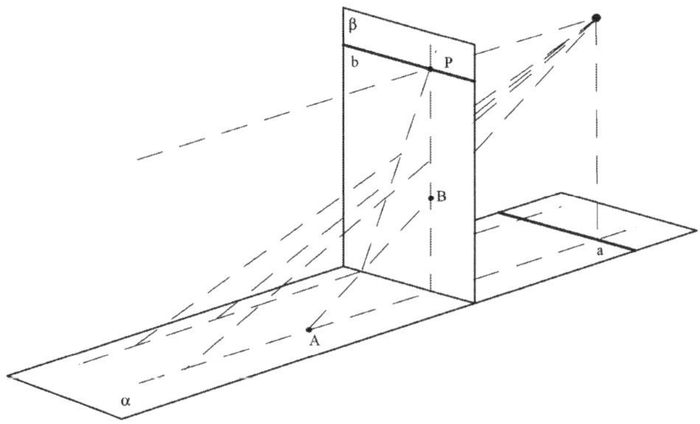

Before we start to consider the central projection of some geometrical objects, we have to clarify what happens with the whole plane, on which these objects are drawn. It is not difficult to see that, with the exception of two parallel planes, the projection of a plane is not the whole plane. On the first plane (plane α – the plane from which we project) there is a line *a* of points for which there are no images. On the other hand, on the other plane (plane β – the plane onto which we project), there is a line *b* onto which nothing is projected. To make the central projection a mapping, Desargues had first of all to supplement both planes with infinitely remote points. After this, the line *a* consists of those points of the plane α which are mapped onto the infinitely remote points of the second plane β. On the other hand, the line *b* consists of the images of the infinitely remote points of the plane α. So by supplementing each plane with the infinitely remote points, the central projection becomes a one-to-one mapping.

In the pictures of projective geometry there is a remarkable point, different from all other points: the centre of projection. As shown above, the centre of projection represents, in an abstract form, the eye of the painter from Dürer's drawing. For Desargues, nevertheless, the point of view is explicitly incorporated into language. Besides this point, the pictures of projective geometry also contain a remarkable straight line. It is the line *a*, which is responsible for many of the singularities occurring by projections. The position of the line *a* on the plane α is determined by the centre of projection, which represents the eye of the spectator. So it is not difficult to see that the line *a* represents the horizon. But it is important to realise one basic difference between the horizon in a perspectivist painting, and in a picture of projective geometry. In projective geometry, the horizon is a straight line, which means that it belongs to the language. It is not something that shows itself only when the picture is completed, as in the case of the paintings. Desargues drew the horizon, made from it an ordinary line, a sign of iconic language. Thus, we can say that he incorporated a form of the pictorial language of Renaissance paintings into the language itself.

2.2.3. The Riemannian analysis situs

For the sake of brevity, we have omitted four stages of development of the form of language of geometry, which are connected with the development of non-Euclidean geometry (Lobachevsky, Beltrami, Cayley, Klein), and turn now directly to Riemann. The omission of these four stages makes it a bit difficult, perhaps, to recognise in Riemann all the aspects of the form of language, which we described in the case of projective geometry. The problem is that these aspects became more and more abstract, and the only way to identify them, in the case of topology, is to trace them back through the intermediate stages to projective geometry. This will be not possible now, because of our omission, but those interested can read more in (Kvasz, 1998). Our central concern here is just to show that Poincaré did, with the form of the language of Riemann's theory of functions, exactly the same thing as Desargues did with the form of the language of Renaissance paintings; namely, he incorporated the form of the language into the language itself. In showing this, it will be made clear that our method of reconstruction can fill in the gap between the geometrical and the topological parts of Lakatos' book.

Riemann developed the basic techniques of the *Analysis situs* in the course of his work in the theory of complex functions. We will show the basic steps of Riemann's construction on the so called Klein's bottle. Let us take a square, and do what Riemann did with the complex planes, namely pasting together its edges. For each such pasting we have to determine, which two edges we are going to paste together, and in what orientation. We will indicate with letters the corresponding edges, and with arrows the orientation. If we paste two opposite sides, which are correspondingly oriented, we get the surface of a cylinder. If we paste two opposite sides of the square, but twist one of them by 180 degrees, we get the well-known Möbius strip.

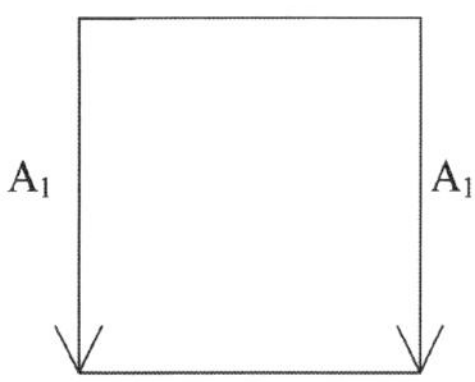

Cylindrical surface

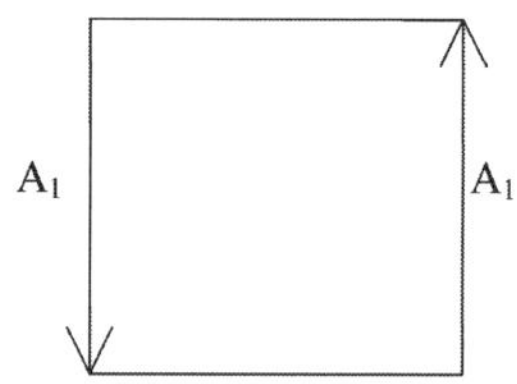

Möbius strip

Both these pastings are easy to perform in our three dimensional space. The next object, which is also easy to construct, is the torus.

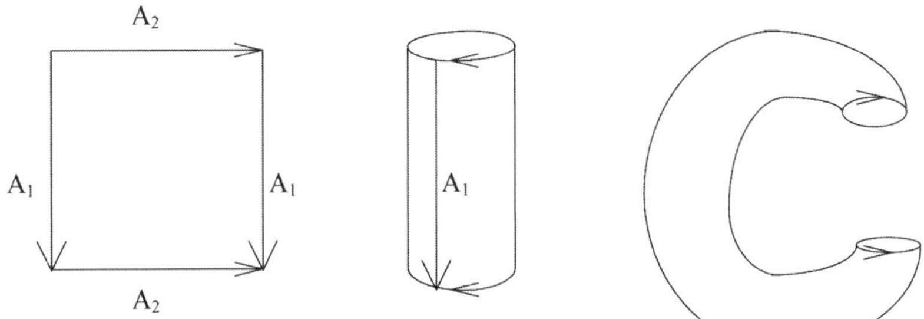

Torus

Let us now imagine a square, which differs from that defining the torus only in our having changed the orientation of one of the edges A_2 to the opposite.

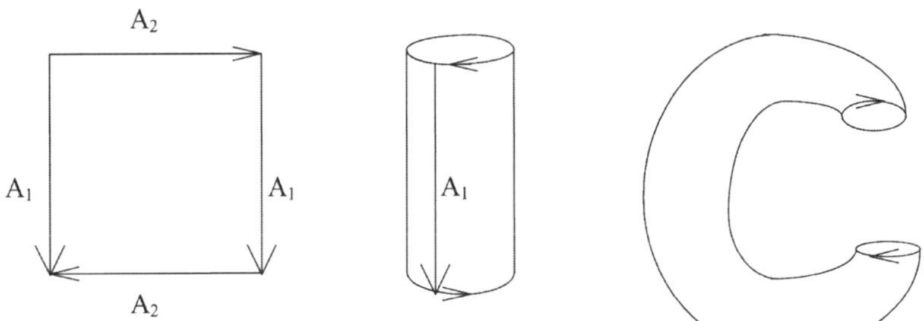

Klein's bottle

We see that, in this case, it is impossible to paste the corresponding circles together, because their orientation does not match. We would need to twist one of them, as we did one edge of the square in the construction of the Möbius strip. In the case of the Möbius strip, we were lucky. The object before the last (and in fact the only) pasting was planar, for which reason we had the third dimension at our disposal, which we used for twisting one edge of the cut. In this way, we obtained the correct orientation of the edges, and could paste them together. In the case of Klein's bottle, we are in a worse position. The cylindrical surface, the one edge of which we have to twist, is a three-dimensional object, so we have no further free dimension at our disposal, into which we could 'lean out' and make the necessary twist. But it is also clear that this is our problem, a problem of our three dimensional space, in which we want to make our construction. Klein's bottle itself has nothing to do with this. Whether we can or cannot construct it in our three dimensional space is not a property of the object, but of our space. Therefore, the diagram consisting of a square (or, more generally, of a polygon with even number of edges), on which it is indicated, which edges should be pasted together (and in what orientation), we can consider as a new geometrical language, which makes it possible to represent the particular objects independently of space. "We should think of space more intrinsically and not as being imbedded in some special way in a particular Euclidean space." (Agoston, 1976, p. 60).

But who can do this? We think that we are dealing here once again with an implicit appeal, which is similar to the appeal of the Renaissance painters. The difference is, however, that this time we are not required to see two parallel sides of a ceiling beyond the two convergent lines in the painting. Now we have to get rid of space. From our previous experience with such appeals, we know that we are dealing with a new form of language, based on a new kind of epistemic subject. Riemann's language describes the constitutive acts, with whose help we can create the surfaces. Therefore, the epistemic subject, which forms the basis of Riemann's language, is the implicit constitutive subject. It is implicit, because Riemann is not able to tell us what, exactly, we have to do in order to get rid of three-dimensional space and see Klein's bottle. And it is constitutive, because it takes over the role which geometry, until Riemann, had assigned to space. Riemann found out how we can, in a technical way, grasp this constitutive function. Something, which was given, which represented the a priori characteristics of space, is replaced by an a posteriori structure.

2.2.4. Poincaré's combinatorial topology

The basic problem of Riemann's language is the implicit nature of its form. On the one hand, it relies on geometrical intuition (we have to see, for instance, that by pasting together the sides A_1 and A_2 we really do get a torus), but on the other hand it requires us to abandon a considerable part of this intuition (namely, its dependence on three-dimensional space). The way out of this dilemma was found by Henri Poincaré. It is the well-known combinatorial topology, which brought into the language in an explicit way an incorporation of the constitutive subject.

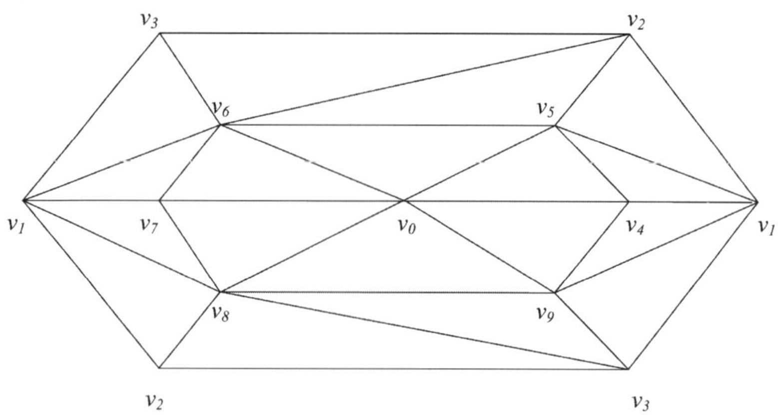

Let us take, for instance, the projective plane, which can be obtained within Riemann's framework by pasting a circle along the side of a Möbius strip. Poincaré's idea was to avoid any pasting (which we are unable follow), and any requirement that we forget about space (which we are unable to do). Instead of appealing to any pasting, we just take a picture of Riemann's language and divide it into triangles. It may seem that we were too generous in this triangulation, and have chosen too many triangles. But the aim is to make sure that no two segments that correspond to different edges, carry the same labels. The internal hexagon ($v_4\ v_5\ v_6\ v_7\ v_8\ v_9$) represents a circle, while the external belt represents a Möbius strip, and thus the whole object is a projective plane.

Now we define a k-dimensional simplicial complex (see Agoston, 1976, p.32):

Definition 1: *Let $k \geq 0$. A k-dimensional simplex is the convex hull σ of $k+1$ linearly independent points $v_0, v_1, \ldots, v_k \in R^n$. We write $\sigma = v_0 v_1 \ldots v_k$. The points v_i are called the vertices of σ.*

Definition 2: *Let $\sigma = v_0 v_1 \ldots v_k$ be a k-dimensional simplex and let $\{w_0, w_1, \ldots, w_l\}$ be a nonempty subset of $\{v_0, v_1, \ldots, v_k\}$, where $w_i \neq w_j$ if $i \neq j$. Then $\tau = w_0 w_1 \ldots w_l$ is called an l-dimensional face of σ and we write $\tau \prec \sigma$.*

Definition 3: *A simplicial complex K, is a finite collection of simplices in some R^n satisfying:*

1. $\sigma \in K$, then all faces of σ belong to K.

2. If $\sigma, \tau \in K$, then either $\sigma \cap \tau = \emptyset$ or $\sigma \cap \tau$ is a common face of σ and τ.

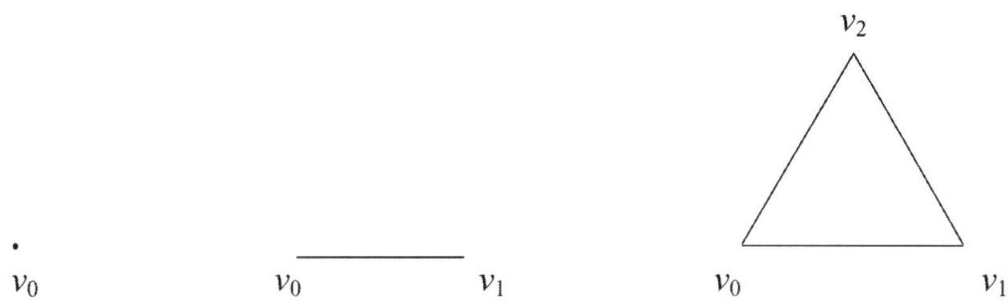

The point v_0 is a 0-dimensional simplex, the segment $v_0 v_1$ is a 1-dimensional simplex and the triangle $v_0 v_1 v_2$ is a 2-dimensional simplex. Thus, the simplexes are the most simple objects of their particular dimension. The simplexes v_0 and v_1 are 0-dimensional faces and the $v_0 v_1$ is a 1-dimensional face of the simplex $v_0 v_1$. The simplex $v_0 v_1$ has no other faces. The simplex $v_0 v_1 v_2$ has three 0-dimensional, three 1-dimensional and one 2-dimensional faces.

The sphere (i.e. the surface of a ball) for instance, which from the topological point of view is an object equivalent to the surface of the tetrahedron, we can represent as the following simplicial complex:

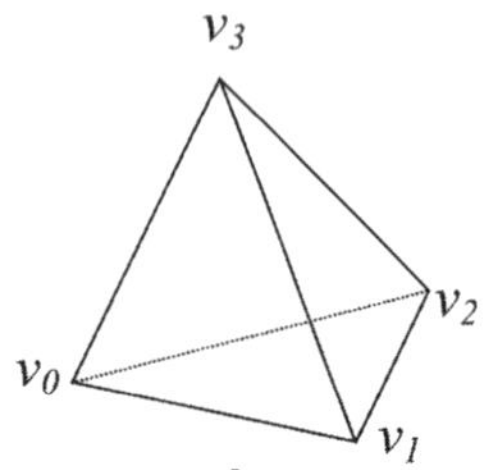

sphere

$$K_S = \{v_0v_1v_2, v_0v_1v_3, v_0v_2v_3, v_1v_2v_3, v_0v_1, v_0v_2, v_0v_3, v_1v_2, v_1v_3, v_2v_3, v_0, v_1, v_2, v_3\}$$

Similarly, for the projective plane we get the following complex

$$K_P = \{ v_1v_6v_3, v_2v_3v_6, v_2v_6v_5, v_1v_2v_5, v_1v_5v_4, v_7v_0v_6, v_1v_9v_3, v_3v_9v_8, v_2v_3v_8, v_1v_2v_8, v_1v_8v_7, v_1v_7v_6, v_1v_4v_3, v_6v_0v_5, v_5v_0v_4, v_0v_9v_4, v_8v_9v_0, v_7v_8v_0, v_1v_2, v_1v_3, v_1v_4, v_1v_5, v_1v_6, v_1v_7, v_1v_8, v_1v_9, v_2v_3, v_2v_5, v_2v_6, v_2v_8, v_3v_6, v_3v_8, v_3v_9, v_4v_5, v_4v_9, v_4v_0, v_5v_6, v_5v_0, v_6v_7, v_6v_0, v_7v_8, v_7v_0, v_8v_9, v_8v_0, v_9v_0, v_1, v_2, v_3, v_4, v_5, v_6, v_7, v_8, v_9, v_0\}$$

Now we can forget about Riemann's picture of the projective plane. The simplicial complex K_P represents the projective plane without any reference to pasting. The edge v_1v_2 in the triangles $v_1v_2v_5$ and $v_1v_2v_8$ is simply the same edge. In the picture it is presented twice, once at the lower right-hand corner and the second time at the upper left-hand corner. But this is a problem of the picture. It is impossible to draw the projective plane without cutting it; thus, the edges which form the cut will be in the picture two times. On the other hand, the simplicial complex represents the particular surface without any reference to cutting or pasting. In the complex, therefore, every vertex, edge and face of the projective plane is present only once (as it is listed only once in the definition of K_P).

The basic achievement of Poincaré was the development of formal techniques and calculation procedures, which only on the basis of the simplicial complexes make it possible to determine the basic topological invariants of a surface. Without any reference to a picture or space, and using only symbols, he managed to compute the Betti numbers (which determine connectedness, dimension, Euler characteristic and many other topological invariants). Every surface is constituted by the way in which the triangles of its triangulation are connected. We need no other information in order to determine the topological invariants, and precisely this information is expressed in the abstract simplicial complex. Thus, we can say that Poincaré succeeded in incorporating Riemann's constitutive acts into the language. Instead of pasting the edges of a square in our imagination (as the real pasting it is usually impossible to carry out, it remains our task to pretend that we know what would result from such pasting) we make the specific triangulation, in which the pasting is already done (or, more precisely, there is nothing to paste, because nothing was cut). The language of combinatorial topology, and topology in general, is thus based on the explicit constitutive subject. It is the language by which mathematics liberated itself from dependence upon space. It makes it possible to speak about objects independently of whether it is or is not possible to represent them in our three-dimensional space. The simplicial complex of the projective plane does not differ in any fundamental way from the complex, which corresponds to the sphere. The fact that it is not possible to realise it in our space is, from the topological point of view, only of secondary importance.

2.3. Returning to Lakatos' proofs and refutations

Our reconstruction of the development of geometry enables us to explain the strange omission in Lakatos' book. The transition from geometry to algebraic topology is a change in the form of language used, consisting in the introduction of the constitutive subject. This transition, nevertheless, did not occur primarily in the theory of polyhedra (i.e. in the theory the development of which Lakatos reconstructs), but in theory of functions of complex variables. Riemann as well as Poincaré were led to make their fundamental changes in the form of the language by problems of complex analysis. Poincaré only afterwards utilised the conceptual progress, achieved in this area and in geometry, and demonstrated the power of the new language with the theory of polyhedra.

A change in the form of the language – even one which had its origin outside the theory which Lakatos was studying – was a rupture, with which his way of analysing change did not allow him to deal. His interpretative tools such as *monster-barring* or *lemma-incorporation* failed, because here we are dealing with a change of the whole conceptual basis of geometry and not just with the assumptions of some theorem. Thus, what appeared to be only a mere omission sheds light onto the boundaries of applicability of Lakatos' method of *Proofs and Refutations*. Lakatos' method of reconstruction can be adopted only in cases where the form of language is not changing. So even if, in the last chapters of his book, Lakatos is writing about concept formation, the fact that he did not even recognise one of the most fundamental conceptual changes in the history of mathematics casts doubts concerning the reliability of this part of his analysis.

After what we have said, it is not surprising that Lakatos misinterprets the epistemological nature of Poincaré's proof. He presents it as a '*translation of the conjecture into the terms of vector-algebra*' (p. 106). But the fundamental epistemological questions, why into the terms of vector-algebra (where is it coming from?) and why into the vector-algebra modulo 2 (why not modulo 3 or 7?) he leaves unanswered. Poincaré made no translation. He only explicitly formulated the structure of the constitutive acts, forming the basis of the concept of space. The fact that these acts are based on arithmetic modulo 2 only shows that we have to do with constitutive acts. The arithmetic modulo 2 expresses parity. Parity means that the chain of acts 'closes itself,' and so constitutes an object (an open chain does not constitute anything).

We saw that the concept of the form of language makes it possible to overcome the barriers, which for Lakatosian methodology remained insuperable. The changes of the form of language he either totally ignores (in *Proofs and Refutations*); or later, in the period of the *Methodology of Scientific Research Programmes*, he is able to describe them only as the 'emergence' of a new research programme (Poincaré's programme in the theory of polyhedra, or Lagrange's programme in mechanics). But one question of fundamental epistemological significance, namely that of what the relationship is between two programmes which follow one another in the development of a theory, – for instance, what is the relation between Riemann's and Poincaré's programme in geometry, or between Lagrange's and Hamilton's programme in mechanics? – this basic question Lakatos did not even raise. Epistemology based on Wittgenstein's concept of the form of language makes it possible to formulate this question (as a question about the relation between two forms of language) and give a clear answer to it (this relation consists in the explicit incorporation of a previously implicit form of language into the language itself, or in the emergence of a new implicit form following the explicit incorporation of the previous form).

2.4. Dialectic versus logic in Lakatos' methodology

The above exposition illustrates the usefulness of the concept of the form of language for epistemology. This usefulness is based on two circumstances. First, the concept of the form of language is closely related to the notion of the subject, and therefore it makes it possible to reconstruct cognitive processes such as heuristics, discovery, interpretation, translation and understanding without the necessity of introducing a subject in the form of an idealised scientist or scientific community. We need not introduce a subject from outside, because the subject is already present in our framework as a constituent of the form of language (cf. Sec. 2.2.1). That means that we are not forced to mix our epistemological considerations with sociological, psychological or historical elements. Secondly, the form of language (the nature of the horizon, background, subject, etc.) is clearly separated from logic (the principle of twovaluedness, etc.). This liberates our approach from dialectics.

Dialectics (in its explicit form, as we encounter it in Hegel or Marxism, as well as in its implicit form, as it is present in the work of Lakatos or Popper) has a tendency to interpret the evolution of concepts, or of knowledge in general, as a process which is logical in nature. It is not by chance that Hegel called dialectics a *science of logic,* and it is not by chance that Popper speaks about the *logic of scientific discovery.* According to dialecticians, knowledge is logical in nature. A necessary consequence of this is that the evolution of knowledge comes into conflict with logic.

Both solutions to the dialectician's conflict between logic and the evolution of knowledge are unsatisfactory. Philosophers who follow Hegel, in the attempt to replace classical logic by some new dialectical one, were unable to offer anything comparable to the successive formal logic, and thus their research programme degenerated. On the other hand, dialecticians like Popper or Lakatos, who were not prepared to sacrifice logic, and thought that logical consistency is crucial to rational discourse, were forced to give up evolution. The fact that Lakatos was unable to reconstruct any deeper conceptual change in history of mathematics or physics is not accidental. As a dialectician, he conceived evolution to be in conflict with logic, but as Popper's disciple he was not prepared to give up logic. Thus he omitted some of the most interesting moments in the history of mathematics. If he had tried to reconstruct them, he would have been forced to violate logic. Therefore he reconstructed only those changes, in which relatively small conceptual changes occur.

Nevertheless, if we interpret the evolution of concepts as development of the form of language, then all of logic remains untouched in the course of the evolution. What is evolving is not logic, not even the separate concepts, but rather the whole conceptual framework, into which they are incorporated. Thus, it is possible to describe the evolution of concepts without any logical inconsistency. What does change in the course of development is the structure of the epistemic subject of language, the horizon, background. Logic remains the same. Epistemology based on the concept of the form of language represents, thus, a passable way between two extremes. The one extreme is *dialectical logic* (of Hegel and Marxism), which for the sake of evolution sacrifices logic. The other extreme is *logical dialectic* (of Popper or Lakatos), which for the sake of logic sacrifices evolution. If we consider evolution to be the development of the form of language, we can develop epistemology as a exact discipline, free of any dialectical inconsistencies and – in spite of this – able to grasp the most fundamental conceptual changes in the history of mathematics and science.

3. THE QUESTION OF THE UNIVERSALITY OF THE METHODOLOGY OF SCIENTIFIC RESEARCH PROGRAMMES

The notion of the "research programme" appears for the first time in Lakatos' *Changes in the Problem of Inductive Logic* (Lakatos, 1968a). Lakatos uses this notion in order to discuss Carnap's research programme in epistemology. Lakatos wrote that

> A successful research programme bustles with activity. There are dozens of puzzles to be solved and technical questions to be answered; even if some of these – inevitably – are the programme's own creation. But this self-propelling force of the programme may carry away the research workers and cause them to forget about the problem background. They tend not to ask anymore to what degree they have solved the original problem, to what degree they gave up basic positions in order to cope with internal technical difficulties. ... One frequently solves very different problems from those which one has set out to solve.

Lakatos presented the Methodology of Scientific Research Programmes in his paper *Criticism and the Methodology of Scientific Research Programmes* (Lakatos, 1968b). According to Lakatos, great scientific achievements are not characterised by an isolated hypothesis (as it was in his *Proofs and Refutations*, where he presented the development of the theory of polyhedra as centred around Euler's formula), but rather by a research programme. A research programme manifests itself in the form of a series of successively developed scientific theories, characterised by a certain continuity which connects the members of the series.

3.1. MSRP – some problems

Lakatos suggests viewing the evolution of science as the development of different research programmes. Nevertheless, he wavered with regard to the possible scale of a research programme. On the one hand, he writes, *"even science as a whole can be regarded as a huge research programme,"* while on the other side we can find the following words: *"But what I have primarily in mind is not science as a whole, but rather particular research programmes, such as the one known as 'Cartesian metaphysics'"* (Lakatos, 1970, p. 47). It is not clear whether these two are the only possibilities for a research programme, or whether there are some programmes of an intermediate scale such as, for instance, quantum physics. Quantum physics, taken as a programme, is not as large as science as a whole; on the other hand, during the development of this programme different metaphysical positions were taken, and so it seems to be a programme of a larger scale than Cartesian metaphysics, which has the same ontology from its beginning to its end. Lakatos did not make clear what should be the core, the protective belt, and the positive heuristics of a programme in the case of large scale programmes, such as science as a whole.

Furthermore, even if Lakatos mentioned the possibility of large scale research programmes, all of his examples operate on the smallest possible scale. On page 50 of his *Falsification and the Methodology of Scientific Research Programmes,* he gives a detailed description of Newton's programme:

> Newton first worked out his programme for a planetary system with a fixed point-like sun and one single point-like planet. It was in this model that he derived his inverse square law for Kepler's ellipse. But this model was forbidden by Newton's own third law of dynamics, therefore the model had to be replaced by one in which both sun and planet revolved round their common centre of gravity ... Then he worked out the programme for more planets as if there were only heliocentric but no interplanetary forces. Then he worked out the case where the sun and planets were not mass-points but mass-balls ... This change involved considerable mathematical difficulties, held up Newton's work – and delayed the publication of the Principia by more than a decade. Having solved this 'puzzle,' he started work on spinning balls and their wobbles. Then he admitted interplanetary forces and started work on perturbations.

In contrast to his detailed description of Newton's programme, Lakatos says nothing about the transition from Newtonian mechanics to the Lagrangean, and later to that of Hamilton. So even if he mentioned the possibility of programmes on larger scales, he never gave any example of such a programme. This seems very strange, if we keep in mind that Lakatos conceived of his theory as an alternative to Kuhn's theory of scientific revolutions, which operates on the large scale of science as a whole. The fact that Lakatos entered into polemics with Kuhn shows that he understood his theory to be universal, but it is incomprehensible why did he not analyse some developments on a larger scale. If he would have done so, he would probably have realised that his characterisation of the core, belt, and positive heuristics was too narrow.

We believe that this vagueness concerning the scale of the research programme on one side, and a very specific characterisation of the programmes core, belt, and heuristics on the other, caused lots of troubles when philosophers tried to apply the basic ideas of the MSRP in mathematics. When G. Giorello tried to use the ideas of the MSRP to characterise the works of Archimedes (Giorello, 1975) and Riemann (Giorello, 1981), he faced the problem of finding the hard cores and protective belts of their programmes. Neither the programme of Archimedes, nor that of Riemann have as a hard core some metaphysical theory. Koetsier, in his book *Lakatos' Philosophy of Mathematics* (Koetsier, 1991) discusses the works of Giorello in length and comes to the conclusion that "notions like protective belt, positive heuristics, empirical versus theoretical progress all concern natural sciences and they cannot be applied to mathematics without modifications. The attempts by ... Giorello to describe certain mathematical developments in terms of the MSRP clearly show the difficulties involved." For this reason, Koetsier comes to the conclusion that it is not possible to use the concepts of MSRP in mathematics, and therefore proceeds to develop his methodology of research projects and research traditions.

We disagree in this point with Koetsier's assessment of Giorello's papers, as well as with his conclusion that the concepts of the MSRP are not applicable in characterising some developments of mathematics. We can take for instance the logicist programme of Gottlob Frege, which can be clearly characterised through the metaphysical belief that there are no mathematical objects as such, and that all mathematical truths are logical. (In addition to Frege's logicist programme, we could also mention the three other programmes in the foundations of mathematics which competed with it. We have in mind Brouwer's intuitionist programme, Hilbert's formalist programme and Cantor's set theoretical programme.) Another candidate for a research programme in the sense of Lakatos (this time, in geometry) could be the Erlangen programme of Felix Klein based on the belief that all geometrical objects can be characterised in terms of invariants of groups of transformations. If one asks what is metaphysical about this belief, we can say that it concerns the very concept of metaphysics, saying that "to be" means to be invariant with respect to some group of transformations. (Aside from Klein's Erlangen programme, we could also mention the structuralist programme of Nicolas Bourbaki.)

Thus, the problem is not that there would be no examples of research programmes in mathematics in the sense of Lakatos. The problem is, rather, that there are many other patterns of development. In our opinion, the fundamental problem with Lakatos' MSRP lies in the tension between the very vague characterisation of the scale of a research programme – which led Giorello to use this concept rather freely – and the very narrow characterisation of the basic concepts of core, belt and heuristics – which led Koetsier to his critical assessment of Giorello's papers. In order to develop Lakatos' theory in a consistent way, we have, in our view, two possibilities. The one possibility is to give up Lakatos' idea, that the whole of science can be seen as a research programme. In this case, we will define a research programme only on a small scale, so that we can preserve Lakatos' narrow characterisation of the core as metaphysical, and of the belt as consisting of auxiliary hypotheses. The other possibility is to stick to Lakatos' idea that research programmes operate on different scales, and to give up his characterisation of the programme's core, belt, and heuristics. We think that the first possibility is less interesting, making our next problem that of characterising the different scales on which programmes operate.

3.2. An alternative strategy of reconstruction – the concept of epistemic ruptures

In our paper *On Classification of Scientific Revolutions* (Kvasz, 1999) we dealt with the problem of characterising changes in science – which occur on different scales – and offered a classification of the scales, on which science is changing. The basic idea behind our classification is the following. It is well known that, when a scientific theory is developed, the new theory has to explain all the relevant facts, which were explained by the old one. This leads to the possibility of embedding the old theory in the new one. Such embeddings have usually the form of a limit (for instance, in the case of the embedding of Newtonian into relativistic mechanics, this limit has the form

$v/c \to 0$). We analysed the development of many different theories and compared the limits, which accompany the embedding of the old theory into the new one. We called these embeddings *epistemic ruptures*, and it turned out that there are at least four different ways in which such embedding can be done. Because of this, we distinguished four kinds of epistemic ruptures.

Examples of epistemic ruptures of the greatest magnitude are the *Galilean rupture*, during which physics was turned into an experimental science, and the *Pythagorean rupture*, during which mathematics was turned into a deductive science. This kind of rupture also separates Aristotelian from Newtonian physics, as well as Egyptian (or Babylonian) from Greek mathematics. We suggest calling this kind of rupture *idealisation*. The basic difference between Aristotelian and Newtonian physics lies in the way in which they idealise motion. Similarly, the basic difference between Egyptian and Greek mathematics could be seen in the way in which they idealise shape.

Examples of epistemic ruptures of the next magnitude in mathematics are the *Cartesian rupture,* consisting in birth of analytic geometry, and the *Leibnizian rupture,* consisting in the birth of differential and integral calculus. These changes are so deep, that it seems as if in the course of these ruptures, quite new universes were created. New universes of curves or formulas. If we consider, for instance, how many kinds of curves were known in mathematics before analytic geometry, we will find out that there were only about ten. Descartes and Fermat discovered a new way of generating curves according to algebraic formulas. In this way, infinitely many new kinds of curves appeared, curves unknown to the Greeks. The Greeks could not grasp them, because they did not have the appropriate analytic language, based on the combination of a co-ordinate system, algebraic formulas and point-by-point construction. So what qualitively changed during this rupture was the scope of the objects present in mathematics. Therefore, we would like to call these ruptures *re-presentations*. They change the ways in which objects and formulas are present in mathematics; the method of idealisation remained, nevertheless, unchanged.

Examples of epistemic ruptures of the third magnitude in synthetic geometry are the *Desargean rupture* (the birth of projective geometry), the *Riemannian rupture* (the birth of Analysis situs), or the *Poincaréan rupture* (the birth of combinatorial topology), which we discussed in the second part of this paper. They all operate in the same universe of objects. But what they change radically is the ontological status of these objects. We would like to call these ruptures *objectivisations*.

The last, fourth type of rupture encompasses epistemic ruptures of the smallest magnitudes. We would like to call this fourth type *re-formulations*. As an example of such re-formulation, we can take Lakatos' description of successive re-formulations in the development of Newton's programme.

We believe that our classification of epistemic ruptures can produce an alternative approach to the problem of the "fine structure" of scientific revolutions, raised by Giorello for mathematical revolutions: "Revolutions are not instantaneous events but long-term processes ... A revolution is not a single break but a complex sequence of breaks" (Giorello, 1992, p. 165). If we consider the development of quantum mechanics, we can see a basic development that is in accordance with the above described distinctions. Planck, in 1900, introduced the hypothesis of quanta as a mere re-formulation of the standard description of black body radiation. He did not believe that they really exist, for which reason he tried to derive his famous formula without the quantum hypothesis. In 1905, Einstein gave the quantum hypothesis a deeper interpretation in his theory of the photoelectric effect, where he considered quanta to be really existing objects. We could say that Einstein carried out an objectivisation of the quanta. Thus, quanta started being considered as real objects, alongside electrons, protons and other particles. Even so, this is not the end of the story. In 1923, de Broglie came up with the idea that quanta were not special objects (particles of light), which exist alongside material particles (such as electrons or protons), but that every particle has the same wave-corpuscular duality, which was characteristic for the quanta. Thus, de Broglie made from the quanta a re-presentation, in the sense that they are not some special kind of object, but rather a universal means of description, in which the whole universe should be represented. We see that our classification of ruptures is not incidental, but is in good accordance with the actual historical development of science. It is so, because scientists are conservative, and try to solve every anomaly at the smallest expense. In this light we can also see Maxwell's theory of aether as an objectivisation, as an introduction of an object of a new kind, with the help of which electrodynamics (and as Maxwell discovered, also optics) could be explained inside of the "mechanical philosophy." Poincaré shared this conservative attitude, and therefore tried to solve the problems which occurred in the theory of aether within this objectivisation. On the other hand, Einstein's view was that, in order to solve these problems, we need a new re-presentation. So even if from the technical point of view, Poincaré and Einstein are close one to another, their interpretations are fundamentally different.

3.3. Returning to Lakatos' MSRP

Our model makes it possible to refine Lakatos' crude alternatives of considering science as a whole to be a programme, or of considering only programmes of the smallest scale. We would like to introduce programmes of intermediate scales, and to clarify the nature of the core and of the belt on each scale. The idea is to consider successions of theories on each of the four above mentioned scales. In this way, we can consider theories which differ from each other with respect to re-formulations, objectivisations, re-presentations, and idealisations. A research programme on a particular scale would have to characterise the unity in such successions of theories.

IDEALISATION | RE-PRESENTATION | OBJECTIVISATION | RE-FORMULATION

In order to find the respective cores of programmes on different scales, let us consider a succession of theories which differ with respect to re-formulations. All examples studied by Lakatos were of this kind, so we can take, for instance, his reconstruction of the Newtonian programme, which we presented above. Lakatos described the hard core of this programme as metaphysical. If we remember that a particular objectivisation determines the ontological status of the objects, we see that Lakatos characterises Newton's programme in terms of objectivisations. So we can generalise his approach in such a way, that we define a programme of a particular scale as a succession of theories which have differences only on that particular scale. The unity of such a succession can be described only in terms of the next larger scale. Therefore, we define the programme on a particular scale by characterising its core in terms of the next larger scale. Hence, for a programme consisting of re-formulations (for instance, those described by Lakatos in the case of the Newtonian programme) the core is determined by an objectivisation. For a programme consisting of objectivisations (for instance, the programme consisting of Newtonian, Lagrangean and Hamiltonian mechanics), the core is provided by a re-presentation. Finally, for a programme consisting of re-presentations (for instance the mechanical, field theoretical and quantum re-presentations of the universe), the core is provided by the idealisation. On the other hand, the protective belt of a programme consists of the use of the techniques of the same scale, to which the succession of the theories forming the programme belongs. Thus, for the programme consisting of re-formulations, the protective belt consists precisely in the use of reformulations; that is, in introducing different hypotheses, but without changing the ontology of the theory, etc. We believe that this is a natural generalisation of the concepts of Lakatos, and that it solves all the problems he was facing.

This generalisation of the MSRP leads us to a more complex concept of the hard core. We propose considering the hard core as consisting of different layers. These layers would correspond to programmes operating on different scales. In the absolute centre of the core of a scientific research programme, the nature of ideal objects is fixed. In physics (i.e. the programme connecting the mechanical, field theoretical and quantum re-presentations), for instance, with the programmatic declaration: *"Try to understand the natural phenomena as dynamic processes, obeying natural laws."* Around this centre, there is a layer fixing the re-presentation. In the case of mechanics (i.e. the programme connecting the Newtonian, Lagrangean and Hamiltonian mechanics), it is fixed by the programmatic thesis: *"These dynamic processes consist in mechanical motion of matter in space. Space is an absolute three-dimensional continuum, and particles of matter interact at a distance.*" The last layer of the core fixes the objectivisation. In the case of Lagrangean mechanics it is fixed by the programmatic instructions: "*Find a system of parameters, which characterise the configuration of the system. Using these parameters, express the kinetic and potential energy, and their difference is the Lagrange function. Substitute this function into the Lagrangean equations, and solve them.*" This layer of the core is surrounded by the protective belt of various re-formulations, which help solve these equations.

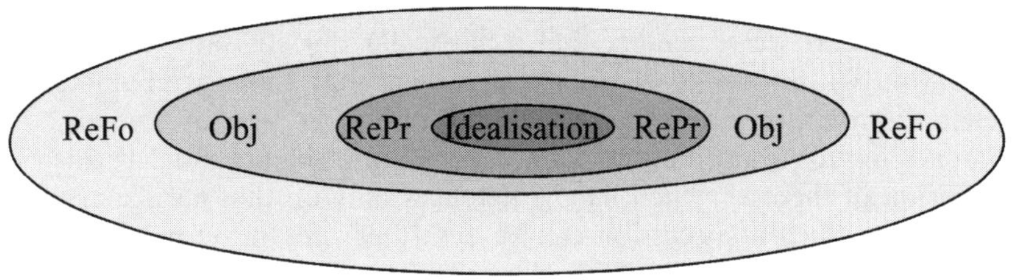

If some anomalies appear, the scientific research programme defends itself by developing a protective belt. At first, the re-formulations are examined. In some cases, it is sufficient to introduce a further planet into our planetary system, and so, by a reformulation of the number of planets, solve the anomaly of Uranus' motion. In this way, the conceptual framework of the programme was saved. In the case of more serious anomalies, for instance those imposed on mechanics by different constraints or friction, it is necessary to broaden the belt and incorporate into it the objectivisation, as well. Thus, we can turn to Hamiltonian mechanics, instead of to the Lagrangean. Nevertheless, the basic re-presentation of physical processes as mechanical motion of matter in space, as well as the absolute character of space and the interaction at a distance was preserved. An even deeper anomaly, as for instance the question of the nature of aether, requires the incorporation of the re-presentations, as well, into the belt. The new theory of motion, the theory of relativity, was in this way able to neutralise this anomaly by changing the absolute three-dimensional space into a relative four-dimensional space-time, and giving up the concept of force acting at a distance. But even such deep anomalies were solved inside the belt, and so the hard core itself, based on the idealisation of motion, could be saved.

We believe that our approach can solve a problem posed by Forrai: "Nor can the concepts of negative and positive heuristics serve as a basis for a general heuristic in the sense of the MPR [Method of Proofs and Refutations], for they belong to individual research programmes, and may have very little in common" (Forrai, 1993, p. 174). We think that our differentiation between programmes of different magnitudes gives us a chance to find something between the general heuristics, in the sense of the MPR, and heuristics belonging to individual research programmes. We have in mind heuristics belonging to each of the particular scales. If we consider idealisations, re-presentations, objectivisations, and re-formulations in more detail (see Kvasz, 1998; Kvasz, 1999; Kvasz 2000), we will find out that there are some common patterns of development on each of these scales. Some of these patterns belong to heuristics.

Lakatos' description of the heuristics in his MPR is not as general as it seems to be. Let us remember that it is based on only one case study. So it is possible that, in areas of mathematics other than geometry, the heuristics might work in different way (cf. the lemma-exclusion method mentioned in the first part of this paper). With the heuristics of the individual research programmes, it is even worse. Lakatos described them only very briefly. Maybe, if one would describe Bohr's or Newton's programmes at the same length, at which Lakatos described the development of the theory of polyhedra (analysing works of 39 mathematicians), some more general patterns of heuristics could be found.

3.4. Some further developments of the MSRP

Our refinement of the concept of the research programme can resolve the conflict between Giorello's reconstructions of Archimedes and Riemann on the one hand, and Koetsier's criticism of these reconstructions on the other. If we consider Archimedes or Riemann, it is obvious that they operate on the level of objectivisations (i.e. they change the ontology of mathematical objects). That is why it was impossible to characterise the hard core of their programmes in terms of metaphysics, and why Giorello was forced to characterise it in terms of re-presentation. If we accept that research programmes operate on different scales, and that only on one of these scales, the smallest one, is the unity of the programme provided for by the metaphysical hard core (i.e. by a fixed objectivisation), whereas on the next scale, the core has the form of a fixed representation, then there is nothing wrong with Giorello's approach.

Thus, in the question of the reconstructions of the programmes of Archimedes and Riemann, we believe that Koetsier's criticism was unjustified. There is no need for the unity of these research programmes to be given in terms of a metaphysical hard core. Also, Koetsier's concept of research projects seems not to fit into the Lakatosian methodology. Lakatos' basic idea was to consider not isolated theories (or projects), but their successions, and to characterise their unity. Thus, we believe that a research project is just one step in the development of a research programme. The different models used by Newton, for instance, could also be called projects.

On the other hand, Koetsier's concept of research traditions can be nicely incorporated into our global picture. It seems that what Koetsier calls research traditions coincides with what we characterised as programmes of the second magnitude – programmes connecting theories and differing in objectivisation, the unity of which is given by a common re-presentation. The characteristic features, with the help of which Koetsier characterised a research tradition (a fundamental mathematical domain and general assumptions about the methods of proof – see Koetsier, 1991, p. 151) as well as the basic examples he gives (the Demonstrative Tradition, the Euclidean Tradition, the 17th Century Geometric Tradition, the Formalist Tradition, the Conceptual Tradition, and the Structuralist tradition) conform well to our description of re-presentations in (Kvasz, 2000). Thus, even if we prefer the terminology of research programmes (because this theory is able to explain the relationships between programmes of different scales), we do have to admit that Koetsier described re-presentations in mathematics in a precise way. He did not study the objectivisations in mathematics, because he restricted the concept of the research programme (in the narrow Lakatosian sense, where the core has to be metaphysical) only to science, and he did not take into consideration, for instance, the logicist or the Erlangen programme. This led him to see research programmes and research traditions as alternatives, rather than descriptions of different scales of development. This means that we, in contrast to Koetsier, believe that there are also programmes (in the narrow Lakatosian sense) in mathematics, and that there are also research traditions (analogous to Koetsier's ones) in physics.

We believe that misunderstandings often root in the fact, that authors speak about changes on different scales, but are not aware of it. In particular, we think that this was the case in the controversy between Lakatos and Kuhn. Thus, at the end, we would like to use our theory to clarify the differences between the Lakatosian and Kuhnian theories of scientific growth. We believe that these theories are based on the analysis of change in science on different scales.

Lakatos' theory is based on the analysis of the smallest of changes (reformulations, on which the original concept of a research programme was based). As we showed, the concept of a research programme can also be generalised for the analysis of objectivisations, and in a metaphorical sense, we can also use it to describe re-presentations. But, as Lakatos' theory is a theory of continuity, and in order to be able to describe a continuous development on a particular scale, it is necessary to have a larger scale, which provides us with tools for the description of this continuity; Lakatos cannot deal with idealisations. Thus, we can characterise Lakatosian theory with the following diagram:

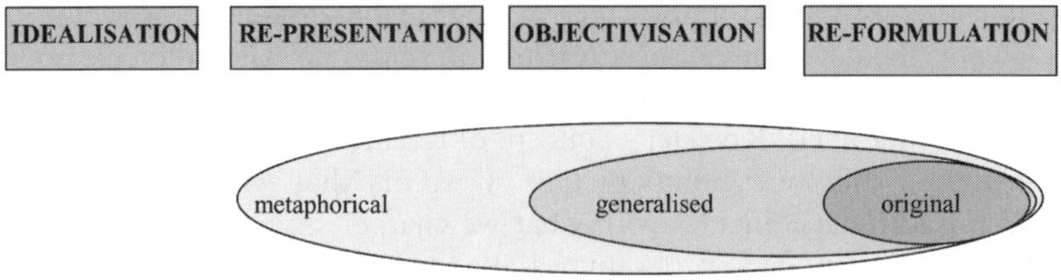

On the other hand, Kuhn's theory is based on the analysis of the largest changes (idealisations, as was the scientific revolution of the 17th century, on which the original concept of a scientific revolution was based). The concept of a revolution can also be used in the analysis of re-presentations (such as the Copernican or Einsteinean revolutions). For the description of objectivisations, nevertheless, we can use it only in a metaphorical way. Kuhn's theory is a theory of discontinuity, and in order to be able to represent a discontinuous development on a particular scale, it is necessary to have a smaller scale, which provides us with a continuous background, against which the discontinuity can appear. Therefore, Kuhn cannot deal with re-formulations as changes, and he describes science on this smallest scale as being mere puzzle solving. Thus, we can characterise Kuhn's theory with the following diagram:

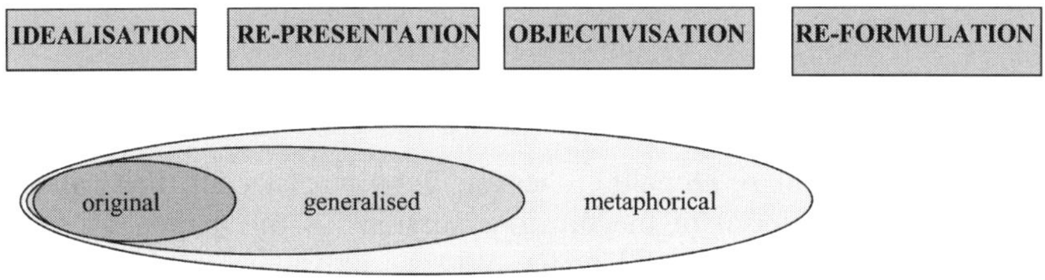

If we compare these two diagrams, it is obvious at the first glance that the re-formulations, on the analysis of which Lakatos based his theory fall outside the Kuhnian framework; on the other hand, the most important examples of change for Kuhn are outside of the scope of Lakatosian theory. On the intermediate scales, where again, one of the theories is generally valid, the other is only metaphorical. So it is obvious that they had not understood each other.

4. FINAL REMARKS

The aim of our paper was to show that the methodology of Imre Lakatos has potential for further development. We believe that it was due to Lakatos' unfortunate combination of methodology with dialectics, which slowed down – or even halted – the development of his ideas. Dialectic is, maybe, as good as heuristics, for the purpose of asking questions about the growth of knowledge. Nevertheless, dialectic is contraproductive as a methodology, as a method of giving answers. We concentrated in this paper on three ways, in which dialectic is contraproductive.

First of all, after disclosing some methods of knowledge-growth (such as *monster-barring*, *exception-barring*, and *lemma-incorporation*), dialectic incorporates these methods into a dialectical scheme, thereby halting further research (for *lemma-exclusion*, and many other possible methods within the MPR).

Secondly, the general dialectical scheme of development is presented in logical terms, what creates a conflict with formal logic and involves methodology in unnecessary controversies, in the course of which the real methodological problems are neglected.

Finally, the dialectical scheme of development is presented in such a general way, that all other patterns of development are ignored. As the above mentioned controversies are usually very vivid, nobody realises that only a small part of methodological problems and patterns of development is discussed.

We hope that by separating the methodological core from the dialectical belt, we can introduce a positive problemshift into the Lakatosian methodology, which will be just as progressive theoretically, as it is empirically.

ACKNOWLEDGEMENTS

I would like to thank to Donald Gillies and Teun Koetsier for their valuable comments and criticism of the previous versions of this paper. Financial support of Aktion Österreich-Ungarn, Wissenschafts- und Erziehungskooperation, Projekt Nr. 25oe11 is gratefully acknowledged.

REFERENCES

Agoston, M. (1976) *Algebraic Topology. A First Course*, Marcel Dekker Inc., New York.

Corfield, D. (1997) "Assaying Lakatos's Philosophy of Mathematics", *Studies in the History and Philosophy of Science*, Vol. 28, No. 1, pp. 99–121.

Forrai, G. (1993) "From the Method of Proofs and Refutations to the Methodology of Scientific Research Programmes", *International Studies in the Philosophy of Science* Vol. 7, No. 2.

Gillies, D. (1992) *Revolutions in Mathematics*, Clarendon Press, Oxford.

Giorello, G. (1975) "Archimedes and the Methodology of Research Programmes", *Scientia* 110, pp. 125–135.

Giorello, G. (1981) "Intuition and Rigour: Some Problems of a Logic of Discovery in Mathematics", pp. 113–135, in: M.L. Dalla Chiara (ed.), *Italian Studies in the Philosophy of Science*, Boston Studies in the Philosophy of Science, Vol. 47, Dordrecht.

Giorello, G. (1992) "The 'Fine Structure' of Mathematical Revolutions: Metaphysics, Legitimacy, and Rigour. The Case of the Calculus from Newton to Berkeley and Maclaurin", in: Gillies 1992, pp. 134–168.

Koetsier, T. (1991) *Lakatos' Philosophy of Mathematics, A Historical Approach*, North-Holland Amsterdam.

Kvasz, L. (1998) "History of Geometry and the Development of the Form of its Language", *Synthese*, Vol. 116, pp. 141–186.

Kvasz, L. (1999) "On Classification of Scientific Revolutions", *Journal for the General Philosophy of Science*, Vol. 30, pp. 201–232.

Kvasz, L. (2000) "Changes of Language in the Development of Mathematics", *Philosophia Mathematica*, Vol. 8, pp. 47–83.

Lakatos, I. (1968a) "Changes in the Problems of Inductive Logic", in: *Mathematics, Science and Epistemology. Philosophical Papers of Imre Lakatos Volume 2,* Cambridge University Press, Cambridge, 1978, pp. 128–200.

Lakatos, I. (1968b) "Criticism and the Methodology of Scientific Research Programmes", in: *Proceedings of the Aristotelian Society*, 69, pp. 149–186.

Lakatos, I. (1970) "Falsification and the Methodology of Scientific Research Programmes", in: *The Methodology of Scientific Research Programmes. Philosophical Papers of Imre Lakatos Volume I,* Cambridge University Press, Cambridge, 1978, pp. 8–101.

Lakatos, I. (1976) *Proofs and Refutations*, Cambridge University Press, Cambridge.

Lakatos, I. (1978) "The Method of Analysis and Synthesis", in: *Mathematics, Science and Epistemology. Philosophical Papers of Imre Lakatos Volume 2,* Cambridge University Press, Cambridge, 1978, pp. 70–103.

Steiner, M. (1983) "The Philosophy of Mathematics of Imre Lakatos", *The Journal of Philosophy* Vol. 80, pp. 502–521.

Wittgenstein, L. (1921) *Tractatus Logico-philosophcus.* Routlege, London 1974.

Faculty of Mathematics, Physics, and Informatics
Comenius University
Mlynska Dolina
84248 Bratislava
Slovak Republic

Olga Kiss

Mathematical Heuristic – Lakatos and Pólya

Amidst the growing interest in actual scientific activity, the process of discovery became a legitimate, sometimes even central topic in philosophies of science. It can be studied from psychological, historical, sociological and philosophical points of view, as well. Instead of emphasising its irrational or unconscious aspect, some philosophers tried to grasp it in rational terms. Among them, Pólya and Lakatos studied the classic problems of the 'art of discovery,' and elaborated their views with respect to the more general problems of knowledge and understanding. In this paper, I compare their approaches to mathematical heuristic.

Ars Invendi

"Heuristic, or heuretic, or 'ars invendi' was the name of a certain branch of study, not very clearly circumscribed, belonging to logic, or to philosophy, or to psychology, often outlined, seldom presented in detail, and as good as forgotten today."[1] This thousand-year-old topic attracted excellent authors like Pappus, Archimedes, Descartes and Leibniz. It was Pólya, who revived this field in the twentieth century, formulating his famous six-step structure of problem-solving.[2]

The aim of heuristic investigations is to find the *method of thinking*, the *rules* by which one can receive results more easily and surely. Scientists sometimes talk about their experience in this field. Archimedes, for example, wrote a letter to his friend, Eratosthenes, in which he described how he had *achieved* his result about the area of the parabolic segment. It says that first he found it in a mechanical way; he performed a thought-experiment, imagining analogous mechanical problems. *Then* he *proved* the result in a (deductive) geometrical way. After reading his letter, one can follow his way of thinking, his method, in similar situations. Pólya did something similar. He showed several different ways of fruitful thinking in mathematical problem-situations through thousands of examples in his books. Lakatos – following Pólya's suggestion – chose one single example, and showed its historical variances, changes and re-formulations.

For Pólya and Lakatos, heuristics did not mean psychological studies. Let me stress this with a counterpoint. When Littlewood wrote about "The

G. Kampis, L. Kvasz and M. Stöltzner (eds.),
Appraising Lakatos: Mathematics, Methodology and the Man, 243–254.

Mathematician's Art of Work," he described the circumstances in which he would get new ideas; he refers to walking, skiing and shaving.[3] These activities have close connections to neither the problem investigated, nor the subject of mathematics. They are general suggestions about how to keep our mental activity high, how to avoid breakdowns and infertility. Littlewood told us stories about the physical, emotional and psychic conditions most suitable for hard intellectual work with high aesthetic quality. Heuristics, for Pólya and Lakatos, was outside of this domain.

PÓLYA'S MATHEMATICAL HEURISTIC

Pólya's works contain "*mental operations typically useful for the solution of problems*" (his italics).[4] Though he elaborated upon a special heuristic applied to mathematics, his aim was more general. As he put it: "*What is the unknown? What are the data? What is the condition?* These questions are generally applicable, we can ask them with good effect dealing with all sorts of problems. Their use is not restricted to any subject-matter. Our problem may be algebraic or geometric, mathematical or nonmathematical, theoretical or practical ... ."[5] His questions and instructions are based on commonsense experiences. His advice is to be simple and natural, as if anyone could find the solution(s), if he really wanted to. They advise you "to do what you would do anyhow, without any advice, if you were seriously concerned with your problem."[6] This common sense is, however, definitely mathematical. His main interest is, of course, mathematical heuristics. As he wrote, "mathematics has two faces: it is the rigorous science of Euclid but it is also something else. Mathematics presented in the Euclidean way appears as a systematic, deductive science; but mathematics in the making appears as an experimental, inductive science."[7]

To speak about mathematics behind the scenes, 'in statu nascendi,' before it receives its rigorous deductive form – this is the aim of Pólya's heuristic.[8] Plausible reasoning that helps us *to find* the final exact forms, has special patterns which do not follow the rules of formal logic. These patterns make rational steps in the process of problem solving.

The most important patterns of plausible reasoning are induction and analogy. Induction, for Pólya, means to start from factual experience in order to find more general relations. We have observations, and they 'suggest' conjectures. An example: we realize that

$$1 + 8 + 27 + 64 = 100.$$

If we are experts in the field of numbers, we easily transform this relation to $1^3 + 2^3 + 3^3 + 4^3 = 10^2$, and if we are interested in mathematical theorems, we pose the question: is it a usual thing that the sum of cube numbers is a square number? And here we find the starting point to a conjecture:

$$1^3 + 2^3 + 3^3 + \ldots + n^3 = x^2.$$

Our next conjecture in this line comes from further observations:

$$1^3 = 1^2,$$
$$1^3 + 2^3 = 1 + 8 = 9 = 3^2,$$
$$1^3 + 2^3 + 3^3 = 1 + 8 + 27 = 36 = 6^2.$$

Not bad! We can feel that we may have found something true. And (again, if we know the world of numbers well enough) we can find $x = 1 + 2 + 3 + \ldots + n$ to be a suitable substitution. In this way, we can express the conjecture in a nice algebraic form:

$$1^3 + 2^3 + 3^3 + \ldots + n^3 = (1 + 2 + 3 + \ldots + n)^2.$$

This is an *inductive generalisation.* Now our task is just *to prove* it. *Induction is not enough* to do so. (Induction here doesn't mean inductive logic. We can say, following Lakatos' later distinction, that induction is used here in a special sense, not *as* a logic of justification, but *as* a logic of discovery.[9]) Pólya does not speak about inductive logic, but rather about the *inductive attitude.* "This attitude aims at adapting our beliefs to our experience as efficiently as possible."[10]

Analogy is the other main pattern: you use it each time when you – to solve a problem – look for similar cases, similar situations to the one which you must cope with. If you have a problem, you can look for other problems, which are analogous to yours, but already solved, or (more) easily solvable. How many vertices does an n-dimensional cube have? In two dimensions: a square has 4 vertices, in 3 dimensional space: a cube has 8 vertices. It suggests that in n-dimensional spaces cubes have 2^n vertices. And so on.

After finding a conjecture, it can be useful to *test* it. We must test our hypotheses by facts; in mathematics we must carry out *quasi-experiments.* If our example is the Euler thesis[11] we can verify some of its consequences pretty easily. The Euler thesis, in its original conjectural form, states that the number of the v(ertices) – e(dges) + f(aces) of a polyhedron is equal to 2. In the case of a cube it must be the same: $v = 8, e = 12, f = 6$, that is $v - e + f = 2$; the conjecture is true. If we test it in more cases of regular polyhedra, it becomes more and more credible. Of course, in this way of reasoning we never achieve truth. Counterexamples may also occur sometimes, like a picture frame, where $v - e + f = 0$.

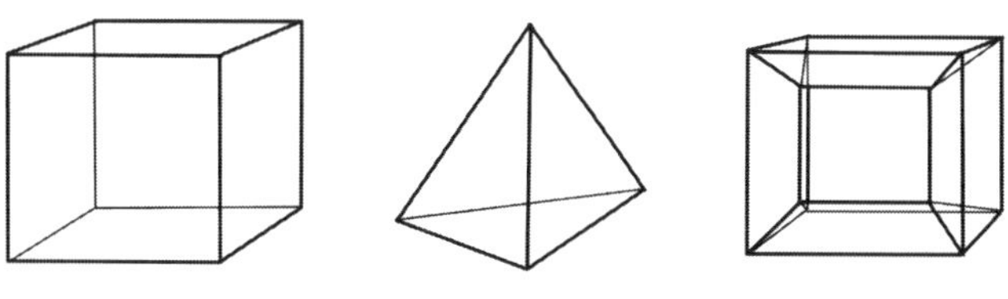

Figure 1. Cube, tetrahedron, 'picture-frame'

In this case, we must change the original conjecture, or change our definition of a polyhedron in order to exclude this 'picture frame.' Pólya's suggestion: lets talk about convex polyhedra. The special cases and tests suggested the modification of the hypothesis, and in this form it seems to be much more credible. *This is induction* again, when we learn from experiences. If our hypotheses become more credible through this process, we can start to prove it (sometimes induction itself suggests ideas for the proof, as well).

Generally speaking: if we try to test hypothesis *A*, and we know that *A implies B*, and we find that *B is true*, then we can conclude from this heuristic reasoning, that *A is more credible. Degrees of credibility* form an infinite set between the *true* and *false*. This is the usual characteristic of plausible reasoning: it is just plausible, having no forcing power. Truth cannot come from plausibility – only a higher degree of credibility can. There is an asymmetry between the truth-values in this respect. While the test cannot ensure the truth of a given conjecture, it can ensure its falsity. Just as in natural sciences, where one can address a question to Nature: "I suspect that this law is true. Is it true? ... Nature may answer Yes or No, but it whispers one answer and thunders the other. Yes is provisional, No is definitive."[12]

The term 'quasi-experiment' became famous from the papers of Lakatos discussing the quasi-empirical character of mathematics. It stems from a citation used by Pólya in his *Induction and Analogy*. Pólya attributes the text cited from Euler's *Opera omnia* to Euler. Anyway, as Lakatos showed, its source is the introduction of the work, which is written not by Euler, but rather his editor. The text runs as follows: "the properties of the numbers known today have been mostly discovered by observation, and discovered long before their truth has been confirmed by rigid demonstrations."[13] – And he adds: "The kind of knowledge which is supported only by observation and is not yet been proved must be carefully distinguished from the truth; it is gained by induction, as we usually say. Yet we have seen cases in which mere induction led to error."[14]

This is why Pólya stresses the importance of intellectual courage and honesty. Courage, to be ready to revise one's standpoint. Honesty, to change it if there is enough reason. His third advice is, however, wise restraint, which tells us not to change it without good reason.[15] Suspicion is not enough, but in some cases a single counterexample would be enough for Pólya to give up his theory. He said, for example, that one counterexample would be enough to refute the Goldbach conjecture.[16] But when he talks about the Euler theorem, the 'picture frame' only inspires him to modify this.[17] What is enough, and what is not, is a matter of *personal taste*. This is one aspect of Pólya's heuristics that sets it apart from logic.

Besides the inference from induction, and analogy, there are other patterns of plausible reasoning. Pólya, in his books, elaborates examples of these patterns, demonstrating several ways of problem solving. In this way a question emerges: *are there rules* for such inferences? *Is there a logic of plausible reasoning*? Discussing this problem, he mentions his talk with a friend, who maintained that "there are no rules, there can be no rules, there should be no rules, and if there were some rules, they were useless anyway."[18] It is important to see that when Pólya talks about rules, he does not mean the sort of rules which would be applicable in a mechanical way. There is no formal logic of plausible inference. Additionally, these rules, applied in the same case by different persons, can lead to different results. His heuristic is not a logic of plausible reasoning.

Plausible reasoning is not the same as demonstrative reasoning. Demonstrative reasoning must be *impersonal, universal, self-sufficient and definitive.* Plausible reasoning is, at least partially, personal: the *degree of credibility* of a hypothesis coming from the tests is up to *personal* opinion. *The fact that it is confirmed is*, however, *impersonal* – according to Pólya.[19] The following also applies to the principle of plausible reasoning: "The verification of a consequence strengthens the conjecture."[20] It is impersonal in this form, but the degree to which it strengthens the conjecture is already personal choice. It is also *universal;* it is admitted in other fields, not only in mathematics. But "in order to judge the *weight* of the evidence, you have to be familiar with the domain; in order to judge the weight with assurance, *you have to be an expert* in the domain."[21]

And what can we tell about self-sufficiency and definitiveness? In demonstrative reasoning, if we accept the premises, and the inference is correct, we cannot help but accept the conclusion. The premises also support plausible conclusions, but some of them can be implicit. New information, new understanding of the situation, may change our opinion about the credibility of the conjecture. That is, the *plausible inference is also self-sufficient, but its strength is not.* In demonstrative reasoning, "if the premises are unquestionably certain, we can 'detach' the conclusion from the syllogism," says Pólya, but "we cannot 'detach' the conclusion of our pattern of plausible reasoning. 'A is rendered more credible' is meaningless, without reference to the premises that explain by which circumstances it was rendered so."[22]

In short, the patterns of plausible inference have two faces. One of them is similar to logical rules: they are impersonal, universal, and self-sufficient. The other facet is *non-logical, personal, contextual, situational and temporal.* This contextual, temporal character of plausible reasoning has been extended so far as to have a historical perspective in Lakatos' philosophy.

LAKATOS' LOGIC OF MATHEMATICAL DISCOVERY

Lakatos' brilliant essay, *Proofs and Refutations,* brings together mathematicians of different historic epochs to discuss the proofs and counterexamples of the Euler theorem. We read the well-ironed dialogues, while the slightly confusing 'real history' takes its course, in the footnotes, as citations from past mathematicians and philosophers.

Lakatos speaks of this dialogue, as if it were a continuation of Pólya's work. "The phase of *conjecturing* and *testing* in case of $V - E + F = 2$ is discussed in Pólya (1954)[23] ... Pólya stopped here, and does not deal with the phase of *proving* – though of course he points out the need for a heuristic of 'problems to prove'... Our discussion starts where Pólya stops."[24] This last remark is true at least in one sense:

One really can follow the process of conjectures and refutations in Pólya's *Induction and Analogy*, so that the two texts can be read together, as if they were two parts of the same story. That is to say, when the dialogue of the *Proofs and Refutations* begins, we find ourselves in the classroom, where the students have already formulated the Euler conjecture (they still do not know whether it is true or false). We enter the classroom after the period of work Pólya described. In the first part (with Pólya), the students formulated, tested and reformulated our conjecture and learned from the experience.[25] In the second part (with Lakatos), they try to prove it or to reject it; to use the counterexamples for rejection or the improvement of both the theorem and the proofs.

The problem-situation is unravelled when the teacher proves the conjecture, and the students are dubious, trying to find explicit ways to refute it, to find counterexamples. The situation became dramatic: proof and refutations of the same theorem. At least one of them must be wrong. A debate starts at this point, and a rational reconstruction of the history of mathematics acquaints us with scientists' usual ways of treating such situations. Here we can join Lakatos in saying: not only the student of meta-mathematics, but I'm afraid, those of any topic "will recognise the patterns described here in his own field."[26]

When faced with this challenge, one has, in answering, several possibilities. Lakatos classifies scientists' usual reactions as the following: defence, exclusion of counterexamples as exceptions, exclusion of counterexamples as monsters, and the adjustment of counterexamples in order to re-interpret them as examples. All of these reaction-types are illustrated in footnotes. All of them lead to different definitions, or re-formulations of theories, and can be traced back to the scientific literature. This is the point where these reactions became interesting for Lakatos: when they determine the *content* of science. Monster-barring, monster-adjusting definitions, exception-barring formulations of the theorem are what results. This is why some years later – in the methodology of scientific research programs – Lakatos can speak of these reaction-types as the positive heuristics of the research programs, as if they were different heuristic rules.

But if that were all he did, we could end this paper here. It would be a perfect step in the Popperian research program; he would stick to the hard core of the 'trial and error'-like conjecture-counterexample-reformulation scheme. Lakatos, instead, found something much more interesting: the method of *lemma-incorporation*. It is at the heart of the method of proof and refutations. It differs from the previous methods by the way in which it uses the counterexamples. While previous methods used them to modify solely and directly the conjecture, or the definition of its terms, Lakatos introduced *proof-analysis* to make the logic of mathematical discovery more sophisticated. Global counterexamples (of the theorem) must be local counterexamples (of at least one step of the proof) too, otherwise the proof fails to prove the theorem. He looks for the guilty lemma, and when he finds it, he incorporates it into the conditions of the theorem. In this way he can exclude the counterexample from the theorem's field of validity. Local – but not global – counterexamples can be dealt with in an easier way: we just have to replace the refuted lemma with a non-refuted one, which can complete the proof again.

Proof analysis sometimes leads to strange results; there are counterexamples that are global, but not local. This means that the proof has not proved the theorem. We can reject such a proof as useless, but we can also modify it by means of the counterexample. "You are interested only in proofs which 'prove' what they have set out to prove" – says the Teacher to his students in the *Proofs and Refutations* – "I am interested in proofs, even if they do not accomplish their intended task. Columbus did not reach India, but he discovered something quite interesting."[27]

No doubt: Columbus did it, but what is interesting in a wrong proof? What can we do with it? A methodological invention can be a good treatment of such situations. If we have a global – but not local – counterexample, it will show a *gap in the proof*. If this proof seemed to be correct earlier, there must be *a hidden lemma*, which we have assumed unconsciously, and which was supposed to be right, but which is falsified by this counterexample. If we find that (or such a) lemma, the gap can be filled. In this way, the situation became similar to the one, where we had a global and local counterexample. Now we can do the same: to incorporate it into the conditions of the theorem, in order to exclude this counterexample from the field of validity of the modified conjecture.

Do we not, by this method, make the field of validity of our theorem narrower and narrower? According to Lakatos, not necessarily. Sometimes the *counterexamples*, which gave rise to this process, are crazy, fantastic, unforeseen enlargements of the previous *world of mathematics*. That is, our defence can be seen as holding the same boundaries in a *widening* conceptual *universe*. Now we can see what those sorts of 'wrong' proofs are suitable for. They can initiate "a decomposition of the original conjecture into subconjectures or lemmas,"[28] resulting in a new classification of the changing world of mathematics. This is the outcome of the method of *proof and refutations*.

This method, in a sense, was not Lakatos' invention. Its logical basis (that if we have a proof for a false theorem, there must be a – perhaps implicit – false lemma in the proof) is a well-known practice among contemporary mathematicians. The method itself states that one can identify this 'guilty' lemma, and that one can, by means of it, modify the conditions of the conjecture in order to exclude the counterexample. Lakatos attributes this methodological invention to Seidel, who had used it already in 1847.[29] Lakatos' contribution to this method was to give it a name and a systematic form. This form came from Pólya's book, entitled *How to Solve It*. Lakatos added his radical approach, based on the view that falsifications can be as important as proofs (Popper).

We have arrived here at the method, which gave the essay its title. There remains only a small step forward. The method of *proofs and refutations* differs from the previous one in terms of its openness towards completely new proofs, as well. It is the radical version of the method of proof and refutations.[30] It allows the replacement of one, or several, or even all of the lemmas. It accepts the guess as a legitimate form of discovery.[31] We can change the whole proof – and the conjecture as well – as we like it.

'PROBLEMS TO PROVE'

The remark of Lakatos, that the dialogue of *Proofs and Refutations* starts where Pólya stopped, is true in a second sense,[32] as well. Lakatos begins not only the story, but also the heuristic of *problems to prove*, where Pólya stopped. If we read Pólya, we can see that he also talks about the heuristic of the 'problem to prove.' Why doesn't Lakatos simply apply Pólya's heuristic to the case of proving the Euler theorem? The answer will be clear only by looking carefully at the sort of 'problems to prove' about which Pólya talks.

Pólya – following Euclid and some of his commentators and translators – makes an important distinction: "We shall consider two kinds of problems, problems 'to find' and problems 'to prove.' The aim of a *problem to find* is to find (construct, produce, obtain, identify) a certain object, the unknown of the problem. The aim of a *problem to prove* is to decide whether a certain assertion is true or false, to prove it or to disprove it."[33] Here, Pólya uses the term 'problem to prove' in a restricted way: it refers to such problems that all of us know well from school. There, the 'problem to prove' is a well-formulated theorem, which one has to prove or falsify. It contains the exact conditions under which the theorem is valid. One does not have to formulate the theorem itself. The only thing left to do, is to substitute the definitions, and find connections between them and the other theorems already proved in the course of the previous classes. In such cases it is a sign of incompleteness of one's proof, if one has not yet used all of the conditions of the theorem. This is why Pólya is right in calling our attention to doing so, asking: "Did you use the whole condition?"[34]

In Lakatos' heuristics, the theorem is not ready when we start to prove it. It is stated in a possibly false generality, and it can be reformulated several times in the process. It is the process of theorem-formation, with which he is dealing. Proofs and counterexamples play a constitutive role in this story, and Lakatos tries to formulate the most suitable rules to guide it.

He states that, "the theorem does not jump out of the head of the mathematician, like Pallas Athene, fully armed out of Zeus' head."[35] It will be armed after the trials and errors of proving and falsifying. The conditions of the theorem, and the concepts used in it, go through a continuous reformulation. The form, which students can meet in their textbooks, is usually the end product of a long (sometimes already historical) process. "The student of mathematics is obliged, according to the Euclidean ritual, to attend his conjuring act without asking questions, either of the background or about how this sleight-of-hand is performed."[36] The deductivist style of proof (and mathematics) hides the process of discovery. Deprived of its history, however, we cannot understand the real process of making mathematics.

In this way, we can say that the two heuristics complete each other. As I see it, Lakatos is too strict in saying of Pólya's *How to Solve It* that "it is wrong to assert that 'the aim of a 'problem to prove' is to show conclusively that a certain clearly stated assertion is true, or else to show that it is false.' "[37] We can accept the existence of different sorts of 'problems to prove.' In the heuristic of Pólya, we deal with those situations, in which we already have the well formulated assertion – that is: someone else already proved it, and our aim is to find a (perhaps different) proof for the same theorem. In the heuristic of Lakatos, the theorem is not yet firmly formulated. The aim of a 'problem to prove' here is its construction. His heuristic rules are there to help mathematicians change conjectures into better-grounded theorems, ones which are better rooted in mathematical knowledge.

None of these problems is '*the* true problem to prove.' We can, rather, say that Pólya's *How to solve It* helps those who want to find a proof to a theorem, and Lakatos' *Proofs and refutations* helps those who want to discover it. Pólya's heuristics may be closer to the students, and that of Lakatos to the working mathematicians. In these cases, Pólya and Lakatos speak of different processes. Pólya speaks of learning mathematics, Lakatos of making mathematics. This is why Lakatos' first rule states: if you have a conjecture, set out to prove it and to disprove it. Both processes belong to scientific research. But this is only a rough comparison.

As we have seen, the method of proofs and refutations belongs to a different problem-situation, not one with which Pólya's heuristics deals with *explicitly*. Sometimes, however, we can find in Pólya's works formulations of theories (e.g. in order not to go far astray, his story about the Euler theorem). His *Induction and Analogy* is about creating general statements about mathematical facts. When he speaks about the patterns of plausible reasoning, he means mathematical problem-solving thought in general.

Moreover, we can see explicit references to the ideas of *Proofs and Refutations* in Pólya's later works. In his "Mathematical Discovery," he refers to that methodology as the 'alternating method'; when first one tries to prove a theorem, and when it is not successful, one tries to find counterexamples.[38] Here, he mentions Lakatos as one who wrote more about it. He also refers to historical examples here, as well as later, when he speaks briefly about the different concepts of mathematical proof and their historical changes. At this point, he also mentions the *Proofs and Refutations*.[39] Anyway, one cannot say that what Lakatos did, was nothing more than to follow Pólya.

We can say, rather, that Lakatos is much more explicit when speaking about theorem formation and theoretical work, while Pólya's aim is to find general patterns for plausible thinking.

THE NATURE OF MATHEMATICS AT BIRTH

To catch mathematics at the moment of its birth – this is the aim of both their philosophies. Their respective work seems to be very close, but there are some crucial, fundamental differences. According to Pólya, mathematics seems to be inductive. His heuristics can be general because there are 'facts' in mathematics; the properties of mathematical objects can be discovered by observations or quasi-experiments. The method of mathematics, therefore, is similar to that of the natural sciences. Lakatos characterised Pólya's approach as follows: Pólya's "stress on the similarities between scientific and mathematical heuristic is one of the main features of his admirable work. What may be considered his only weakness is connected with this strength: he never questioned that science is inductive, and because of his correct vision of deep analogy between scientific and mathematical heuristic he was led to think that mathematics is also inductive."[40]

If one reads these words with Lakatos' early papers in mind, one tends to suppose that Pólya belongs to the Inductivist Program. In his "Infinite Regress and the Foundations of Mathematics" Lakatos mentions three rational enterprises which attempt to achieve firm foundations of knowledge: the Euclidean, the Empiricist and the Inductivist programs.[41] All of them take a justificationist position, and Lakatos sets the critical fallibilist program of Popper against them. According to this classification, Pólya would seem to adhere to the Inductivist Program, but this is not the case. The aim of the Inductivist Program is to build inductive theories, whereas the truth of the basic empirical statement could flow up to the theoretical ones through the channels of inductive logic. That is to say, we speak here about induction in the context of justification, while Pólya talked about it in the context of discovery. This is why we do not find Pólya's place among the classical approaches to the foundations of knowledge (and why Lakatos did not mention him in this paper at all, nor in his later paper on inductive logic[42]). The context of discovery (while it is the very field of Pólya's interest) seemed to be outside their domain.

Inductive reasoning, as we have seen, plays an important role in Pólya's heuristic even if he cannot be counted among the adherents of the Inductivist Program. Lakatos, following Popper, cannot accept that. As he put it: "*Mathematical heuristic is very like scientific heuristic – not because both are inductive, but because both are characterised by conjectures, proofs, and refutations.*"[43] This is what he wanted to show in *Proofs and Refutations*.

NOTES

1. George Pólya: *How to Solve It. A New Aspect of Mathematical Method*. Princeton: Princeton University Press, 1945. p. 102.
2. "Understand the problem. ... Find the connection between the data and the unknown. ... consider auxiliary problems, if an immediate connection cannot be found ... You should obtain eventually a plan of the solution. ... Carry out your plan. ... Examine the solution obtained. ..." (Pólya: *How to Solve It*, inner cover).
3. John E. Littlewood: The Mathematicians Art of Work. in: *Mathematical Intelligencer*, 1 (1978) n. 21, p. 112–118.
4. Pólya: *How to Solve It*, p. 2.
5. Pólya: *How to Solve It*, p. 2.
6. Pólya: *How to Solve It*, p. 3.
7. Pólya: *How to Solve It*, p. vii.
8. Pólya: *How to Solve It*, p. vii.
9. This distinction came from Lakatos: "Changes in the Problem of Inductive Logic" (*Philosophical Papers*, Vol. 2. Cambridge: Cambridge University Press, 1978 p. 136), written in 1968. In the *Proofs and Refutation* – first published in 1963–64 – Lakatos still did not refer to it.
10. Pólya: *Induction and Analogy (Vol. 1 of Mathematics and Plausible Reasoning)* Princeton, Princeton University Press, 1954, p. 7.
11. As with Lakatos in the *Proofs and Refutations. The Logic of Mathematical Discovery* (John Worrall and Elie Zahar, eds.) Cambridge: Cambridge University Press, 1976.
12. Pólya: *Induction and Analogy*, p. 10.
13. Euler: *Opera Omnia*, ser. 1, Vol. 2. p. 459, quoted by Pólya: *Induction and Analogy*, p. 3.
14. *Ibid.*
15. Pólya: *Induction and Analogy*, p. 8.
16. It says that *any number that is neither a prime nor the square of a prime is the sum of two odd primes* – see: Pólya: *Induction and Analogy*, p. 4.
17. Pólya: *Induction and Analogy*, p. 42.
18. Pólya: *Patterns of Plausible Inference*, Princeton: Princeton University Press, 1954, p. 109.
19. What is confirmation and what is falsification will be relativized by Lakatos.
20. Pólya: *Patterns of Plausible Inference*, p. 114.
21. Pólya: *Patterns of Plausible Inference*, p. 114.
22. Pólya: *Patterns of Plausible Inference*, p. 115.
23. Pólya: *Induction and Analogy*.
24. Lakatos: *Proofs and Refutations*, p. 7.
25. See Lakatos: *Proofs and Refutations*, p. 73 n2.
26. Lakatos: *Proofs and Refutations*, p. 5.
27. Lakatos: *Proofs and Refutations*, p. 14.
28. Lakatos: *Proofs and Refutations*, p. 9.
29. The question here is what counts as a methodological discovery. Is it enough to do something, or does it become a discovery only if it received its systematic form and name? Koetsier holds that proof-analysis was only a mean to achieve something else, and Seidel was unaware of 'his' methodological discovery. (See Theun Koetsier: *Lakatos' Philosophy of Mathematics*. Amsterdam: North Holland, 1991, p. 54.) It is, however, not unusual in the history of mathematics to give a name to a theorem or concept not according to historical data (e.g. a correct formulation) but rather mathematicians' respect toward the master who was his predecessor.
30. Lakatos: *Proofs and Refutations*, p. 64.
31. Lakatos: *Proofs and Refutations*, p. 73 and 76: Rule 5.
32. The first one was discussed at the beginning of the previous chapter, on p. 7.

33. Pólya: *Induction and Analogy*, p. 119.
34. Pólya: *How to Solve It* (inner cover).
35. Lakatos: *Proofs and Refutations*, p. 145 n1.
36. Lakatos: *Proofs and Refutations*, p. 142.
37. Lakatos: *Proofs and Refutations*, p. 41.
38. Pólya: *Mathematical Discovery*, § 9.11.
39. Pólya: *Mathematical Discovery*, Vol. 2. § 14.10.
40. Lakatos: *Proofs and Refutations*, p. 74 n.1.
41. "Infinite Regress and the Foundations of Mathematics", Lakatos: *Philosophical Papers*, 2, p. 4.
42. "Changes in the Problem of Inductive Logic" in: *Philosophical Papers*, 2, p. 128–200.
43. Lakatos: *Proofs and Refutations*, p. 74.

Department of Philosophy
Budapest University of Economics
Fővám tér 8
H-1093 Budapest
Hungary

Hans-Christian Reichel

Lakatos and Aspects of Mathematics Education

Amongst mathematicians and philosophers, Imre Lakatos is well known as a scientist who linked mathematics, philosophy and epistemology perhaps more closely than did any other mathematician during the second half of our century. But Lakatos also had a strong influence on mathematics education. This influence regards everyday mathematics education at school, as well as the scientific background theories. Particularly his well known book *Proofs and Refutations* (which has been published in German, as well), influenced the theory and practice of mathematics education (or "didactics" as German speakers prefer to say).

In this note, we try to shortly describe and specify some of the respective influences and trends which have their origins in Lakatos' work.

With his book, *Proofs and Refutations*, and through his general investigation of the "nature" of mathematics and the particular sort of mathematical thinking, he more or less indirectly contributed to the change of several petrified views of teaching mathematics and proofs in mathematics education.

Of course, we recognize influences upon daily classroom-work, too, but his major influence is through his scientific writings and – as already stated – background theories, which, of course, influenced most of the authors of schoolbooks and classroom notes, as well. Here we primarily mean new ways of discussing (and proving) in school math courses, and a new way of developing concepts (often "copying" and including historical items and developments). Below, we shall present some concrete items. But first, we must clearly state that perceptions of the nature and role of mathematics held by our society have a strong and general influence on the development of school mathematics, curricula, instruction and research.

In this respect, Lakatos' view of mathematics as a quasi-empirical science had an impact on the whole field. According to more recent views of mathematics education which, of course, were influenced by *Proofs and Refutations*, instruction and education should be quasi-empirical in any case, whether or not this label is, in general, justified for mathematics itself.

This means that mathematics should not be taught as a matter of absolute truths, as a matter of structured sets of axioms and mere algorithms,[1] but that instruction should start with problems, bold ideas and strategies. Mathematics as a process, not as a ready-formulated bunch of theories.

G. Kampis, L. Kvasz and M. Stöltzner (eds.),
Appraising Lakatos: Mathematics, Methodology and the Man, 255–260.

Advancement, progress and permanent changing of views is more important than the understanding of logical foundations or collections of formulae and the impression of eternal truths.

In other words, mathematics should be seen a bit more as an experimental science (as is done in physics or biology, for example.[2]) In his *Proofs and Refutations*, Lakatos defines a proof as a "thought experiment," a decomposition of the original conjecture into subconjectures, or lemmata.

In light of this contemporary view of mathematics education, it is clear that Lakatos' epistemology and philosophy of mathematics should be important to mathematics educators. In education, the investigation of the balance between methods and heuristics – as is typical for Lakatos – seems more important to me than the provision of theoretical or axiomatic foundations at school, as was intended by the so-called New Math of the seventies, as well as by Bourbaki, of course.

The main influence of Lakatos on education in this sense concerns the role of proving and the role of proofs in school mathematics and curricula (formal proofs, pre-formal ones and heuristic arguments). Proofs as a constant fluctuation and interdependence between trials, counter-examples, reformulations, generalizing, discovering errors and so forth. Proving (arguing mathematically) as a *process* rather than a system of ready-made concepts and "figures of argument." (Process versus product, so to speak.) Proving and mathematically arguing is an "open" process, as several math-educators note (Steiner 1983, to give but one example).

Another important feature in this respect is the problem of teaching how concepts are developed in order to provide for strict proving and arguments. Most pupils (as well as university students) think that concepts come first, and that mathematics consists of working with given concepts. From Lakatos' work, amongst others, we know that mathematics is not a "solid building" consisting of given concepts (bricks), above all. Since Lakatos' work, this fact is – at least partly – reflected in school-mathematics.

All these points are surely positive ones, but Lakatos' ideas were often misinterpreted, as well, and mathematics education was ill-served by incorporating these items too heavily. John Convay, for example, remarked recently that *Proofs and Refutations* is, of course, an interesting book, but that it may be misleading with regard to mathematics in general (see Hanna 1997, p. 79). He writes:

> It is misleading to take the example of Euler's theorem as typical of the development of mathematics. Most mathematical theorems do get proved and stay proved; the original proof may not be quite satisfactory to later standards of proof. But that is a fairly trivial matter. In many cases there has been a significant omission or error in the first attempt of proof, which later had to be corrected. There have been very few cases like Euler's theorem in which the discussion continued for several centuries.

Although many similar statements critical of *Proofs and Refutations* exist, it is certain that Lakatos' book and later work is of eminent importance to the whole of philosophy and to mathematics education. That is clear to everybody, who is informed of recent development of the field. (Compare for example Davis and Hersh, 1981, and others.)

Let me come back to my original theme: Lakatos' influence on mathematics education, instruction and curricula. In Lakatos' view, proofs do not reveal eternal truths like they do to the Platonist or Euclidean eye, but – at the same time – proving in mathematics is not just playing with language and establishing truths as mere conventions, as nominalists may perhaps see it.

But, according to many mathematics educators, this is exactly what proofs should do for learners – especially younger pupils. Proofs should show how the language of mathematics and its whole system works formally, but at the same time establish absolutely reliable formulae and truths – especially at the *end of a process* or of a dialogue. This, in my estimation, is the most important thing that Lakatos had to tell us about, when he discussed Euler's theorem and the evolution of the concept of the polyhedron: the step-by-step refinement of everyday language and proof as a *process*. What may get lost somewhat is the "proof as a *means of insight*." When my students ask whether the exam following a lecture will contain proofs, too, I usually answer that proofs are the only thing I shall ask for – since the terms and concepts of a theory, formal methods and the theory itself are not paramount in mathematics, but have been invented "after the fact," to turn fuzzy reasoning and rough ideas into formal arguments, which then can be communicated and understood independently of subjective views and believes. Such ideas in mathematics education are partly based on Lakatos' ideas!

The word *quasi-empirical*, used to indicate the interplay of example and generalisation, is clearly an essential part of the mathematical enterprise, and it is surely due (at least in part) to Lakatos that this interplay is now much more central to mathematics education that it was in the past. There were, of course, mathematics educators who espoused similar ideas earlier. Most of them, however, described the quasi-empirical nature of mathematics learning as a *methodological* issue, whereas after Lakatos, is was a question of the *philosophy of mathematics education*. Lakatos may have thought that mathematics lies at the centre of the philosophy of mathematics (many authors contradict this claim), but concerning mathematics *education*, anyway, he is right.

In Hanna (1997), Lakatos' credo – and her criticism thereof – are described as follows:

> It is not surprising that such a fascinating new way of looking at mathematical discovery diverted attention from its weaknesses. The method of proof analysis is admittedly engaging, but the case for it as a general method rests upon two examples mainly, one of which is the study of polyhedra – an area in which it is relatively easy to suggest the counterexamples required. This method does not even begin to explain some important cases of mathematical discovery, however. It has nothing to say about the emergency of non-standard analysis, of numerical mathematics in fact the many mathematical discoveries that did not start with a primitive conjecture.

In the interest of fairness, we must remember that Lakatos himself brought up the example of group theory, in order to show that his philosophy of mathematics is not described completely if one only has in mind his discussion of Euler's theorem!

There are two other – and more general – influences of Lakatos on mathematics education. The first one concerns those parts of its background theory, which are called "Stoffdidaktik" in German (and in the English literature, as well): "Stoffdidaktik" (after Lakatos) may be described as summarizing by providing "rational reconstructions" of mathematical theories and concepts, including the process of their genesis and (historical) development.

A second fact resulting from the widespread reception of Lakatos' work (among others, of course) is that the psychology of learning had become a major part of recent mathematics education ("didactics"). Moreover, studying the process of students' mentally developing and/or incorporating mathematical ideas contributes in an important way to the theory of present-day mathematics education.

Aside from the issues described above, there is another aspect – to be seen particularly in the United States – that has to be mentioned when speaking about Lakatos and educational and curriculum aspects of school mathematics: over the last two decades, many math educators in North America (less in Europe, interestingly enough) have suggested that the proof be relegated to a lesser role in the secondary mathematics curriculum.[3] A significant reason (amongst others, of course) for that was a misinterpretation – or say, "over-interpretation" – of Lakatos' thinking. After Lakatos, many mathematics educators stated that *discussion* itself is more important in mathematics than formal reasoning and proofs. Interestingly, European mathematics educators did not participate in this shift. In Europe, many papers continue to be published, which discuss various aspects, and the value, of proofs and the proving of facts in school mathematics.

Very recently, one has been able to discern a new movement towards the use of proofs in North American school mathematics. A prominent example of this is the already cited paper of Gila Hanna (1997) on "The Ongoing Value of Proof," or the call for papers on this subject by the panel of the NCTM journal "The Mathematics Teacher" for a special issue on the role of proofs, which appeared in 1998.

In conclusion, it can be said that many of the well-known, and sometimes misleading, proposals to downplay so-called "formal" mathematics in curricula all over the world, were influenced by, or even based on, Lakatos' ideas which he published in *Proofs and Refutations.* Although these requests had many positive aspects and consequences (think of the decline of the so-called New Math mentioned above), proofs do remain central to mathematics, and this fact should somehow be reflected in school mathematics. A good method of doing so is the concept of so-called *pre-formal* proving, developed by W. Blum and A. Kirsch in their 1991 paper. There are also many European school-books showing that it is nevertheless possible to incorporate ideas compatible with Lakatos' philosophy and still maintain the spirit of "classical" mathematics, characterized by the kind of rigorous proving used in most areas of the mathematics field.

NOTES

1. Remember that (in England for example) geometry was taught strictly following Euclid's "Elements" up to the the first years of our century.
2. A similar request, incidentally, is made and justified by examples by G. Chaitin in connection with his algorithmic information theory (Chaitin 1993). – On the other hand, S. MacLane states that "mathematics does not need to copy the style of experimental physics." Mathematics, he says, rests on proof, and proof is essential; in Hanna 1997: MacLane 1996, p. 193.
   Other recent suggestions divide mathematics into two camps, into so-called rigorous and non-rigorous mathematics ("theoretical mathematics"). This suggestion caused a long discussion; compare, for example, the sixteen counter-statements of well-known mathematicians in the Bulletin of the AMS in 1994, including papers of Atiyah, Thurston and others. Those discussions on the nature of mathematics and mathematics research nowadays have strong impacts on mathematics education, curricula and research, of course. And all of these recent discussions are somehow influenced by Lakatos' work, too. The interplay between his work and the field of didactics should be studied in as many respects as possible in the years to come. The well-written paper of Gila Hanna (Hanna 1997) is an initial example which should be followed by others.
3. Curiously enough, Lakatos himself did not really overestimate the role of proofs. (Compare just the title of his *Proofs and Refutations*.) He only had another *view and concept* of proofs.

REFERENCES

Blum, W. and Kirsch, A. (1991) Preformal Proving, Examples and Reflections. *Educ. Stud. Math.* 22 18; 183–203.

Bürger, H. (1979) Beweisen im Mathematikunterricht – Möglichkeiten der Gestaltung in der Sekundarstufe I und II, in: *Beweisen im Mathematikunterricht*, ed. by Dörfler, W. und Fischer, R. Verlag Hölder-Pichler-Tempsky, Wien.

Chaitin, G.J. (1993) Zahlen und Zufall – Algorithische Informationstheorie, Neueste Resultate über die Grundlagen der Mathematik, in: Prat de la Riba, E. and Reichel, H.-C. (eds.) *Naturwissenschaft und Weltbild*, Verlag Hölder-Pichler-Tempsky, Wien, 30–44. (English: Driessen, A. and Suarez, A. (eds.) (1997) *Mathematical Undecidability, Quantum Nonlocality, and the Existence of God*, Kluwer, Dordrecht).

Davis, P.J. and Hersh, R. (1981) *The Mathematical Experience*, Birkhäuser, Boston.

Feyerabend, P. (1975) Imre Lakatos. *British Journal for the Phil. of Science* 26, 1–8.

Jaffe, A. and Quinn, F. (1993) "Theoretical Mathematics", Towards a Cultural Synthesis of Mathematics and Theoretical Physics. *Bull. of the AMS* 39/1; 1–13.

Lakatos, I. (1976) *Proofs and Refutations*, Cambridge University Press, Cambridge.

Hanna, G. (1997) The Ongoing Value of Proof. *Journal für Mathematik-Didaktik* 18, 171–185.

Lehmann, H. (1980) An Examination of Imre Lakatos' Philosophy of Mathematics. *The Phil. Forum* 22/1, 33–48.

MacLane, S. (1996) Despite Physicists, Proof is Essential in Mathematics. Paper presented at the Boston Colloquium for the Philosophy of Science; Proof and Progress in Mathematics. Boston University.

Reichel, H.-C. (1995) Hat die Stoffdidaktik Zukunft? *Zentralblatt für Didaktik der Mathematik* 95/6; 178–187.
Stein, M. (1986) *Beweisen.* Verlag Franzbecker, Salzdetfurth.
Steiner, M. (1983) The Philosophy of Imre Lakatos. *The Journal of Philosophy* 80/9; 502–521.
Tietze, U.-P. and Wolpers, H. (1997) *Der Mathematikunterricht in der Sekundarstufe* II (vol. 1). Vieweg, Braunschweig and Wiesbaden.

Department of Mathematics
University of Vienna
A-1090 Vienna
Austria

III.

LIFE AND POLITICS

Lakatos in Moscow, 1949
(By kind permission of Éva Pap)

Lakatos with his wife Éva Pap at Balatonvilágos, Summer 1955
(By kind permission of Éva Pap)

Jancis Long

The Unforgiven: Imre Lakatos' Life in Hungary†

Everything Happened So Quickly

When Imre Lakatos left Hungary in November 1956 following the Soviet crushing of the short lived Revolution, he wrote to a friend from the border "Everything happened so quickly. There was no time to say goodbye." To another he wrote "I believe I shall come back because one cannot live without a country."[1]

Everything continued to happen quickly. Within two months of walking across the Hungarian Austrian border with his wife and her parents Lakatos was at King's College Cambridge on a Rockefeller Fellowship pursuing a Ph.D. in the philosophy of mathematics with R.B. Braithwaite. Within 3 years he was an assistant lecturer at London University, working with Karl Popper, and within 13 years Professor of Logic and Scientific Method with world wide fame among mathematicians, philosophers and historians of science, but with an irreparable rift with Popper.[2] His work in the philosophy of mathematics made a strong impression and received the highest praise. His philosophy of science, specifically the ideas about scientific research programs and rational reconstruction of the growth of theories were considered less original than his mathematical philosophy, but have also remained part of the discourse of the field.

Lakatos never did return to Hungary. His death, too, came quickly, of a heart attack, in 1974 at age 51. Though he had made a number of enemies, most notably Popper, colleagues around the world, mourned his loss. P.T. Bauer said of him: "He [had] the utmost intellectual integrity and the greatest moral and physical courage ... he was an outstanding exponent of that fundamental belief that there exist objective standards of truth."[3]

Despite being famous as a non-stop talker, raconteur and lover of gossip, Lakatos' was virtually silent on his early life. His associates in England knew only a few anecdotes from his Hungarian years. When encouraged to write his life story, Lakatos told his colleague John Watkins. "I shall only do so when I understand how my history fits with the history of Hungary in those years." I have taken this as a research program for considering Lakatos' first 34 years, in Hungary.

G. Kampis, L. Kvasz and M. Stöltzner (eds.),
Appraising Lakatos: Mathematics, Methodology and the Man, 263–302.

The first part of the paper takes him through his university years in the Second World War, his underground teaching of Marxism and eight months of hiding from the Nazis. I make a first attempt at linking the person with his times and his philosophy by analyzing a tragic act he committed in the summer of 1944 at the height of the Hungarian Holocaust. The second part traces how Lakatos helped the communists establish totalitarian power in Hungary between 1945 and 1948, and concludes with a discussion of what kind of communist he was. Part three tells what is known and what may be conjectured about Lakatos' fall from influence, arrest, and imprisonment. Part four takes him from Hungarian prison camp to Cambridge University, from politics to mathematics, from Stalinism, through reform communism, to anticommunism. In a final section *The Unforgiven* I try to get nearer to understanding the data of his life by some psychological conjectures, by considering his virulently bad reputation in Hungary, and by asking what persisted and what changed in the two halves of his life. Who was this man, this famous professor, theorist of truth known as a liar, passionate spokesman for Reason given to irrational outbursts, described as *kind, warmhearted humane* and *fiercely loyal* by some of his western colleagues and students, but for whom many people in Budapest, 40 years after his emigration, keep finding the words *truly satanic, demonic, diabolic, Mephistophilian,* and *not fully human*? How did he, his times and his ideas fit, or clash together?

1922–1944 SURVIVING FASCISM

Childhood

Imre Lipsitz was born in Debrecen, eastern Hungary, on November 9, 1922, only child of Jewish parents Jakab Márton Lipsitz and Margit Herczfeld. His father, originally from eastern Hungary, was a wine merchant in Budapest. Márton Lipsitz spoke Hungarian and German from childhood, read Hebrew and taught himself French and English. He told his son that he kept himself awake for learning at night by putting his feet in iced water. Deep into his 70's he got pleasure from music and extensive thoughtful reading.

Less is known of Margit Herczfeld. She lived in Debrecen with her mother and brother, the latter also being in the wine business. She worked from time to time, if money was needed, as a cosmetician. Margit lived for a while with Márton in Budapest but returned with Imre to her family in Debrecen after a short time. Márton visited Debrecen at times and attended to Imre's education by writing to him "every day" in German, requiring a German response, and by "trying to excite his curiosity about mathematics."[4] After Márton Lipsitz left Hungary for Australia in 1947 father and son met only once, in Australia, in 1966. They exchanged only a few letters before 1956, but wrote weekly, in English, after Lakatos' emigration.[5] The relation between Imre and his mother was said to be close but not comfortable. She looked after every detail of his life, and he gave vent to demands and criticisms as he felt inclined.

Imre met his lifetime friend Gábor Vajda around age six during German lessons taught in the Debrecen city park. From age 10 to 18 they attended the (boys only) Jewish Reál Gymnasium in Debrecen emerging as the two top students of the school, Imre's record being only a shade better than Gábor's. Gábor became interested in communism first. Imre's uncle claimed Gábor had brought two evils into their house, communism and chess.[6] Imre appears in the published memoirs of a fellow high school student, György Magosh, as "the genius student."[7]

Debrecen University

In 1940 Imre Lipsitz entered the University in Debrecen. Most of Imre's class were kept out of the university by the Horthy government's "Numerus Clausus" restrictions on Jewish education. Even Imre, the top student of the year, applied to the law faculty because that was the best chance of entry. Once inside, he was able to move to his preferred subjects of physics, mathematics and philosophy. By at least the second year of university he was reading Marx, Engels and Lenin and becoming committed to communism. Around that time he fell in love with Éva Révész, an attractive Jewish student excluded from the university. Though Éva had an uncle in the Soviet army, it was Imre who brought the communism to her house. They talked of it from the start of their relationship which, Imre said, following Lenin, was conditional on their continuing the class struggle together.[8]

During 1943 Imre started his own illegal Marxism study groups, embarking on his life long avocation of teaching, and his six years as a communist teacher and activist. Though communism had been illegal in Hungary since 1920, there were organized underground groups operating throughout the country, frequently subject to raids and arrests of members. Imre visited several of them in the Debrecen area before starting.[9] No-one who attended Imre's groups has forgotten the intensity and brilliance of the atmosphere. "He opened the world to me!" a participant said.[10] Even those who were later disillusioned with communism or ashamed of acts they committed remember the sense of inspiration, clear thinking and hope for a new society they felt in Imre's secret seminars.

These groups were not open societies. Imre ran them rather rigidly and firmly rejected suggestions either that their members should engage in anti-fascist resistance, or make an alliance with factory workers. Either course would increase their risks and take them from the task of preparing themselves, as Lenin prescribed, for the coming revolution. Imre was criticized by a number of people for such caution and theoretical purity. Others criticized him for his lack of caution in his exuberant Marxist references around the university.

The third major development of 1943 was Imre's arrival in Árpád Szabó's seminar on Plato. Szabó was 10 years older than Imre, the youngest professor to have been appointed by the university, and already with a name in his field of Greek and Latin philology.[11] Szabó's future wife Klára was also a student in the class. She remembers Imre sitting exactly opposite Szabó, and the two of them engaged in dialogue of a speed and intensity that excluded the rest of the students. Szabó, who was engaged in clandestine antifascist resistance, tried to caution Imre to keep his Marxist enthusiasm more hidden.

Like all of Lakatos' teachers from Debrecen to Cambridge, Szabó perceived Imre as having come to class full of ideas on the subject. He was always more interested in arguing positions than being "taught" in any conventional sense. Szabó felt that though his argumentation was brilliant, his knowledge was always quite superficial.

Survival

On March 19, 1944 to forestall Hungary's attempts to negotiate a separate peace with the Allies, Germany invaded and occupied the country. Eichmann supervised the ghettoization, forced marches and transports to concentration camps of Hungarian Jews. Six hundred thousand died in the next nine months. In early April, Imre's mother and relatives were forced into the Debrecen ghetto. Imre, Éva Révész and some other young people fled at this time with false papers[12] to nearby Nagyvárad (Oradea).[13] His friend Gábor Vajda had many contacts and was able to make arrangements for several families to take in his friends. He himself was in hiding in Miskolc north west of Debrecen. Tibor and Zoltán Molnár, young working class men who had attended Imre's seminars helped them obtain false papers. Imre's were in the name of Molnár. After a few weeks in Nagyvárad Imre moved to the house of a family of Horthy supporters, unknowing, it is said, that Imre was Jewish and communist.

As soon as they were settled Imre restarted the Marxist group. Its character changed markedly. From the rather authoritarian seminar of his Debrecen meetings, it became a quite tightly organized communist cell along lines laid down by Lenin, with Imre (and to some extent Éva Révész) assuming the right to control and criticize group members, require permission for love affairs, open mail etc.[14]

In May they were joined by a 19-year-old Jewish girl, an anti-fascist activist named Éva Izsák. She was from another Romanian town, also living with false papers and requiring the hospitality of a non-Jewish family. I discuss her story in more detail below. The disturbing facts are these. Following some difficulties at the place where Éva Izsák was living the difficulty finding her another lodging, and dangers of discovery if she had to "register" for this, Imre decided there was a special likelihood she would fall into Nazi hands and be forced to betray the rest of the group. He decided her duty to the group and to communism was to commit suicide. He found fault with or ignored suggestions to remove her from Nagyvárad. The group, including a young man who had become her lover, met without her and assented to the plan. When told, Éva asked "Is there no other way?" Imre decided it would be dangerous for her to die in Nagyvárad since she would have been seen with the group. A non-Jewish young member who visited them from Debrecen to take her back with him and obtain poison.[15] He took her to a wooded park at the edge of the city, where she took the poison. A report of a child finding her body appeared in the Debrecen paper of August 3, 1944.[16]

Later that month in Budapest Éva's sister Mária received a letter, without stamp or address dropped through an open window at her work place . It said "Don't write for a while, take care, I'm fine at all times. Éva."

Discussion: "Is there no other way?"

Several dimensions of Lakatos' character and roots of his thought emerge in these early years, the brilliant mind, passionate embrace of a philosophy, avocation for teaching, dialectical structure of interchange, but a threatening manner that suggested control rather than argument, and a willingness to do harm. To make a first attempt at deciphering the interplay of Lakatos' character and his particular history and ideas, the tragic story of Éva Izsák requires a closer look.

This story is known mainly because Mária Izsák Zimán, Éva's next oldest sister, comrade in activism since their early teens made it a lifetime mission to discover the truth of her death. She made the difficult journey from Budapest to Nagyvárad immediately after liberation in February 1945. She found the man who had been Éva's boyfriend who told her in a rather harsh impersonal way that Éva had been killed at the Hungarian Romanian border. Mária did not believe this. She enlisted the help of a friend of the family who had become head of the post-war local security police. He held a formal hearing in which Éva's boyfriend and another member of the group were witnesses. The transcript of this hearing was later obtained by Mária, and taken with her when she emigrated to Israel in 1977. She used it in a book *Betűsírkő É vának* "A Memorial of Words for Éva" which she privately published in Hebrew and Hungarian in 1989.[17] Litván (1995) reviewed the book in print, and Congdon (1997), has also retold the story.

The report of the hearing was sent by the Romanian police to the Hungarian police. Additionally, at least three members of the group, including Lakatos gave an account of the death in the "autobiographies" that applicants to the, now legal communist party were required to prepare. The point to remember for further events in Lakatos life is that the Party knew of this from the beginning. It did not affect Lakatos' rising influence in the early years, but it remained in his file and was used against him at times until he left Hungary.

This story of Éva Izsák, commanded to suicide by the group in which she sought survival, still hangs scandalously on Lakatos' name in Budapest. I have heard three distinct theories to explain her death. They may be labeled rational, romantic, psychological. The rational (situational logic) version is similar to the only one Lakatos ever made on the rare occasions he is known to have spoken of this story. Éva Izsák presented a security risk to the group. She was likely to be arrested. If arrested she was likely to be forced to betray the group. If she did this many people would die. Éva's death would prevent this. The rational argument rests on the likelihood a) that Éva was more likely than other members of the group to be caught, and/or b) that, if caught, she was more likely to betray the group, and c) that there was no other way than her forced suicide to protect them.

Before examining these criteria it is worth noting two things. First, group leaders who make tough decisions in dangerous circumstances, are usually treated leniently even if the decisions are later considered wrong. Furthermore semi-voluntary sacrifice of a group member to spare the others has been recorded in numerous Resistance situations. Several people have suggested that this is how Lakatos' decision and the assent of the group, including Éva herself should be viewed. On the other hand most endangered groups hiding from the Nazis did not consider killing one of their members an acceptable option. This can be seen in hundreds of first hand accounts of the Holocaust. The rational theory implies the acceptability of forced suicide among the solutions to the problem.

Was Éva Izsák more likely than others to fall into Nazi hands? In the account given by group members to the police investigation the suggestion was made that, no good new hiding place was available, and the necessity to "register" for any that might be found would increase her chances of discovery. It was also said that she looked more Jewish than the others, a suggestion that cannot be evaluated without photos of them all. And finally that her false papers were not adequate. Even in the interrogation this suggestion was withdrawn. It was admitted her papers were good enough to allow her to travel the length of the country. *Not* mentioned in the enquiry was a possibility that the Gestapo might look especially for Éva because of her activist past. Of the three rational criteria her greater likelihood of her coming to Nazi attention is the one that seems the most plausible, though not very plausible.

Was she more likely than the others to weaken under pressure and endanger the others? This was clearly Imre's fear, but Éva's history suggests it as unlikely. She had had a strong family background of activism. From the age of 13 she and her sister Mária had been engaged in clandestine groups. Her brother had been captured, tortured and had not betrayed anyone. It is not known how much of all this she conveyed to the group, but I conjecture her manner gave them little reason to suppose she would be specially lacking loyalty and courage. Both participants in the investigation noted how calmly and bravely she dealt with the group's "decision."

Was there no other way? This seems to be the weakest part of the rational argument. A letter to the group was received offering that a friend of Éva's family would be at the Nagyvárad railway station at a given time to take her to another city. Imre rejected it because the new host family (anti-fascist activists) were "under surveillance." But no one appears to have asked their non-Jewish contacts in Debrecen, or Miskolc to find a new city for Éva to hide in. Several people later said they were mystified that they had not been asked for the solution they believe they could have found. This seems the strongest case for rejecting the rational theory as the only explanation for Éva's death.

If the fear of capture and betrayal and the perception that there was no other way is not a sufficient explanation, what should be added? People hearing this story often independently mentioned Dostoevsky, "revolutionary romanticism" and a book named *Chocolate*. In the romantic explanation Imre was inspired by ideas drawn from revolutionary literature (including of course Lenin) in which the beleaguered revolutionaries voluntarily submit to tight discipline, and develop the value of sacrifice to the "will of the group" often revealed through the leader.

The book *Chocolate*, published in the Soviet Union in 1924 by Alexandr Tarasov-Rodionov, foreshadows the show trials by telling the story of a Bolshevik accused of betraying the communist party. He submits to accusation and punishment far in excess of any mistake he made for the sake of an example to strengthen party discipline. This book in Hungarian translation was widely circulated in the communist underground at that time, and read by the group.

For Mária Zimán, Lee Congdon (1997) and some of the interviewees, the key factor in Éva Izsák's death was Imre Lakatos' demonic nature, and her death the best evidence for it. Translated into psychological terms this includes his drive for power, his disregard for human beings in the working out of ideology, his demand to have his way, and obsessional holding on to an argument. In this, I believe plausible, view the suggestion is that even if it was misguided rationality that gave Imre the idea of forced suicide as a solution to Éva Izsák's housing problem, it was his own obsessive, "possessed" psychology, buttressed by Lenin, and perhaps by *Chocolate* that kept him impervious to alternative solutions.[18] Thus while the plan may have appeared to Imre merely rational, the psychological view supposes that there were moments in its enactment when he had other motivations for wanting to carry it out. Consciously there was Leninist ideology supporting strong discipline, and sacrifice to the group.

Historian György Litván (1995) reviewing Mária Zimán's book under the title "Victim of Two Madnesses" emphasized that Imre was correctly in tune with the psychology of Marxism-Leninism in making this decision, even though later the party censured him for it. We may conjecture that Imre's already formed character habit of demanding his way was augmented by a wish to overcome his recent helplessness to save his family from the Nazis.

Acting powerfully to protect his group may have been his way of dealing with the enormity of the times, just as Mária's insistent uncovering of the truth about one of her murdered sister was hers. The difference was that he "identified with the aggressor"[19] in his attempt, and she with the victim.

For reasons of space I shall not discuss here the responsibility of others in Éva Izsáks' death, or the different ways ideology, rationality and personal considerations may have been involved. While there is no doubt that Lakatos proposed the suicide and pressured all parties into carrying it out, Éva Révész who took several leadership roles in the group, the members who assented to the plan, and the young man who spent two days with Éva between leaving Nagyvárad and giving her poison in the Debrecen wood, and even Éva herself played their part.

The missing piece for this discussion is of course Lakatos' own assessment. To the few people he told the story he presented the "rational" view and indicated no regret.[20] He did cite his decision in a discussion of ethics with a friend in 1955 as an illustration of how complicated ethical decision making can be.

Liberation

The year 1945 set Imre's life course for the next eight years. He returned to Debrecen with Éva Révész late in 1944. The Red Army had entered Hungary's eastern border in November 1944 and "liberated"[21] the area after a massive tank battle near Debrecen. A Provisional Assembly was convened in Debrecen in December. The city, therefore, became the center of post war activity until the liberation of Budapest in February 1945, and its citizens well represented in post war national activities.

Both Imre and Éva Révész returned to ransacked houses and the absence of all family members.[22] Éva wept, for the first time since the hasty farewell as she escaped to Nagyvárad, when she found a note from her father with a disguised reminder of some hidden money. Her parents eventually returned, but Imre's mother, grandmother, uncle and many family associates perished in Auschwitz.

Almost immediately on his return Imre changed his name from the Germanic Jewish Lipsitz to the Hungarian, proletarian Lakatos.[23] He also made a mysterious "conversion" to the Calvinist Church taking as "god-father" the head of his Ph.D. dissertation committee, literary historian Sándor Karácsony. The mystery is because Marxism is antireligious, and no one ever heard or saw Imre representing Calvinism.

At any rate Lakatos at once became active in the new, legal communist party. Vigorous recruitment of members was clearly one of his activities. István Király, later professor of literature and lifelong spokesman for the regime, claimed that "in February 1945 [Lakatos] convinced me to be a communist and took me into the party."[24] The Szabós remember his relentless, ultimately successful, push for them to join, and his insistence that Klára remedy her inadequate knowledge of Marxism by studying texts. He told both Szabó and Király that their being non-Jewish made them more important to the cause since the party was full of Jews.

By mid 1945 when he left Debrecen he was known at high levels and scheduled for important party work. Undoubtedly his recruitment efforts, brilliance, dedication and energy were enough to get him noticed in those earliest months of communist organization in Hungary, and at that time, in contrast to later, his recent underground activities added to his desirability. We do not, however, know exactly how and via whom his career developed in those months between Oradea and Budapest, or at what point he became known to József Révai, chief party theorist of propaganda and cultural matters.

Another person who may have been important at that time was László Kardos, Imre's Gymnasium teacher. Kardos moved at the same time as Lakatos to a job in the ministry of education, and his son[25] at the same time to Eötvös College. That Lakatos considered Kardos, Karácsony and Szabó his particular mentors is indicated by his inviting them before he left for Budapest to a meeting to "advise" him on his future. Szabó says they were all rather surprised and wondered what they could say. Lakatos had clearly been directing his own life and had jumped into the heart of the party. He had already rejected Szabó's advice that he should finish the doctorate (docent) he had registered for at Debrecen University in 1944 before moving. But they met, talked a little and assented to his plans.[26] Perhaps he was seeking some sort of parental blessing, the inventing of a home and family from which to leave for his new life after the tumult and losses of the past. Presumably he had learned by this time that his own father had survived and was awaiting him in Budapest.

1945–1948 BUILDING COMMUNISM

Overview

Between May 1945 and December 1948 Lakatos helped establish Hungary's communist totalitarian state. Brief histories of the post war era often do not indicate a time period between the Soviet occupation of Eastern Europe in 1945 and the establishment of fully communist "satellite" dictatorships in the countries occupied. In Hungary Stalin's plan was for a gradual communist take-over, overtly complying, for a while, with the post war agreements for democratic process and rule of law. Hungarian communists were told to contest elections, insist on coalition governments if they did not win enough votes outright (which they did not), and gradually assimilate the other political parties (the infamous "salami tactics").

Thus between 1945 and 1948 the communists were simultaneously engaged in winning over key sectors of the population in the context of a multi party system and still fairly free press, undermining independent segments of society, and building up the state apparatus that would later replace enthusiasm with fear as the basis for one party rule. Lakatos was involved from the beginning in these three big tasks. His brilliance, humor and relentless intensity were well suited to arguing with the intellectuals. In addition to recruiting them for party membership, already started in Debrecen, he published articles, eagerly engaged in informal argument, and was used on several occasions for public debates. At the same time he entered the restructuring program in the field of education. This put him under the aegis of József Révai, editor of the Party newspaper *Szabad Nép* (Free People), and generally agreed to be "the real ruler of Hungarian communist cultural policy."[27]

A grey eminence

Lakatos arrived from Debrecen in mid 1945 to a Budapest devastated by the two month bombing siege and street fighting at the end of the war. Seventy per cent of the buildings had suffered some damage, many were reduced to rubble. The departing Germans had blown up all the Danube bridges linking Buda and Pest. In many ways, however, recovery was astonishingly rapid. By early summer essential services had been restored, streets more or less cleared, food supplies organized, and a temporary bridge put across the Danube. Despite the trauma the nation had just been through, family members lost to genocide, war and Soviet army round ups, despite massive dislocations of careers and households, most accounts of the immediate post war period stress the energy and hopefulness of Budapest. The communists were seen by many as leading the way out of Hungary's shameful legacy of feudalism and fascist alliance. Lakatos at 23 embodied the times' energy and ideology, as well as the losses and dark deeds of the recent past. Like Budapest, he immediately embarked on a very full schedule of activity.

Éva Révész came to Budapest after a few months, but they did not marry for another year and a half. Before Éva arrived Lakatos had met the beautiful Vilma Balázs, at local party headquarters. She became a life long friend, whom Lakatos many times asked that she leave her husband and marry him. She was one of the few people who consistently saw him as attractive, human and vulnerable, as well as brilliant and impossible. But she always knew marriage to him would not be a good idea. In 1945, however, he was still in love with Éva, despite their frequent quarrels.[28]

In the Party Lakatos was seen from the start as one of the upcoming young elite of its education strategy. He had an office in the Ministry of Education. Though this was still a conservative ministry, his work was central to the new order – evaluating credentials of university teachers to see whose class background and political affiliations were appropriate, and drawing up lists of people to be fired or hired to move education increasingly into communist hands.

He made a good friend of a colleague at the Ministry, psychologist and educator Ferenc Mérei. For 13 years (except when he was in jail) Lakatos was a regular visitor to Mérei's home. His wife Vera always enjoyed the excitement of the conversation when Imre visited and the way he played with and taught their children.[29] Lakatos and Mérei were also part of a small "inner circle" of the Party who met regularly to discuss strategy in matters of ideology and education. A member of this group Miklós Szabolcsi called Lakatos "a grey eminence" whose brilliance and intensity gave him a far bigger influence on policy making than his place in the party hierarchy would suggest.[30] Révai was either present at these meetings or was briefed on them.

Lakatos also attended a small seminar given by György Lukács the world famous Marxist literary critic who had returned from exile in Moscow with the other "Muscovite" Hungarian communists. Other members of this seminar were József Szigeti, philosopher, István Király, literature historian and György Lázár (formerly Leó Lám who later became secretary to Party Chief Rákosi.) All were part of the young communist elite. Szigeti (who later married Éva Révész) became Lukács' chief assistant, but on two occasions followed the Party line by publicly denouncing his mentor. Both he and Király remained Party supporters for as long as there was a Party.

Lakatos also helped the Party court the intellectuals and the new generation by at least eleven book reviews and articles published in journals during 1946 and 1947, and probably numbers of commentaries in *Szabad Nép.* Most of these were strongly political, exposing bourgeois thought even in authors that appeared radical or scientific. But Lakatos' writing was always well ordered, philosophical and problem oriented. His lifelong interest in teaching and the "real" history of scientific theories was already at times apparent. In 1947 he published a review of Károly Jeges' *I Learn Physics.* Lakatos approved its aim of popular scientific education but deplored its introduction of concepts "in a scholastic manner, without making them real in terms of experiments" or giving "the historical dialectics of theories." "It is incorrect to give the impression that physics is an eternal science" (Lakatos\Jeges 1947).

Two of his three main papers in 1946 and 1947 were published in *Valóság,* a journal newly revived as part of the communist youth organization (MADISZ) and a focus for intellectual communist writing. Its authors were sometimes referred to as the "*Valóság* circle." Its editor István Márkus admired Lakatos' clarity of thought ("in a discussion he was always several steps of ahead of everyone else") but found "diabolic" his "total disregard for people" in the working out of strategy.[31]

Lakatos' writing career brought him another good friend, Miklós Gimes, journalist at the party newspaper *Szabad Nép* who was tried and hanged with Imre Nagy after the 1956 Revolution. A friend of both of them said Lakatos became close to Gimes in the late forties. "They were both brilliant. This was the bond between them. Gimes was very different, anything like Nagyvárad would be unimaginable for Gimes. But they could argue endlessly over fine points of doctrine, philosophy, hair splitting, quite Talmudic in fact."[32] Gimes was probably influential in turning Lakatos from hard line to reform communist in the mid '50's.

But of all the ways Lakatos contributed to the communist takeover none is remembered more dramatically than the downfall of Eötvös College.

Lakatos was not a gentleman

Among the problems for the communists in Hungary was a strong tradition of excellent, internationally oriented education. In 1945 lively political debates were in full swing, with teachers again letting young minds roam among free thinkers from the Greeks to Aldous Huxley. No place better exemplified this spirit than Eötvös College, a place characterized as being in "eternal intellectual ferment."[33]

Eötvös College was separate from the University, though all College students were enrolled in both places. It had been founded in 1895 in conscious emulation of the French "École Normale Supérieur," to provide high level classical education, including science, languages and literature, to future Gymnasium teachers. But many would-be intellectuals with other careers in mind found their way there at 18. Its pride was a culture of intensive Socratic teaching in small seminars with intensive and extensive use of the large library in between. A rival, Györffy College, was less exclusive and intense, transforming itself after 1945 into the vanguard of the Peoples Colleges movement (NEKOSZ).

In the summer of 1945 Lakatos applied for entry to Eötvös College. Keresztury has written of this in his memoirs. Lakatos at 22 was 4 years older than the typical entrant, and was applying to study sociology, which was essentially not a subject offered. He was probably choosing a topic that would not involve much time, but may also have been trying to prepare himself for his doctoral degree from Debrecen University which was to be on the sociology of scientific concept formation. Keresztury was hesitant to admit him but was swayed by knowing that Lakatos was well connected in the Communist party. This might have some benefits for Eötvös, and make rejection of Lakatos inadvisable. Also he felt Lakatos seemed "an abject character" who might be benefited by the Eötvös spirit, and furthermore might himself benefit those younger students whose last year at school had been violently disrupted by occupation and war.

Lakatos lived for a while at the college, but even after he left and despite his Ministry work, he was a frequent visitor remembered well by almost everyone there. One third year student, József Vekerdy, on his way to an eventual mastery of 10 languages,[34] remembers Lakatos as mainly sitting in various places in the attractive old buildings and courtyards arguing intensely. With one student[35] in particular there were a whole series of arguments. Other students would gather around for the show. Vekerdy saw them as Naptha the ideologue and Settembrini the humanist from Thomas Mann's "The Magic Mountain."[36]

Lakatos was not the only communist at Eötvös College. Many of those entering in 1945 believed in the cause and quickly formed a "communist faction." Many others supported different parties in the November elections of 1945, or were totally unpolitical. For another year and a half they were free to voice their opinions and often did. Lakatos was of course urgently trying to capture their opinions and actions for the Party.

The reputation he made for himself at the college and his own humour was reflected in the 1946 year end cabaret. These famous year end events always interspersed comic skits of college life with sophisticated parodies of such as Mozart arias and Greek plays. In the skits students usually played the parts of professors, public figures or other students. In one skit in 1946 a student impersonated a professor famous for his complaints that students did not work hard enough. On to the stage bounced Imre playing himself: "Just give me their names professor – I'm from the Party's Department of Denunciation!"[37]

Lakatos used the Eötvös library to prepare his publications of 1946 and 1947, and his 1947 doctoral dissertation, and attended the Wednesday evening lectures and debates. Lukács and Révai among many others spoke at these initially lively occasions. At one of them in 1947 Lakatos made some sharp criticisms of a speech by Révai.[38] This may have been significant for later events.

Lakatos' denunciation of Eötvös College came in February 1947 when *Valóság* published his long paper *Eötvös Collegium – Györffy Collegium*. The problem for the paper was "the education of the new Hungarian intelligentsia" meaning both the new principles of (Marxist) education and the new (working) class from which they would be drawn. The paper's main thesis was that Eötvös College was still governed by its original principles of producing "scientists and educators for Hungary's liberal bourgeoisie" and had carried its isolationist position with which it combated fascism into an "isolation" of "rigid opposition to the vanguard" (of the working class). Györffy College which had allied itself with the peoples college movement were now in the vanguard. Their resources, library and tradition, however, were insufficient to train the new intelligentsia. Logically they needed the resources of Eötvös to help them in their task, but they were rightly suspicious that Eötvös' fine education and "scientific absorption" was "inseparably linked" to its elitism and isolationism. To train the new working class vanguard, and help Györffy turn itself into "the Eötvös of the peoples democracy." Eötvös would have to abandon its elitism and isolationism. Lakatos concluded that Director Keresztury and most teachers were unlikely to help Eötvös fulfill its role with the new intellegensia. He offered Keresztury the chance to emulate past great directors of the college by "daring to create a new college spirit," but ended with the threat that if he did not take this role the college would come to an end.

This paper was correctly seen as a stab in the back to Eötvös and Keresztury, and created a furor at the college. The press was still free enough for rebuttals to be published and a long series were. Eventually, however, Keresztury was called on to dismiss the students who had signed their names to counter arguments. He resigned instead.

Keresztury was replaced by Tibor Lutter, an English literature scholar, already known to Lakatos, and well known to have put himself entirely at the service of the communists. However, he was a good teacher and students continued to find the College an exciting place.[39] Lutter had the same vision for Eötvös as Lakatos had outlined – to embody *both* educational excellence and the aims of the "people's democracy."[40] But the aims of totalitarian politics and independent thought are incompatible. Three years later in 1950 Eötvös College, Györffy College and the NEKOSZ peoples college movement all came to an end.

Among the non-communist students of that era, the attack confirmed the demonic image they had already formed of Lakatos, "the Mephistopheles of Eötvös."[41] His name has lived in infamy among them for 50 years. Many others, however, do not put his Eötvös role among Lakatos' sins. Later the ideology proved false and the hope for educational excellence and Soviet model communism seemed naive. But at the time it represented Lakatos' struggle in good faith for a better society. Congdon (1996) has linked the attack with Lakatos' conservative role against student demands in England in 1968, suggesting that he learned that to turn an educational establishment into a center of social revolution can be but a step into turning it into an instrument of government policy. Midway between these two rounds of educational politics was a speech he gave in Hungary during the calls for reform that preceded the 1956 Revolution.[42] It is interesting to compare the three positions.

In 1947 Lakatos saw, (to use today's term) political correctness as fundamental to education. He did recognize the danger of "politics and rote learning" degenerating education, but thought that could be solved by converting fine educators to the communist cause. And at that time he trusted the (communist) state to be both politically and educationally progressive. In 1956 Lakatos' problem was the tedium and narrowness that state control, political correctness and censorship had in fact brought to education, especially for gifted students. He concluded that education required certain freedoms, and that the Party should be guided by scholarship, rather than the other way round. In 1968 the problem in England was preserving academic excellence from students demands for control. He called for separation of students' justified demands to criticize and advise on their education, from unjustified demands to participate in decisions about the content of teaching and research. He spelled out the dilution of quality and increase of governmental control that would follow. It is of interest that in 1968 Lakatos attributed his "return to politics" against the students' demands to having seen the harmful and ludicrous effects on education of ideology and Government control under the Nazis and in the Soviet Union.[43] He never mentioned his own role in Hungary.

Postscript

When I started this search Dezső Keresztury, was still alive. I phoned to request an interview, but when I mentioned Lakatos he said "The mathematician? Oh no I couldn't tell you anything good about *him*." He was not swayed by my noting that I was looking for history, not goodness. "It would be too agitating for me." His voice got weaker as we spoke. Finally he murmured into the phone "Lakatos was not a gentleman."

In action

In early 1947 Márton Lipsitz, Lakatos' father, emigrated to Australia. Not long afterwards Imre found time for the long delayed and short lived marriage to Éva Révész. Árpád Szabó and Tibor Balázs, husband of Vilma, were the witnesses. In February the Eötvös College article came out. He published seven other articles and reviews that year and finished his doctoral thesis for Debrecen. In July he passed his Oral examination for the doctorate degree which was awarded the highest possible commendation. It was entitled "*On the Sociology of Concept Formation in the Natural Sciences*." Two of his recently published articles, *A Critique of Physical Idealism: Stebbing Vs Jeans*, and *Modern Physics, Modern Society* were cited as part of the doctorate. The readers noted that the work was based on "dialectical materialism, but modern, not orthodox, and even that was only a basis for Lakatos' own 'independent ideas'." The thesis itself has vanished from Debrecen university and any other place a copy might be expected to have been placed. Even at the time Lakatos said it was "worthless." He never showed it to the Szabó's. A faculty member at Debrecen questions whether he might have removed it himself from the library.

During 1947 as the Party continued to build up its pervasive control of daily life, and weaken all forums for dissent, disagreements about György Lukács began to surface. Lakatos seems never to have been an enthusiast. Several people remember his calling Lukács "too bourgeois" and his thinking "silly." In December 1947 "about 50 members of the *Valóság* circle met at the Opera Cafe to discuss their mutual criticisms of Lukács."[46] According to István Király, Lakatos "shone" at that meeting. Over the next eighteen months "official" Party opinion appeared to blow hot and cold on Lukács (participants of the December meeting received party warnings of disapproval), but in May 1949 a major attack was printed on Rákosi's behalf by another former "Muscovite" László Rudas. Lakatos had private conversations with Rudas at his house on several occasions,[47] and it is rumored that he contributed to selecting Lukács passages for criticism.[48]

József Szigeti had by this time become Lukács chief assistant. This may have given Lakatos extra impetus for a Lukács criticism, since at the end of 1947 Éva Révész had just left him for Szigeti. Lakatos was distraught. Throughout his life being left by women put him into a state of acute agitation and despair. This loss of his youthful love, and the insult of her leaving for Szigeti felt particularly intolerable. "I am as alone as an abandoned old dog" he told Vilma Balázs. Within a few months he was trying to persuade her to leave *her* husband, but settled for being her lover and friend. He moved into a single flat in a block reserved for medium level Party officials.

1948 was "the year of the turn" to complete dictatorship. The communist party's name was changed in June to the Hungarian Working People's Party. Six men held most of the political power. First Secretary of the Party Matyás Rákosi, Finance Minister Ernő Gerő (said to have the most power in running things), Defense Minister Mihály Farkas, Education/Culture minister József Révai, Interior Minister Lászlo Rajk, Police Chief Gábor Péter. Of these all except Rajk were Jewish, and all except Rajk and Péter had spent the war in the USSR.

That summer saw the beginning of forced collectivization of agricultural land, the end of press freedom, and the suppression of, among much else, psychoanalysis. György Palóczi-Horváth[49] had written a Marxist critique of Freud. Révai took the opportunity to get a team of the young faithful to "debate" publicly the respected psychoanalyst Imre Hermann. Lakatos, Szigeti, Király and Lazár were the Révai team. Ferenc Mérei chaired the meeting. A few weeks later psychoanalysis was banned in Hungary.

The "turn" also signaled the Party's final plans for replacing unreliable teachers . Lakatos prepared new "B-lists" of unreliable teachers to be fired and their reliable replacements. He also prepared an overall plan for university education. Lakatos was Secretary of the Party's Science Committee, and Assistant Secretary of a scholarships committee headed by Tibor Lutter, and including also his friend Mérei.

No longer needing numbers to prove its popularity, the Party became more concerned with the reliability of its members. Everyone, including former Communist Party members, were reviewed by a committee headed by Gyula Kállai for their eligibility for the new Party. According to Lakatos' police record he "passed" his review and was recommended for a scholarship to Moscow.[50] There are indications that this represented a removal of someone who, former Prime Minister András Hegedűs reports, was coming to be seen as too explosively outspoken and unreliable to fit the "turn."[51] But whether intentional at the time or not, Lakatos' departure for Moscow ended his period of influence in Hungarian politics.

Discussion: "a natural born Trotskyite"

How important was Imre Lakatos in Hungary's communist takeover? How far did he rise in the system? What kind of communist was he? What was his own experience of those years? I will consider these questions briefly.

The preceding section provides the first answer. Lakatos was an enthusiastic imaginative contributor to the communist takeover in the field of education, working in the three major tasks of the time, attracting people to Marxism, undermining the old structures, and building the new system. He was more influential in policy decisions than his age and relatively low official position would suggest. However, it would be false to assume that without him the advancing tide of communism in Hungary would have been different. Eötvös College would not have survived the Stalinist era, and the liberal teachers would still have been swept away. At the level of influence over individuals Lakatos of course held considerable power. Many teachers would have remained at their jobs, and others not have got them, if their names had not appeared on Lakatos' lists. Unquestionably he caused harm to some through individual denunciations, and saved from others whom he could have but did not denounce. One of the ironies of totalitarianism is that the citizen, who has no power to change the rulers and severely constrained hegemony over his own life, has extraordinary power over fellow citizens to harm by denunciation or protect – by silences, helpful lies, warnings etc.

How high did Lakatos rise? Officially not nearly as high as has been written in western summaries of his life. He was not "number three in the Ministry of Education" or really "a senior official," as was claimed after his death, probably based on his own words.[52] His last title was that of "Secretary" which meant a person in charge of the administration of a committee or bureaucratic department. Probably "medium level official" would better describe him. He was important enough to be granted a Party flat, and to be given the rare authorization to travel by plane to a conference in Prague and his studies in Moscow. Unofficially, we know he had influenced policy debates, was known by the rulers, and had a high enough profile for his explosive critical style to be a cause for concern.

What kind of communist was he? Among the possibilities were the "true believers" who managed to keep their faith through all the changes of doctrine and obvious difficulties of the system; the "cynics" (in party terms "opportunists" or "careerists") who would stay with the party for their own gain without concern for ideals or theories; and the "critical believers" who kept trying to make the real party fit their ideals, but sooner or later, as Klára Szabó put it "woke up." Everyone agrees that, though ambitious, Lakatos was not a cynic or "careerist." He was as likely to endanger his career by impulsive faith driven words and acts as to promote it. In style he was a "critical believer," always willing to argue and joke about party theory and practice, but he appears to have in no way "woken up" to problems in the system until after his release from jail in 1953. His criticisms of Révai were that Révai was not Stalinist enough.

György Litván, however, calls him "a natural born Trotskyite," better fitted to the agitation of Trotsky's "permanent revolution" than to Stalin's iron hand. Hence his good fit with the party's needs in 1945–48. Litván's diagnosis also referred to Lakatos' incautious questions to comrades that could be dangerous whatever way they did or did not answer. Was this calculated party business, impulsive truth telling with thoughtless disregard for the vulnerability of others or sadism? Sometimes it was obviously calculated. At other times it may well have been more thoughtless, or even in his own mind a joke. He told Szabó at one point that of course all bourgeois professors like himself would have to be shot and was surprised to learn later that this had disturbed the Szabós.[53] In 1956 he asked Litván openly what he thought of a "multi-party political system." Such comments were certainly experienced as those of an *agent provocateur*, and contributed to his demonic reputation. Their lasting sting came as a surprise to Lakatos when told about it later. He often claimed no memory of the event or of any animosity intended. Psychologists would certainly discern unconscious sadism in such acts, but he did not fall into that category of sadist who when given power by an inhumane government develop an enjoyment of causing and witnessing pain.

As for Lakatos' subjective experience of his years of influence, we again know very little. We do not know what his immediate career hopes were in 1948. We know he was always expecting to get back to science and philosophy, and sometimes complaining of having no time for this. We know from his writing that he cared about teaching science with a historical, dialectic, non-reverend approach. We do not know what he saw as problems in Marxism, how he solved them, and how he reacted internally to the party's changes of direction, to the increasing constrictions of freedom of speech, and forced adulation of party leaders. He did say after his emigration that at that time he saw the world in terms of "angels and devils"[54] and that he "moralized too much and hence landed in jail."

1949–1953 SURVIVING COMMUNISM

Overview

During these years Hungary and Lakatos were plunged into a Stalinist nightmare. Not only was Eötvös College finally closed in the summer of 1950, but the peoples colleges as well. Cardinal Mindszenty was tried and sentenced to life imprisonment for "conspiring against the state." Foreign Minister László Rajk was arrested in May 1949 and hanged in October after a show trial at which he confessed to a lifetime of imperialist espionage. Rudas' attack on György Lukács accused him of "idealism, cosmopolitanism and blasphemy against Lenin."[55]

Those whose lives most fully partook of the Stalinist nightmare were the ones in jail. Teenagers were in jail in those years for penciling a mustache on a Rákosi poster. Parents were in jail because their children had denounced them. Though one could be in jail for nothing at all, everyone had got there along some pathway. This section sets out as much as I know (in early 1998) about Lakatos' path to Recsk, the infamous labor camp where Rákosi again tried to earn his title as "Stalin's best pupil" by reproducing the gulag in Hungary's Matras hills. Where there are only conjectures I have set them out and indicated the evidence to support them. Any attempt to make sense of this segment of Lakatos' life has to take into account the profound irrationality of the system.

Moscow

The nightmare had not begun for Lakatos when he left for Moscow in January 1949. Whatever the motivation for his Moscow candidacy, the Party would still appear to be investing in his scientific education, and presumably he was considered safe enough not to embarrass Hungary in the USSR. He left his many books with a colleague, sociologist Sándor Szalai. Szalai, however, was arrested a few months later, charged as an "anti state collaborator" and imprisoned until 1956. Lakatos never saw his books again. In the summer of 1956 the government began to pay him a small sum of money in compensation for what was claimed to be a 1000 book collection.

Physicist Pál Román, a candidate in Moscow three years later has written of the student hostel where Lakatos lived. It was a huge building, eight kilometers from the center of town with "endless dismal corridors lit by bare light bulbs burning day and night." Students slept many to a room on "simple metal frame beds" and ate "poorly cooked food" in a "huge unfriendly badly lit cafeteria." From a contemporary of Lakatos' in Moscow we know that the hostel was host to numerous rats.

Lakatos formally studied physics in Moscow but later told a friend he had also used the time to research the writing of József Révai for "incriminating" quotations to be used against him if necessary.[56] While he was there much blatantly false science was "declared" true by the Soviet government and scientists punished for teaching contradictory positions. Later in life Lakatos frequently cited this as impressing him with the danger of allowing too much government control of science. At the time, however, he acted with passion in service to the regime.

Before his departure a colleague had shown him a letter received from the colleague's wife Kamilla who had arrived in Moscow in the same "candidates" program several weeks before. The letter complained of poor food and living conditions in the Moscow dormitory. On arrival he had a friendly talk with Kamilla who was ill in hospital, told her to come to him if she needed anything, and cautioned her against writing such complaints as he had seen in her letter to her husband. Later at a meeting of the students he made the warning general "for those people not developed enough such reports might promote anti-Soviet feelings." He cited Kamilla's letter by name, and she was required to engage in "self criticism."[57] Another student there was the daughter of a high level official. Her father's ability to send her packages created envy among the other deprived Hungarians. Gábor Vajda believes that Lakatos criticized her for these packages, that she complained to her father, and that this may have been a contributing factor to Lakatos early recall from Moscow. We do know that Lakatos' police record notes that he was recalled from Moscow for "un-Party-like behavior."[58]

In July 1949 he telegrammed Vilma Balázs, that he was coming home unexpectedly for a short holiday. He came back by plane, claiming later that he had persuaded a Soviet military attache to carry a large sheaf of papers through customs for him It is not known whether he left possessions in Moscow, as the "short holiday" plan would suggest.

Limbo

One of the most mysterious chapters of Lakatos' life is the nine months following his recall from Moscow. I shall first set down the facts I know, and then discuss indicate factual conjectures that have been made, and leave for the discussion the possible reasons why events unfolded as they did.

Soon after his arrival in Hungary Lakatos was told that he was not permitted to return to Moscow, could not return to his job in the Ministry, receive back his party card (surrendered automatically when party members went abroad) or return to his old flat (which had been let to a friend of Vilma's). He was directed to teach at a Gymnasium. It is not clear what, if any, reasons were given him, either for the recall, or the punitive treatment. None of his friends recall the shock, depression or anxiety one would expect after such an abrupt exclusion from his former life, though he told Vilma he hated the teaching job. Tibor Lutter, the successor to Keresztury as Director of Eötvös College, close to Révai, and part of the Party's cultural inner circle, gave him a room in his own apartment at the College where he lived for nine months. It is said Lutter occasionally asked Révai if there was any reason why he should not house Lakatos, and Révai said he did not know of any.[59] But in December 1949 Rákosi's secretary Jolan Simon, who was also the wife of police chief Gábor Péter wrote a letter to the Hungarian embassy in Moscow requesting the immediate recall of a number of students because of their behavior ("immoral lifestyle") or attitudes ("nationalism," "chauvinism," etc.). For two of them their undesirable characteristics included "friend of Imre Lakatos."[60] This indicates a new infamous status known at the highest party levels, and raises the question of why he was not already arrested, and why the Party's faithful servant Lutter was housing him. Editors at that time were nervous enough (or so ordered) to published some book reviews he wrote only under a variety of initials taken from his names.

Emotionally, too, this must have been an intense time for Lakatos. While the Lutter flat in the lovely old college building provided an incomparable improvement over the Moscow hostel, Lakatos had to contend with the fact that his nemesis József Szigeti and former wife Éva Révész were in another flat a few steps down the hall. At the same time he was living in close quarters with another Éva, Tibor Lutter's teenage daughter. Éva Lutter remembered Imre while he was a houseguest as "a shining figure" wonderful at helping her with her mathematics, bringing to life the Party ideology, and creating a sense of fun.[61] Everyone who knew her in her youth remembers *her* as a shining figure, beautiful, sexy and intelligent. Lakatos was said to be "educating her to be his wife"[62] although she was only 14 at the time. Certainly it was an intense relationship and adds to the mystery of Tibor Lutter being willing to house him for so long. Éva Lutter herself had a tragic history and played a role in Lakatos' life after his emigration, but that is another story.

In August 1949 just after Lakatos' return from Moscow and before the Rajk show trial, the World Youth Festival was held in Budapest. Eötvös College was taken over by the Chinese delegation, with all students, faculty and users of the library under strict orders not to enter. A music student Dénes Zoltai had left some work in the library, and persuaded the janitor to give him permission to get it. He was surprised to see in the library a table covered with papers. Glancing at them he saw they were mainly articles by Révai. Later he asked Szigeti who might have been there despite the orders. Szigeti said it would be Lakatos who was living in Lutter's flat and "always in the library." When Zoltai mentioned the Révai articles, Szigeti showed great interest. The following spring Zoltai was called in by the secret police and asked about Lakatos' work in the Eötvös library the preceding summer. He realized their police knowledge of this could only have come from Szigeti.

In mid April 1950 Gábor Vajda who was still living in Miskolc was also called in for questioning about Imre, primarily whether he considered Lakatos a "careerist."

On April 20 Lakatos and the man who supervised the forced suicide of Éva Izsák were separately called before the same committee and reprimanded for their roles in this. In the official account[63] of this hearing Lakatos' biography was given a negative slant, indicating that in his underground communist group he had misrepresented himself as having the authority of the official underground party. It also noted his recall from Moscow for "un-Party like behavior." There was no mention of his former positions. The decision was taken to expel Lakatos from the Party. The other man was given probation in view of the fact that he had been misled by Lakatos into thinking that the orders for her death came from the "official" party, and in view of his continuing good Party work.

That night six police officers arrived at Eötvös College, went through the customary rapid brutal searching of Lakatos' room and took him away to the secret police headquarters. The next morning Éva Lutter stared at the ransacked room, mattress cut, drawers overturned, photos and papers everywhere. Some college students on learning of the arrest asked Tibor Lutter whether the college or any students would be in danger as a result. He said there was nothing to worry about.

Hell

The "Mephistopheles of Eötvös College" entered hell as a frightened prisoner in a basement cell of the secret police headquarters. He was kept there six weeks. Many prisoners have written their descriptions of that infamous place in the center of town. The cells were tiny and damp. Bare lightbulbs shone day and night. Guards were trained to demean, insult, threaten, insinuate. Prisoners were summoned to spacious rooms on the upper floors for interrogation, where they might be subjected to torture, sleeplessness, and forced typing of confessional life stories. Police Chief Gábor Péter presided over this particular piece of Stalin's vast hell.

We do not know what exactly Lakatos was accused of, or what kind of a confession, if any was demanded of him, or what associated "criminals" he was linked with. His police record enters his accusation as "Suspected of Espionage." It notes that he was transferred to Kistarcsa Internment camp for political prisoners in July and to Recsk labor camp in September.[64] There was no trial.

In September Lakatos started three years in the Recsk forced labor camp. The camp was five kilometers from the village of Recsk in an area of gravel pits and stone quarries. This was opened early in 1950. When Lakatos emerged from jail 3 years and 5 months after his arrest, he told Vilma Balázs "Don't ever ask me about it, because I shall never tell you." We know from several published accounts that the camp was run entirely by the State Security Police with brutal conditions, physical abuse and arbitrary humiliation. Witholding the insufficient and revolting food, forcing consumption of too hot liquids, or tipping cooked food on to the floor were frequently used as humiliation. There were no reading or writing materials whatever. Unlike the prisoners, the guards had the luxury of using newspaper as toilet paper. Prisoners would seek news by fishing this paper out of the sewer.

Prison appears to have been the only time in his life that Lakatos kept a low profile. Only a few ex-prisoners I have talked to remember him at all well. Someone from the same barracks said he was a "good worker" and "kept his head down." Another said he was a "bit isolated at first because of his Party background, but later seemed better."[65] Some group allegiance was necessary for improving one's chances of food and necessities. Lakatos naturally allied himself with the communists (rather than the social democrats and others who had been sent to Recsk because of suspected dissident tendencies). He may have received some small rewards for extra work done for the authorities, but he was not an informer.[66] At his release he had broken and missing teeth, and a healed arm wound with considerable loss of muscle tissue.

Lakatos did not appear to change his Party allegiance or ideology much as a result of this experience. He said later that while he knew it was a mistake for the Party to have sent him there, he assumed many of the others were in for "good reason." He did not participate in the nightly "seminars" run in the huts by the prisons' many intellectuals. Lakatos said he survived mentally by telling himself a joke a day, and by going over one by one all the mathematical proofs he knew. It was here he said that he recognized the falsity and dullness of the formalist presentation of mathematical proofs as though they formed an eternal edifice of a-historical truth.

Shortly after his return he told a friend[67] with some pride of a cart of apples that was delivered to the camp, the first fruit inmates had seen in months. Somehow the cart tipped, spilling the apples on the ground. Prisoners rushed to pick them up. Lakatos called the guard to halt "this theft of state property." Only in theory did he identify with the oppressed.

In the wake of Stalin's death, struggles for power in the Kremlin and riots in Berlin, the Soviet Union decided to reduce Rákosi's power. His successor as Prime Minister, Imre Nagy, quickly moved to start agricultural reforms, close all Soviet style labor camps, and allow writers a little more freedom of expression. Lakatos was released on September 3, 1953. He had spent as a victim of the system precisely the same time as he had been influential in building it. He was amazed to have his watch, confiscated at his arrest returned to him.

Like all Recsk discharged prisoners, he was offered first an apology for the "mistake" of having been sent there, and second a warning that to speak about it to anyone would be considered a criminal offense. Many released prisoners, including Lakatos, were offered the "opportunity" of some form of future informing for the police. What threats or rewards were offered is not known.

Discussion: why did Lakatos go to jail?

As mentioned before, this question is not fully answered. The following summarizes the structure of the situation and likely elements involved.

The situation: (1) There are three questions, which may or may not be closely linked, and may or may not involve the same "authorities." Why was Lakatos recalled from Moscow?[68] What was the Party's intention in not returning him to his former work, but allowing (requiring?) him to live with Lutter who was high in the Party's educational hierarchy? What were the intentions of the arrest? (2) Especially high numbers of arrests were made in 1949–1951 among medium level communist party officials, from those who had been in the illegal underground before 1945, and among those with international connections. Lakatos belonged to the first two categories, and relatives in Australia and Israel might have qualified him for the third. It seems unlikely that his path to Recsk was only a statistical chance, but these factors could have weighed in decisions made on other grounds. (3) His antipathy to Révai, and being seen as unreliable may not have been enough to explain any one of the three questions above (particularly the "limbo" period), but again could mean that at crucial points along the way he had more enemies than friends.

The recall from Moscow: First it should be noted that there is a little confusion about the length of time he was supposed to be in Moscow. His police record states that he was cleared to be an aspirant in Moscow for two years, the usual length of time. There is general agreement among his friends that he came home for an "unexpected" holiday in the summer of 1949 and was "unexpectedly" refused permission to return. In the records of the Ministry of Education, however, there is a notation that he had received a "seven months" scholarship to Moscow, and was requesting an advance on his salary as secretary of the scholarships committee for that amount of time.[69] Shortly after his arrival in Moscow he wrote to Vilma Balázs suggesting "let us not write for seven months." However in July 1949 he did cable her saying he was coming home in August "for a short holiday." Two official documents say he was recalled for "Un-party like behavior."

There is a persistent rumor that the whole path to Recsk, and not just the excuse given for expelling him from the party, was the result of Éva Izsák's death catching up with him, perhaps because the Romanian communist party specifically requesting action and there were political reasons to agree to this. Éva Izsák's sister Mária and Mária's husband had positions of influence in the Romanian Communist Party. A further rumour is that he made friends with a Hungarian military attache at the Moscow embassy who was recalled to be arrested and Lakatos was seen as connected.[70]

The most interesting suggestion that has emerged from the study of Lakatos' path to jail is that it may have been connected with a possible new show trial being prepared with Révai as its target. There are many reports of Rákosi's and Farkas' dislike for Révai, of Révai, like Lukács being too slow to move from the concept of "people's democracy" to "dictatorship of the proletariat," of Révai being watched by the police (Aczél 1966), and of Soviet reservations about him. Specific links to Lakatos' Moscow and post Moscow activity are: Lakatos telling Ágnes Heller that he had researched Révai's works in Moscow; Zoltai's observations of Révai material in Eötvös College; István Király reporting that Lakatos had told him he was to have played a role in a possible Révai trial;[71] an editor named Zoltán Zsámboki in written testimony to the Oral History Archives of the 56 Institute reporting that Lakatos had researched the evidence and prepared a written paper on the "Crimes of József Révai," and had been arrested the day after he handed it in.[72]

Ultimately Révai had on his side his heart condition (they did not want him to die in jail)[73] and, perhaps, his being a "Muscovite" (there was claimed to be a loose agreement that Muscovites would not arrest each other). At any rate there was no trial, and we are left with guesses that Lakatos was recruited (by Hungarians or Soviets) or recruited himself to work for others against Révai; that his recall from Moscow was to continue this clandestinely; that Tibor Lutter was told to house and watch him (without necessarily knowing why); that the arrest came when the Révai trial possibility had been abandoned; that when Lakatos' conspirators had lost power, other reasons to arrest him (behavior in Moscow, his unreliable characteristics, Éva Izsák, or a forceful denunciation) resurfaced, or he just knew too much.

After his arrest, he was caught up in police process – of which at that time virtually *no* rational reconstruction is possible.

1954–1956 A Scholar Reared

Catching up

Lakatos returned to Budapest in September 1953 thin, unhealthy, with missing teeth and only a watch as possession. He had no job, or place to live. This was a problem because it was illegal to be without either, and despite Party directives that those released from internment camps were not to be discriminated against, many people were nervous of ex prisoners. In the first weeks he visited the Lutters where he was given a little money but told no further help would be possible. Gábor Vajda helped him get some translating\editing work. The Szabó's and Mérei's gave him clothes, and food.

Once again he was in a hurry.[74] He had decided in prison that mathematical proofs as presented in text books were unsatisfactory, and that his future work would lie in this direction. He felt he had wasted precious years in politics and prison. He must do everything quickly, including getting married. Again he asked Vilma. When she again said no, he told her he would be married within a year.[75]

He found his way to Tamás Lipták, a young mathematician who had been expelled from Eötvös in the 1949 wave of dismissals, and had for a while been engaged to Éva Lutter. They worked together through a book by Alexandrov, *Introduction to the General Theory of Sets and Functions.* Later there were other books, the start of a collaboration, eventually dropped, on Lakatos' translation to Hungarian of George Polya's *How to Solve it* and an intensive informal seminar. Lipták found Lakatos' ideas of mathematical theorems and proofs as problem oriented, synthetic, developed dialectically and with reference to ideas held at the time, transforming to his own thought and the work with Lakatos stimulating. Emotionally he felt that Lakatos was "not fully human," "his drives and his mind in place, but the rest missing."

Lipták introduced him to Alfréd Rényi, director of the Mathematical Institute of the Hungarian academy, and to his friend Éva Pap. Rényi gave Lakatos a job in the Institute Library, and later a research position. This enabled him to work, to pursue his studies in mathematics, and to read philosophy, including, he claimed, Popper. Éva Pap married him. She was a teacher of biology, and daughter of Endre Pap, a well known researcher in maize development. Their marriage in early 1955 was almost within the year he had specified.

He moved into her room in her parents' house. Imre appeared to be very content, not only to have a wife but also a family, and to be cared for while he spent most of his time studying mathematics. This did not, according to the diary Éva kept at the time, prevent him from getting angry or upset many times a week.

Speaking out

Lakatos' two interesting journeys of the mind during these years were his changing views of education, and his development as a mathematician. He had emerged from prison with the Stalinist language and mindset of 1949. Though only months had passed since Stalin's death, Imre Nagy's partial thaw and the "waking up" of many intellectuals, were causing revisions to the ideology, and a cautious return to language less laced with ideology. Éva Pap was amused by Imre's old lines, and by his sitting her down to study *Szabad Nép* in their early meetings.

Of Lakatos' friends, Árpád Szabó had already been turned out of the Party. In 1954 he refused an invitation to return. Ferenc Mérei (expelled from the party on the same day as Lakatos) was among those privately discussing the problems of the system, and the possibilities for reform. In 1955 Imre Nagy's reforms were thought by Moscow to have gone too far. He was relieved of all his posts and replaced by a young hardline Rákosi associate, András Hegedűs, a product of Györffy College, the same age as Lakatos. Journalist Miklós Gimes moved into the circle of those around the deposed Nagy, actively seeking change. Lakatos paid many visits to Gimes, who still worked at *Szabad Nép* during 1954. These discussions and the many lively evenings at the Mérei's house helped move him toward his reformist position of 1956.

Lakatos' most serious political act of his last year in Hungary the defining moment of his reformist position came just before the Revolution in early October 1956. Despite hard line Hegedus being put in to slow the reforms of Nagy, it was impossible to quell the rising will for more normal conditions of life. Kruschev's "secret speech" on Stalin's crimes in February 1956 was a major impetus for change. Moscow struggled to find a balance between some reform and continued control in the face of the Polish uprisings of June. In Hungary the writers continued to experiment with less censorship and more debate. One rising issue was the rehabilitation of Rajk and those executed with him.

To let off steam the government allowed "Petőfi Circles"[76] to meet and discuss reforms. Despite heavy police presence at these meetings, they turned into overflowing debates, far more outspoken than was intended. These were held throughout the summer of 1956 each focusing on a different topic. In July, Lakatos invited Gimes, Mérei and a number of other people to a meeting to discuss education policy. In September he gave his speech "On Rearing Scholars" at the pedagogy Petőfi Circle meeting. In it he asked again, as he had at Eötvös how best to educate the elite, but this time instead of focusing on the need for the elite first to join the proletariat, he addressed the problem of party education dulling good minds with censorship, propaganda and fearfulness to think and speak. He ended with a call for scholarship to guide the Party instead of the other way round.

Mostly, however, between 1954 and 1956 Lakatos was training himself as a scholar. In addition to his private studies and work with Lipták, he attended a mathematics seminar given by László Kalmár and came under the influence of a special Hungarian school that specialized in explaining mathematics to the general public with clear ideas, and without recourse to complicated formulas. Central to this school were George Polya (Hungarian born, but settled in the USA), Kálmár himself, his boss Alfred Rényi and Rozsa Péter. He found Polya's "analysis of patterns of intuitive and plausible thinking in mathematics" and view of theorems as the outcome of a "struggle to solve problems" particularly helpful to his own developing ideas on the real history of mathematics (Ravetz, 1974). Polya's book *How to Solve It* was available only in English and German. Lakatos translated it into Hungarian, as well as Chincin's *Mathematical Foundations of Information Theory,* and Faddeeva's *Computational Methods in Linear Algebra.*

In December 1956 waiting in Vienna after his escape he wrote to Richard Braithwaite in Cambridge that he "would like to pursue two lines of activity: a historico-comparative study of the modern schools of probability theory, and the question of heuristics in mathematics." Applying for his entry to doctoral study after five months at Cambridge he wrote: "In my last year in Hungary I became interested in the history of mathematics. I found that the formal exposition of mathematics in university text books, with postulates dropped from heaven, and sometimes with proofs established by quasi-magical devices is rather harmful for educating mathematicians with real insight."

Getting out

Between October 23 and November 4, 1956, the Hungarian Revolution became the most dramatic attempt of any country in the 45 years of the "Eastern bloc" to throw off Soviet domination. Lakatos was not very active in the Revolution, but co-authored a statement from the National Committee of the Hungarian Academy of Science calling for "the freedom of science in written and oral form, in all locations, and without restraint or moral pressure."

On November 1, at the height of the euphoria (for those who wished the Revolution to succeed) Éva's brother Gábor apparently committed suicide. He had been taken into the Party while still a schoolboy in recognition of his intense devotion to the communist cause, and had attempted to repudiate his bourgeois background by factory work, spartan living and active Party work. Among the many events of the Revolution was the freeing of political prisoners. Among those freed were a group who had spent eight years in jail following a denunciation when they were schoolboys by Gábor Pap. When they came to the university to join the revolution, they saw him distributing guns. They demanded that he be sent out to them. The head of his unit refused, kept them talking, told him to exit the back way and not come back. He went to his flat and, it would appear, shot himself.

Gábor's death and the crushing of the Revolution by the Soviet army in early November convinced the Paps they must leave the country. Éva would not let them go alone. Imre was very doubtful. He did not know what they would face, what work he could do. Éva went right on with the plans for their departure. It had to wait until they had buried Gábor. This was delayed because of the gravediggers backlog of work.

Even up to the last night, November 23, Lakatos was uncertain about going, though he had no hope for things getting better in Hungary.[77] A friend reported him saying "Nothing good will happen here" and looking like "a man from whom the scales had fallen from his eyes." Still, without the Pap's insistence on getting out and Éva's organizing he would not have gone. Then, Éva notes, "there would have been no Imre Lakatos." The Paps turned over their flat to Sándor Szalai the sociologist with whom Lakatos had left his books before going to Moscow. Szalai had only that year come out of jail. Lakatos prepared to leave behind his third library of books.

His last visit was to the Méreis, friends and colleagues since 1945 with whom he had moved from Stalinism to reform to revolution. He particularly loved their small daughter Anna. He took a walk with Ferenc. On the walk he told his friend that from the time of his discharge from prison he had had an agreement with the police to provide information on the Méreis. Later he told the Szabós that they too were part of the arrangement. Mérei was astounded, furious and sad. He waited until Imre had left to tell Véra. To Imre he said about his emigration "You should go!"

What did he tell? Árpád Szabó believes "everything," that he reported conversations, opinions, people who came to the house. Imre later told him that he would go through his wastebasket for evidence of their lives. But said he never told damaging information.[78] Véra Mérei and Klára Szabó believe he may have kept the information relatively undamaging since they are not aware of any bad consequences that befell them as a result. To many people, however, the point is not exactly what he told, but his willingness to keep the authorities informed of the mood and mindset of the intellectuals, and his friends in particular. Why did he make this arrangement? He may have felt forced when he left Recsk, though many others declined to cooperate. It may have been the condition for not being kept at Kistarcsa after Recsk for a further term. Several people were. Or for living in Budapest. Some think it more likely Lakatos sought out the role after Recsk as a way of getting back into Party circles. Why did he tell Ferenc before leaving? Véra believes it was to help him overcome his feelings against going. After he was established in England, Imre was active in getting Szabó get some overseas invitations and publications, and the relationship continued to be warm.[79] He had a little correspondence over the years with Véra Mérei, but the long friendship was over. "My name is still anathema in Budapest" Lakatos cheerfully wrote to his father in the late '60's. This story and the Éva Izsák story are what most firmly keep it so.

The next morning a lorry came early to pick them up. They were supposed to cross the border that day, but there were several delays, and the guide was drunk. Imre had time to write to Vilma Balázs ("I though it was a joke, but now I am at the border I am frightened") and to Gábor Vajda. Gábor's letter included a sealed one to send to Szabó.

On November 25 they walked 12 miles, carrying bags, especially hard on Éva's sick mother and lame father. They heard gunfire (the border was rapidly being resealed by that point). The first snow began to fall. In Vienna friends of the Paps arranged for them to stay in some elegant offices in the middle of town. Éva's mother immediately got a job translating between the refugees and western organizations. Endre Pap had job offers from Argentina, the USA and England. Gábor Dirac, son of the physicist, one of many students who had come to Vienna to help, met them, and spoke of England. Imre made an instant decision to try to get there. They queued for meals, funds and donated clothing and discussed scholarship possibilities. Imre spent days in the library writing proposals and letters to find a place to study mathematics. Szegő and Victor Kraft wrote him recommendations. By mid December Richard Braithwaite wrote from Cambridge accepting his letter of application, and the Rockefeller Foundation awarded him a fellowship.

On January 7 Imre and Éva arrived in Cambridge ... and everything happened very quickly.

THE UNFORGIVEN

This then is the outline of Imre Lakatos' life in Hungary. He was a theatrical person, and these 34 years do fall, theatrically, into four acts. Each ends with a journey and a life change. At 23, survivor of war and Holocaust, he left Debrecen for Budapest, his relatives dead, his possessions looted, armed with an infallible scientific ideology and vision of changing the world for the better by any means necessary. At 26, he left Budapest for Moscow having helped establish a state that was worse in more ways than it was better, whose public utterances were systematically untrue and which frequently punished devotion with the same ferocity as dissent. At 30, he left prison having spent over three years a victim to this state, itself victim to the puppet masters in Moscow, but still not questioning its premises. At 34, he left Hungary, having in a small way joined his country's struggle against the system he had helped build. When it was clear the struggle had failed, he left them both and declared his system false. At 38, in England he described himself as a man "who loves truth, who is curious, and who does not accept authority over his own judgement, neither of god, king or party."[80]

Each of the four acts were structured by equally dramatic chapters in Hungary's intense history, and each contains a paradigmatic story in which Lakatos' character, philosophy and circumstances come together in acts that are easily labeled demonic. So, again, "Who was this man?" Drawing on the documents (few) and memories and opinions (many) of people who knew him, this paper has provided some answers and some conjectures about the facts of his Hungarian life. Only here and there has it been possible to report Lakatos' subjective experience of those years. He was not introspective. He rejected psychology both as a Marxist and a Popperian, and he had much to forget and hide. In this final discussion I try to improve our sense of his "internal history" by a few psychological conjectures.

A true intellectual history or rational reconstruction of Lakatos intellectual journey is still missing. Feyerabend (1976), Congdon (1996), Larvor (1998) and Kadvany (2001) have made major contributions to constructing one. Palló (1998) has made an interesting analysis of the parallels and divergences between Lakatos and Polányi. In claiming Polányi's "more radical opposition to Marxism" he notes that "Lakatos' dialectically competing research programs which are evaluated by their progressive character sound quite familiar to Marxists." Motterlini (1998) has traced the intellectual friendship between Lakatos and Feyerabend, and has shown how Hegel's dialectics, Polya's heuristics and Popper's fallibilism (the three influences Lakatos acknowledged) were in a sense a progressive sequence in Lakatos' views of the historical embeddedness of scientific theory, and his unwavering defense of rationality against Feyerabend's scientific anarchism. Hacking (1979), Larvor (1998) and Kadvany (2001) have traced the influence of Hegel and philosophical problems of interest to Lakatos because of the influence on his early thinking of Hegel and continental philosophy. I have tried to indicate some problems that preoccupied him during his Hungarian years. But Lakatos' intellectual history is not at all complete.

Angels and devils

For at least two decades of my intellectual life I did think of history as of an apocalyptic struggle between angels and devils, even if my criterion of demarcation between good and evil changed immensely. This led to many dramatic, or rather violent scenes with my friends when I suddenly 'realized' that they were slipping into 'evil'... I lived for various reasons, in constant guilt feelings throughout my childhood and youth; my last superego being the Party. In 1956 I decided that I am *not guilty* and was amazingly successful in getting rid of my superego. Perhaps as an overreaction to my youth I ceased altogether to think seriously about or even to give attention to the reaction to my behavior by others. (IL letter to Paul Meehl July 1971)[81]

Many people have done me harm but only one – Lakatos – was evil (Agassi, 1981, who wrote about IL both as "my friend," "my nemesis").

He was an impossible infantile monster, and the most brilliant man I ever met. He was completely unable to understand other people. He walked through life like a 3 year old (Forster, girlfriend briefly in 1954).

Imre Lakatos was above all a kind and warmhearted human being, deeply concerned about the increasing irrationality and injustice in this world, about the almost insurmountable power which mediocrity possesses in all fields and among all people, prepared to defend his ideas and his attitude even under uncomfortable and dangerous circumstances (Feyerabend, *British Journal for the Philosophy of Science*, 26, 1975).

People did not matter to him (Markus, his editor in 1947).

He was fiercely loyal to his graduate students (Musgrave, his student and colleague in London).

Everyone has his dark and his light side; sometimes we see one, occasionally the other. But in Lakatos both were simultaneously and brilliantly displayed at every instant. (Berkson, 1976).

Let us move from what can be seen to what can not.

Style: He had a brilliant, ordered, exciting mind, particularly good at setting out the elements of an argument in logical form. His papers are easy to read because of the clear way he ordered them As a teacher he was histrionic, a showman, a comic, a preacher. He loved to argue, did it all his life, treated ideas with passion, none more passionately than the ideas of rationality and fallibility. But angry outbursts, or the apparent need to overpower, ridicule or even destroy the opponent frequently overcame his rationality. After arguments, whether he had won or lost, he expected to continue a relationship as though nothing had happened, and himself made a habit of forgetting emotional events. He was capable of doing harm to others and conveyed that sense chillingly to them. The harm might be intentional because a person had been construed as an enemy, tangential because people's sensibilities had been ignored, or a mixture of the two when people had been treated as pawns in a political plan. He was capable of warmth, of caring acts, of loyalty and long term friendship, but he could betray even those he cared for. And when his demonic side showed, people never forgot, and rarely forgave.

Diagnosis: Psychologists do not like to diagnose in absentia, but it may be useful as a commentary on some of his characteristics. Lakatos exactly met the criteria for *Type A Personality*, explosive, volatile, impatient, hard working, and with the special hostility features that most closely associate with heart attack. Donald Gillies who knew him at the height of his fame said he "got angry at least once every hour."[82] A heart attack was the cause of his death in February 1974, a fact worth stating because I have seen it attributed in print to brain tumor, suicide and "mysterious circumstances." He also met the criteria for *Anti-social Personality Disorder* in the current DSM IV. Personality Disorder diagnoses in general are given to people with "an enduring pattern of inner experience and behavior that deviates markedly from the expectations of the individual's culture, is pervasive and inflexible, has an onset in adolescence or early adulthood, is stable over time, and leads to distress or impairment."[83] Anti Social Personality Disorder is part of a subgroup characterized by "dramatic, emotional or erratic behavior" and has as its own essential characteristic of "a pervasive pattern of disregard for and violation of the rights of others." It should be noted that this pattern though recurrent was not continuous, and there may be doubt as to whether his personality did bring him distress. Impairments were less than many circumstances of his life would suggest, except for the persistent sense that between the genius and the three year old something was missing.

Lakatos' defensive style of coping with anxiety by "acting out" – transforming unconscious anxiety into immediate action, and "identifying with the aggressor," becoming that which you fear, are most notable, and again are typical for this type of diagnosis. His more mature coping styles of humor and sublimation into intellectual activity and a culturally supported natural warmth may have been the forces fostering his non pathological and relational capacities. The intellectual system of Marxism would seem to have matched and exacerbated his "immature" defenses, while his life long muse of science and later life adoption of Popper's critical rational fallibilism were a mitigating "mature" force.[84]

It is worth noting here when the pathological "inner experience and behavior" of the personality disorder instead of deviating from cultural expectations find a culture that rewards them, very dangerous people result. I conjecture that herein is the answer to why Lakatos remains so virulently unforgiven in Budapest. The *combination* of Lakatos' characterological "patterns of disregard," explosiveness, obsessiveness and brilliance, and an inhumane obsessive society structured to allow citizens to do harm to each other, is what makes him seem so frightening even at a long distance. Many of his contemporaries with a similar record of helping establish Stalinism who, like him, called for reform communism in the 50's, and resistance in 1956, are respected citizens now. Others at least live quiet lives. To some extent the demonization of Lakatos may represent a projection to quiet other peoples' troubled memories of humanity abandoned in the service of ideology. However, his demons are for the most part his own. The memories of Lakatos in England also turn up words such as "evil" and "diabolic" and deeds to match, but they are mitigated with more appreciation for his strengths, including his humane and human qualities. One of the key differences both in what he was and what was perceived of him was the unlimited power to harm granted by the otherwise restrictive totalitarian state and the limits on this power in a lawful "free" society.

What about angels?: Any profile of pathology (a diagnosis) should note the strengths that will modify its dynamics (an antidiagnosis). These are fairly clear. Lakatos did relate, he attracted women, he kept friends, he was never self-pitying, and he had the well known qualities of mind and playfulness with ideas to fill the gaps that emotional and personality difficulties bestowed on him. He did not find comfort in music, art or (mostly) literature. But his life was rich with enjoyment of philosophical reading, letter writing, gossip, jokes, sex, being looked after, politics, teaching and ideas, and he was more able to enrich others than his pathology diagnosis would suggest. I will make the wild conjecture that the traditions of warmth and emotionality in Hungary helped tame his devils, just as the intense cruel ideologies of his day exacerbated them. He loved England but deeply missed the Hungarian emotionality. Four days before his death he wrote to Paul Feyerabend from the hospital "they decided that I am completely Anglicized and that my heart has, therefore, vanished."

What happened?: I shall not here conjecture as to what happened and when to set the parameters for Lakatos' personality. It may be noted that I have supposed his intense history to have augmented his personality traits, not caused them. There is evidence that most factors were present before 1944. Whether Margit Herczfeld failed to help Lakatos build a psychic structure to give him more peace from his furies and fears, or keep people from slipping into abstractions, whether Márton Lipsitz absenting himself from the family laid down a permanent need to find and fight with fathers, and how genetics and culture structured his inner world need more than a few lines to sort out.

I will, however attempt one psycho-philosophical hypothesis. I conjecture that in his deep psychic structure Lakatos feared both a descent into mental chaos, *and* the constrictions of rules. At the mental level his exceptionally good cognitive processes could always build structures to make order from chaos. He valued such structures, ideologies, programs immensely. He hated muddle. He wrote to Vilma (and similarly to Éva Lutter) of the importance of having a firm framework for their work so that "everything does not fall into barbarism."[85] On the other side he feared the constriction of rules that shut down the problem solving, probably calming, ability of thought. Argument, anger and criticism may have been perpetual attempts to break out of this possibility. He was perpetually arguing and valuing its power, as well as perpetually seeming to seek structure to shut down such freedom for others. Agassi has pointed out the contrast between the "lovely utopian classroom" of *Proofs and Refutations* and the authoritarian tone of his proposals for countering students demands in the 1960's.[86]

Emotionally, however, for whatever reason, the usual "structures" of the experience of others and experience of the continuity and goodness of the emotional self through relationship to others was not well developed. He could find others in a way and be comforted through sex and caring, through argument, jokes and teaching, and he suffered terribly when a woman or friend left, but the experience did not build a lasting emotional structure, or an internal realization of the reality of that other. In Klein's terms he had not reached the (weirdly phrased) "depressive position" for being reliably responsible for his actions vis-à-vis others, and taking their feelings into account. The alternative "paranoid position" solves certain problems by judging, disregarding, manipulating, "splitting" perceptions of people into angels and demons. But it produces the greater problem of absence of continuity in the experience of the other or the self. Both Lakatos' wildly erupting angers, and obsessive fixing on a course of action may respectively represent failures of, and attempted solutions to the problem of the vulnerable self. Happily for him a rather strong sense of identity that in later life included some of his vulnerabilities helped him overcome these problems. ("I have been restored to my aggressive self" he wrote to a colleague after an illness.) It was not always happy for others.

How many Lakatoses?: The question is cliché (following his famous criticism of Popper, and its echoes in evaluations of him),[87] but inevitable given his own life that was in his words "broken in two." What changed and what did not in the two halves of his life? One question appears to have been answered. The rumor, fueled by his being twice denied British citizenship, that Lakatos had never "really" abandoned Marxism, and was even an agent sent by the comrades to undermine Popper, has little evidence. His letters indicate his own strong rejection of Marxism and unwavering appreciation for the critical rationalist fallibilism of Popper, even when rejecting aspects of his philosophy. His Hungarian police record, reviewed periodically after his emigration, lists him as "someone to be watched" if he ever returns to Hungary, and is stored in a different archive than would be likely if he had been an agent.[88]

There are many unifying themes between the two halves of his life, in character, in the strong themes of dialectic as a mode of thought, from Hegel to Marx to Popper, dialogue, gossip and argument as modes of learning, researching and teaching, and an unswerving interest in the real history of ideas. But the dramatic introduction of Popper's fallibilism, as a non-restrictive structure of thought, and the change from Hungarian politics to British academia as an arena for action was almost great enough to say there were two Lakatoses. I prefer to conclude, however, that there was only one, a highly problematic, Lakatos. But there were two philosophies. The second, represented a distinctly progressive problem shift even at the psychological level.

First and last Lakatos was a teacher and a joker. It is therefore, perhaps appropriate to end this section with an examination question. A. Imre Lakatos at heart was a seeker of truth, a crusader to change the world for the better, with anti-authoritatian, rationalist instincts and warmth towards others. Unfortunately his impulsive lying, anger, manipulations, obsessiveness, willingness to do harm and dogmatism often got in the way. B. Imre Lakatos at heart was a dogmatic, controlling, power seeking dictator, rationalist only in the narrow sense of his own interest. Fortunately his impulsive truth telling antiauthoritarian beliefs, scientific mind, humor, identity as a teacher and warmth sometimes got in the way. Decide.

EPILOGUE

When Lakatos died at age 51 in 1974, John Watkins phoned Árpád Szabó. Szabó had already heard. He was weeping. "The tragedy of Imre." Szabó began, but could not continue. Watkins wondered what was tragic about Imre. It was not a word he had thought to associate with him.

Many years later Szabó finished the sentence. The tragedy of Imre in his eyes is that he always rushed into political activity and neglected directing his brilliant mind toward the deep development of theory. With his penchant for fast moving gossipy dialectic as a mode of operating and learning, Lakatos might well have rejected Szabó's regret. But I think these stories do indicate several dimensions of tragedy in Lakatos' life; in his times, his character, some of his decisions and some turns of fate.

But it was not surprising that Watkins found "tragic" an unexpected word. Whatever Lakatos' sins and despite all he had seen and suffered, no one who knew him ever heard self pity and few witnessed his occasional sadness and frequent nightmares. So let us leave him on a more typical note. On one of Szabó's visits to England in the late sixties when Lakatos was already well known and speaking out against what he saw as the danger to education of the demands of left wing students, Szabó commented on the remarkable changes in Imre's thought and situation they had both witnessed. With one of his customary quick sweeps of the head from side to side ending with an intense stare at the speaker Lakatos said "My dear Árpád! But you know I haven't changed at all!" Too preposterous to be true? Too preposterous to be false? Either way the essence of Imre Lakatos.

ACKNOWLEDGEMENTS

I am grateful to many people for their interviews, letters and contributions but would like to acknowledge with special thanks many conversations with Joseph Agassi, Tel Aviv University, Gábor Palló, Institute of Philosophy, Hungarian Academy of Science, John Watkins, London School of Economics, and Éva Pap Davies, Gábor and Anna Vajda, Árpád and Klára Szabó, Vilma Balázs, György Litván, Tamás Lipták, Budapest, for letters from Gillian Page, Alex Bellamy and Drs Donald Gillies and Timothy Smiley in the UK, and to John Worrall for permission to work in the BLPES Lakatos Archive. I am particularly grateful to Alex Bandy, Associated Press Budapest, for shared interviews with people I could not have talked to alone, for discovering, extracting and translating material in the archives, and an immense amount of thinking, encouragement and help. I have tried to check the information here in as many ways as I could, and to indicate when dates or facts are in doubt. But there are certainly errors and omissions that I would be most grateful to have pointed out.

NOTES

†This paper is adapted from "Lakatos in Hungary" Journal of the Philosophy of the Social Sciences, Summer 1998, and appears with permission of Sage Publications

Interviews marked * were conducted jointly with Alex Bandy, and the contents made available to Dr. Lee Congdon.

1. Letters to Gábor Vajda and Vilma Balázs written near the border on 25.11.56. BLPES.
2. See Watkins (1997) for a description of the Popper\Lakatos quarrels, and Lakatos (1970, 1977) and Gillies (2002) for Lakatos' philosophical criticisms of Popper.
3. P.T. Bauer comments at Lakatos' funeral. BLPES
4. Letter from J.M. Lipsitz to J. Watkins 1977.
5. Information on IL's early childhood comes from scattered references in the Lipsitz correspondence in BLPES, interviews with Gábor Vajda and his letter to Michael Hallett in BLPES, interview with György Magosh, and letter from J.M.L. to J. Watkins 1978.
6. Gábor Vajda, who became a journalist, has been a major source for details of Lakatos' early life. Dr. Kantor Sándorné (Tünde Varga) of the Mathematics Department of Debrecen University has also contributed documents and information from her work on Lakatos' life in Debrecen.
7. Magosh, György *Vadfa*.
8. Interviews with Éva Révész (1995, 1996).
9. Interview with Zoltán and Tibor Molnár September 1996.
10. Interview with Levente Sóos September 1995.
11. Later his principle publications were on the history of Greek mathematics.
12. False papers saved the lifes of many Hungarian Jews. A good set of papers contained not only required identity documents for a non-Jewish person of approximately the same age as the one in hiding, but also evidence of living or working in a certain place (cafeteria receipts, membership of a club, etc.) While the predominant holocaust story in Hungary is of indifference, extortion or betrayal on the part of the neighbors as Jews were hunted by Nazis, many thousands took risks to hide and help Jews. "Everyone," the Molnárs report of their young friends, "was making papers in those days."

13. I use the current Romanian names for towns currently in Romania, although they were known to Hungarians described by their Hungarian names (e.g. Nagyvárad, and Szatmár).
14. Six years later when Lakatos was being accused by the Party it was said that his "cell" was unauthorized, but it is not at all clear who was supposed to authorize such groups at such a moment.
15. His profound anguish about his role in Éva's death made me decide to follow his wish to withold his name.
16. The information for this story comes from Zimán 1989, interviews with Éva Révész, Zimán, Vajda (who was not there but discussed it with IL and group members afterwards), Central Control Committee Document, and Debreczeni Ujság-Hajduföld August 8 and 11, 1944.
17. Zimán, Mária 1989, *A Memorial of Words for Éva.* English translation by Hajnal Csatorday 1997.
18. Lee Congdon's recent biographical article on Lakatos is entitled "Possessed" (See Congdon 1997).
19. Freud, Anna,1937.
20. E.g. Interview with Edit Erki, 1997*
21. "May you be occupied by the Germans and liberated by the Russians" goes a not so old Hungarian curse.
22. Imre had already begun his lifetime habit of collecting books. Among the other losses in the looting was his first library. Twice more his books would vanish. His final library was donated to LSE.
23. It is sometimes assumed that the name change and conversion was from the Nazi period. But this was not so.
24. Király I. in Csaki (1990).
25. Kardos' son, who changed his name to Pál Pándi, became a leader of the student communist faction at Eötvös College, later a literary historian and lifetime spokesman for the party.
26. Interview with Á. Szabó 1996.
27. Aczél, T. and Méray, T. 1966.
28. Interview with Vilma Balázs*, April 1997.
29. Anna Mérei, one of Ferenc's children is currently making a documentary film about Lakatos.
30. Interview with Miklós Szabólcsi*, March 1997.
31. Interview with István Márkus*, 1997.
32. Interview with Miklós Molnár*, April 1997.
33. Jenő Major in article on Eötvös College in *Valóság*, 1946.
34. Interview with J. Vekerdy, September 1995.
35. Árpád Brusznyai, executed for his role in the 1956 Revolution.
36. Naptha was in fact said to have been modelled on György Lukács whom Mann had known during the years of Lukács German exile.
37. Interview with Barnabás Csongor*, July 1997.
38. Interview with József Vekerdy, September 1996.
39. Interview with Tamás Lipták, November 1997.
40. Interview with Éva Lutter, May 1997* and Gábor Mihálly, March 1998.
41. Fodor, 1991.
42. (In Hungarian) Petőfi Kör proceedings Budapest, 1956 Institute. (In English) BLPES.
43. Lakatos, Letter to the Director of LSE in Lakatos (1978).
44. Theologian sociologist Sándor Karácsony and mathematician Ottó Varga.
45. Commentary on Ph.D. dissertation, University of Debrecen, 1947.
46. Király in TV interview (unavailable) from which his widow selected passages concerning Lakatos and permitted Alex Bandy and myself to view.
47. Interview with Miklós Molnár*, June 1997.
48. Interview with Ágnes Heller, July 1995.
49. Paloczi-Horváth, like the poet György Faludy, and writers Paul Ignotus and Béla Szász had returned voluntarily from the west, to build the new society after the war. All were imprisoned for many years with varying degrees of deprivation and torture. All returned to the west in 1956 and wrote remarkable accounts of their experiences in Stalinist Hungary. Faludy, who was imprisoned with Lakatos at Recsk, returned to Budapest in 1990 at the age of 80.

50. Security Police record in Archives of Interior Ministry.
51. Interview with former Prime Minister András Hegedűs, February 1997*.
52. Ref. to #3 in the Ministry. Barrow.
53. Interview with Á. Szabó, December 1995 and Letter to Paul Meehl, 1971.
54. Letter to Paul Meehl 1971.
55. Kadarkay 1991, p. 406.
56. Interview with Ágnes Heller*, 1997.
57. Telephone interview (by Alex Bandy) with Kamilla Lányi.
58. We also know that Lakatos told William Bartley that he moralized too much noting "I used to do that in Hungary and I ended up in prison" (Bartley, 1976). This of course was Moscow not Hungary, and Lakatos may have been thinking of some quite different incidents.
59. Interview with Éva Lutter*, May 1997.
60. Letter obtained from National Archives, Hungarian Embassy to USSR documents by Alex Bandy.
61. Interview with Éva Lutter*, May 1997.
62. Interview with Éva Pap, February 1998.
63. Report of party's Central Control Committee's decision to expel Imre Lakatos from the Communist Party, 20.4.50, citing Éva Izsák case as the reason. Hungarian National Archive Communist Party Documents. 101 Jegyzőkönyv. Obtained by Mária Zimán who sent it to me.
64. Ministry of Interior, Lakatos Police Record. Obtained by A. Bandy.
65. Interview with Tibor Zimányi, president of the Recsk Prisoner's Association.
66. When I asked this question of former prisoner György Egri, he knew Lakatos so little that he was not sure, but since it was such an important matter not to get wrong in either direction he made phone calls to former Recsk prisoners in Canada and California as well as Budapest.
67. Phone interview with Vera Forster, March 1997.
68. I have earlier speculated on the possible reasons for him going to Moscow.
69. Ministry of Interior, Police record.
70. Interview with Pál Réz (forceful spokesman for the Izsák-jail connection), October 1996. Letter from Eugen Szabó, Israel, former Oradea resident who requested Romanian party to follow up on the 1945 hearing into the death of Izsák. Interview with Gábor Vajda and Paul Roman for possible connection with Hungarian attache.
71. Interview with Ágnes Heller*, 1997 and video tape of István Király.
72. Zsámboki Zoltán in Oral History Archive, Budapest. Researched by Alex Bandy.
73. Interview with Gábor Bognár*, secretary to Révai for one year, 1997.
74. He persuaded a dentist to pull all his bad teeth at one time and fit him with replacements.
75. His final proposal to Vilma came when she visited him in London after his divorce from Éva Pap.. They travelled to Amsterdam for an international conference. Imre arranged that she sit next to Karl Popper. She was surprised when he turned to her saying "Why don't you marry Imre?"
76. The name was from the Hungarian poet who had been a leader of the 1848 attempted revolution against the Austrians.
77. Éva Pap said recently "I have suddeny realized my role in Imre's life. I got him out of Hungary. Without that there would have been no Imre Lakatos!"
78. Interview with the Szabós. 1996.
79. There was one seminar they gave together when students were mystified to find them start quarreling in Hungarian and actually come to blows.
80. Letter from IL to Éva Lutter, 4.5.61, BLPES.
81. IL to letter to Paul Meehl, July 1971, BLPES.
82. Interview with Donald Gillies, Kings College London, 1996.
83. Diagnostic and Statistical Manual of Mental Disorders, 1994 4th edition (DSM IV), American Psychiatric Association.
84. This paragraph draws on Vaillant's (1992) analyses of defense mechanisms.
85. Letters to Balázs and Lutter, BLPES.
86. Agassi, 1976.
87. See e.g. Berkson and Grene in Cohen, 1976. Popper had developed a schema for categorizing reality as World-1, material facts, World-2, psychological states and World-3

constructions of stored knowledge such as theories, language, etc. Lakatos criticized Popper with a parody Popper-1, 2, 3. Berkson and Grene echoed this with evaluations of Lakatos 1, 2, etc.

88. Ministry of Interior, Police record.

BIBLIOGRAPHY

Aczel, T. and Meray, T. (1966) *The Revolt of the Mind*. New York, Praeger.

Agassi, Joseph (1976) *The Lakatosian Revolution*, in: Cohen RS et al.

Agassi, Joseph (1981) *The Philosopher's Apprentice*. D. Reidel

Barrow, John (1992) *Pi in the Sky*. Oxford University Press, Penguin.

Berkson, William (1976) *Lakatos One and Lakatos Two: An Appreciation*, in: Cohen RS et al.

Cohen, R.S., Feyerabend, P.K. and Wartofsky, M.W. (eds.) (1976) *Essays in Memory of Imre Lakatos*. (Boston Studies in the Philosophy of Science), Dordrecht, D. Reidel

Congdon, Lee (1996) *Imre Lakatos: The Passions of Reason*. Paper given at Conference on 1956 Emigration from Hungary. Trondheim, Norway, 1996.

Congdon, Lee (1997) Possessed: Imre Lakatos's Road to 1956. *Contemporary European History December 1997*.

Csaki, J. and Kovacs, D. (eds.) (1990) *Rejktőzködő legendarium* (essays on Pandi Pál) Budapest, Szépirodalmi Könyvkiadó & Szemtanu Kiadó (selected translations by Gábor Palló).

Erdey Sándor, 1995 *A Budai Kitöréstől A Recski Táborig* (From the Buda Siege to the Recsk Camp). Budapest, Literator Kiadó.

Faludy, György (1962) *My Happy Days in Hell*. London, Deutsch.

Fodor, Andras (1991) *A Kollégium: Napló, 1947–1950*. Budapest: Magvető Könyvkiadó.

Gati, Charles (1990) From Liberation to Revolution 1945–1956, in: Sugar P, Hanák P, Frank T, *A History of Hungary*. Indiana University Press.

Hacking, Ian (1979) Imre Lakatos' Philosophy of Science. *British Journal for the Philosophy of Science* 30 381–410.

Huszár, Tibor (1995) *A hatalom rejtett dimenziói: a magyar tudományos tanács*. (The secret dimension of power: Hungarian science council) Budapest, Akadémiai Kiadó.

Kadvany, John (2001) *The Guises of Reason*. Duke University Press.

Keresztury, Dezső (1989) "Emlékezés az Eötvös Kollegium utolsó éveire" *Valóság* XXXII, 3

Kardos, Lászlo (1992) *Bortonírásai 1957–1963*. Budapest, Gondolat.

Kadarkay, A. (1991) *Georg Lukács: Life, Thought, and Politics*. Cambridge, Mass, Blackwell.

Kun, Miklós (1989) The First Novel about Show Trials, Postscript to Modern Hungarian Edition of Tarasov-Rodionov, *Chocolate*.

Gillies, Donald (2002) *Lakatos' Criticisms of Popper* (in this volume).

Lakatos, Imre (1946), Citoyen és Munkásosztály *Valóság* II 6–9.

Lakatos, Imre (1947a) Eötvös Kollegium – Györffy Kollégium *Valóság* III.

Lakatos, Imre (1947b) "Modern fizika, modern társadalom" (Modern Physics, Modern Society) *Továbbképzes és Demokrácia* Budapest University Press (File 1.8 BLPES).

Lakatos, Imre (1947c) Huszadik Százsad (20th Century) *Forum* II, 316–8

Lakatos, Imre (1947d) Review of Károly Jeges "I Learn Physics" *Tarsadalmi Szemle*

Lakatos, Imre (1963–4) Proofs and Refutations *British Journal for the Philosophy of Science* 14, Reprinted in revised form: 1976 *Proofs and Refutations*, edited by J. Worrall and E. Zahar, Cambridge University Press.

Lakatos, Imre and Musgrave, Alan (eds.) (1970) *Criticism and the Growth of Knowledge*. Cambridge University Press.

Lakatos, Imre (1977) *The Methodology of Scientific Research Programmes* Philosophical Papers vol 1, edited by J. Worrall and G. Currie. Cambridge University Press.

Lakatos, Imre (1978) *Mathematics, Science and Epistemology*, Philosophical Papers vol. 2, edited by J. Worrall and G. Currie. Cambridge University Press.

Larvor, Brendan (1998) *Lakatos – An Introduction*. London, Routledge.
Litván, György (1996) *The Hungarian Revolution of 1956*. London, Longman
Litván, György (1995) Két Őrület áldozata (Victim of Two Madnesses). *Élet és Irodalom* 6 January.
Magosh, György (1994) *Vadfa*. Budapest, Mult és Jövő Könyvek, 30.
Motterlini, Matteo (1998) *For and Against Method*. Correspondence Between Imre Lakatos and Paul Feyerabend. Chicago University Press.
Palló, Gábor (1998) Antitotalitarian Research Programmes and Their Tacit Components: Michael Polányi and Imre Lakatos *Polányiana*.
Popper, Karl (1972) Epistemology without a Knowing Subject, in: *Objective Knowledge*. Oxford, Clarendon Press.
Rényi, Alfréd (1967) *Dialogues on Mathematics*. San Francisco, Holden Day.
Replika (1996) 1998 Hungarian Social Science Journal two recent series of articles on Lakatos.
Szász, Bela (1971) *Volunteers for the Gallows: Anatomy of a Show Trial*. London, Chatto and Windus.
Szász, Imre (1985) *Ménési út: regény és dokumentumok*. Budapest, Magvető Könyvkiadó.
Tarasov-Rodionov, Alexander *Chocolate,* USSR 1922, (in Hungarian 1940's, 1989) (in English Westport, Hyperion Press).
Watkins, John (1997) Karl Raimund Popper 1902–1994. *Proceedings of the British Academy* 94 645–684
Zimán, Mária (1989) *Betűsirkő Évának (A Memorial of Words for Éva)*. Privately Published. Raana Israel in Hebrew and Hungarian. (English translation 1997 by Hajnal Csatorday.)

UNPUBLISHED WRITTEN MATERIAL

Letters to the author from: Bellamy Alex, Goss (Page) Gillian, Román Paul, Szigeti Éva, Gillies Donald, Smiley Timothy.
Vajda Gábor, letter to Michael Hallett 1980. BLPES File 11.2.
Román, Paul, *I Remember Hungary* A Memoir of Hungary (unpublished).

DOCUMENTS

Report of Party's Central Control Committee's decision to expel Imre Lakatos from the Communist Party. 20.4.50, citing Éva Izsák case as the reason. Hungarian National Archive Communist Party Documents. 101 Jegyzőkönyv. Obtained by Mária Zimán who sent it to me.
Debreczeni Ujság=Hajduföld 8 and 11. 8. 1944.
Report of finding Éva Izsák's body.
Ministry of Education document on reform. 1948 (Obtained by Alex Bandy).
Ministry of Interior: (a) Lakatos, Police record on a set of cards, some of which review his history. Updated notations continue throughout the 60's (Obtained by Alex Bandy), (b) Document issued July 27 1956 certifying that he had been interned without just cause (BLPES, unfiled, box of personal papers).]
National Archives. Documents concerning Hungarian Embassy in Moscow. Letter from Rákosi's secretary Jolán Simon, who was also Police Chief Gábor Péter's wife. December 1949 (Obtained by Alex Bandy).

Visiting Lecturer, Behavioral Science Institute
Semmelweis Medical University
Budapest, Hungary

LÁSZLÓ ROPOLYI

LAKATOS AND LUKÁCS

Lakatos constructed his major contribution to the philosophy of science, the methodology of scientific research programmes (MSRP), in the late sixties and early seventies in England, after he had already become estranged from the Popperian philosophy of science. In this paper, we attempt to show that the MSRP was motivated by his philosophical and political ideas from the forties and fifties in Hungary, when he was imbued with the communist ideology and was influenced by the philosophy of Georg Lukács. From this point of view, the MSRP can be considered to be a special representation of Lakatos' earlier political values and practice in the field of history and philosophy of science.

* * *

Lukács and Lakatos were very different philosophers. However, their biographies have a lot in common: both of them were Hungarians, they came from Jewish families, their extraordinary talent were already manifest in their boyhood, both of them were active already in the literary and debating societies of their high schools, they were brilliant and enduring polemicists, they started at the university as students of law, which neither of them liked, they wanted to transcend the limits of the usual academic intellectual life and become creators and active players in a new, truly humanistic society, both of them played some role in the intellectual and political life of Hungary, and once they got into situations where they thought they would have to leave the country, they were arrested and imprisoned without trial for shorter or longer periods. The list goes on and on.

It can be argued that these similar aspects of their life-histories do not represent much more than the natural prerequisites for any Eastern European intellectual destiny, and that despite these similarities, their thinking, their ideology, their philosophical and political preferences could evolve along very different paths. Really, it is enough to recall the pure fact that Lukács returned to the socialist Hungary while Lakatos left it, or that Lukács was an absolutely committed Marxist thinker until his death while Lakatos became a vigorous critic of Marxism, or furthermore, that Lukács concentrated his activity on the aesthetic and political problems, whereas Lakatos

G. Kampis, L. Kvasz and M. Stöltzner (eds.),
Appraising Lakatos: Mathematics, Methodology and the Man, 303–337.

investigated the problems of science; the list of their dissimilarities is long, as well.

In this paper, however, we will elaborate on some similarities of their philosophical thinking. We will derive some inspiration from a social constructivist approach, but, fortunately, our findings can be based on concrete details and facts of their biographies, as well as on careful analysis of their writings. Taking into account the fact that Lukács was Lakatos' senior by about 40 years, the similarities of their ideas appear in the form of a Lukácsian influence on Lakatos' thinking. It is, perhaps, interesting that this influence had a variable intensity and a life-long duration, especially in those problem-fields in which Lukács and Lakatos might have collected similar experiences about social reality. In this way, Lakatos accepted some aspects of the Lukácsian world-view and applied them in his actual works. In some cases he radically criticized Lukács, perhaps hoping to overcome and substitute him, but in the end, he returned to some very basic ideas of the great, old master.

First, we will recall the relevant biographical facts about Lukács and Lakatos and discuss Lakatos' early writings. In the second part of the paper, two aspects of the Lukácsian influence on Lakatos will be analyzed in more detail. In the case of the Lakatosian MSRP, its Lukácsian components will illuminated.

1. BIOGRAPHICAL REMARKS ON LUKÁCS AND LAKATOS

There are numerous biographies of Lukács – including his autobiographical sketch –, so we think it is unnecessary to reproduce the details here.[1] In contrast to this fortunate situation, we can find in the literature only very few – and sometimes incorrect – descriptions of Lakatos' life, especially of its earlier part, which took place in Hungary. Fortunately, in this volume Jancis Long published a paper [31] which is full of interesting data and facts about this period of his life. This development releases us from the obligation to present here Lakatos' whole life-history in Hungary. However, in order to support a better understanding of our paper, it seems to be worthwhile to partly recount those aspects of their lives which possess a strong relevance to their intellectual and personal relationship. In the case of Lakatos, we try to fill out Long's analysis with some additional details and a short review of Lakatos' early Hungarian papers.[2]

1.1. Biographical remarks on Lukács

Lukács was born in 1885 and died in 1971 in Budapest. He grew up in a wealthy Jewish family, in which Hungarian and German culture were equally represented. His world-view was fundamentally influenced by the deep social contradictions of the time, which were rooted in the coexistence of a feudalistic heritage and the early stage of free-market capitalism in that country. He rejected categorically the political practice and ideology of the ruling classes, and searched for a better system of values in the literature of his age. The poems of Hungarian poet Endre Ady played a very important role in his intellectual development. First he studied law, then philosophy and literature at the University of Budapest, and later, sociology and philosophy in Berlin with Simmel and Dilthey. In his first important work about the history of the development of the modern drama, he correlated this process with historico-social relations. In his mid twenties, he was influenced by some mystic masters and Kierkegaard; later he went to Heidelberg and had intensive contact with the sociologist Max Weber and some members of the neo-Kantian school. He started to prepare an aesthetics of Kantian style, he was shocked by the inhuman imperatives of the age embodied in the escalation of the First World War; this diagnosis provoked a major turn in his intellectual life. He was radically opposed to this inhuman ideology; first only theoretically, later practically, as well. At the beginning of this process – according to his recollection – his position was a mixture of "left wing ethics and a right wing epistemology," but later on he became more radical and drew on the practical consequences of his standpoint, accepted the idea of the social revolution, and joined the Hungarian Communist Party. He played a political role in the Hungarian Soviet Republic which ruled the country in 1919; following its decline, he had to leave the country.

He moved to Vienna for ten years, during which he tried to rethink the perspectives of the proletarian revolution. As a result, his famous book entitled *History and Class Consciousness,* was published in 1923 (in German) [32]. (Later he called the intellectual position of his own book messianic Marxism, and criticized and rejected most of its basic statements.) This book became popular world-wide, especially amongst left wing intellectuals. In order to send an important progressive message, he reconstructed and presented the essence of Marxism as a method and, analysing it, he identified the philosophical constituents of Marx's (and some Marxists') worldview and their close, but critical, relationship to the Hegelian dialectical philosophy. The concepts of totality, alienation and reification, among others, played a central role in his analysis. He argued for the idea that the essence of Hegelian (and Marxian) dialectics is the category of concrete totality which can only represent complex reality. Applying the dialectical view of totality, we can comprehend reality as a social occurrence. The rationalism cannot be described as an abstract, formal, ahistorical principle. The Hegelian dialectic was depicted, in this important book, as a version of rationalism. On the basis of its sectarian attitude, compromises of any kind were rejected because they would prevent the achievement of the proletariat's final goal.

As a result of the criticism of his book and the political struggles between the fractions of the (illegal) communist party, his position became less sectarian, and in 1928 he suggested a democratic dictatorship as the political program for the party. In 1930 Lukács went to Moscow, where he was able to study Marx's *Economic and Philosophic Manuscripts of 1844* and some other unpublished works. This was very important for his gaining a better understanding of the real difference between Hegelian and Marxian philosophical views, and it enabled him to do away with his Hegelianisms – which had appeared essentially in his *History and Class Consciousness* – and elaborate upon an authentic Marxian position for the first time in the fields of aesthetics, social and political philosophy and the history of philosophy.

From Moscow, Lukács returned not to Vienna, but to Berlin – for about two years – and he quickly became prominent among progressive German writers. On the basis of his freshly clarified and strengthened Marxian view, he took part in *political and literary* discussions. He started to elaborate upon his idea of *realism* in literature – an activity he pursued for many years thereafter in Moscow and Budapest, as well. It is very interesting and important that during this work he considered the products of literature together with their socio-historical contexts. The quality and the value of a novel or a drama could be evaluated only in this context. From this position, he criticized the avant-gardism, the expressionist attitude of some proletarian writers. (This criticism was a more or less indirect attack on Brecht, as well. This conflict between the two groups of German left wing writers in 1932 is sometimes called the Brecht – Lukács debate [4].)

His critical position with regard to the avant-garde movement was closely connected with his renewed political philosophy. In his view, avant-gardism expressed a sectarian political attitude, the dangers of which were clear to him because of his own development. Moreover, in the shadow of the emergence of fascism he gradually realized the significance of the collaboration of progressive societal actors to the formation of an antifascist political front. (His idea harmonized with the political strategy of the Popular Front policy accepted by the Comintern in 1935.)

During his longer stay in the Soviet Union between 1933 and 1945, Lukács maintained his alignment with the neccssity of the socialist revolution, and because of the historical situation, he tried to analyze the social, cultural, and political environment of the rise of fascism. In this analysis, he wanted to save the progressive aspects of German intellectual development from the realm of fascist ideology and to present the achievements of Hegel, Marx, Goethe and Thomas Mann as the proper basis for social development. In his historical, political, philosophical and aesthetic works during this period, he elaborated many details of this problem. The central idea of his analysis was a distinction – of a Hegelian origin – between the two groups of citizens, the *citoyens* and the *bourgeoises*. The citoyens form the progressive part of the civic classes; they are the representatives of the classical civil values of civil revolutions such as freedom, equality and fraternity, they preserve and present the most important positive human values, and they are interested in human progress. Opposing them are the bourgeoises, the reactionary class; these can accept only their own selfish interests, are interested in the increase of exploitation, imperialist conquests and war. Fascism, of course, is the product of bourgeois dominance. On the other hand, in the building of a new, socialist society, and especially in the struggle against fascism, the proletariat can form an alliance with the citoyens without endangering its revolutionary goals. In this way, the citoyens serve – not fully aware of their role – to support the socialist revolution, but they play only a temporary role in the development of the revolution, and the historical process will roll right on over them. (This type of alliance-policy was, at the time, called the Popular Front Policy.)

In the Lukácsian view, the citoyen's attitude was related to the values of *rationalism* in thinking *and realism* in literature and art. On the other hand, the bourgeois attitude is expressed in various forms of *irrationalism and anti-realism*. According to his position, the rationalist view is a decision in favor of progress, while the irrationalist standpoint is the result of an inability to chose progressive solutions to problems. The relationship between realism and anti-realism has ideological characteristics similar to those expressed in the arts. (He criticized, from this standpoint, the regressive aspects of naturalism and avant-gardism, as well.) These ideas appeared, for example, in his works on Goethe, Keller, Thomas Mann, Hegel, and in his manuscript on Germany as the center of fascism, which he started to write in Tashkent in 1942.

When, after the war, on 28th August, 1945, Lukács returned to Budapest, he was a 60-year-old, world-famous Marxist philosopher. He received some – not very important – official and political positions, and was awarded a professorship in aesthetics and cultural philosophy at the University of Budapest. Between 1945 and 1950, he lectured on aesthetics, Hegel's *Phenomenology of Spirit*, and the history of irrationalism in philosophy. (Within these courses, he presented some chapters of his Tashkent manuscript, too.) In the first years of his teaching activity in Budapest, his lectures were an important event for the younger Hungarian intellectuals. The lecture rooms were overcrowded with several hundred listeners; many people wanted to become directly acquainted with the authentic Marxist approach to the problems of the age [45]. Lukács was aware of these expectations, and the possibilities resulting from them: parallel to his public lectures, he promised seminars for smaller circles and welcomed students and other intellectual partners into his university office and his own flat for more detailed discussions [7].

On the other hand, a great number of his works were translated into Hungarian and published very quickly during these years. More than fifteen books, containing different collections of his studies, appeared within a few years. (Until these years, his works were essentially not published in Hungarian with the exception of his books from the 1910s, whose philosophical positions were radically rejected by the older Lukács.) He published regularly in periodicals, first and foremost in those which expressed communist political values (Szabad Nép, Társadalmi Szemle), but also in other journals (Forum, Valóság, etc.). In these books and papers, he discussed problems of philosophy, politics, history, ideology, literature and art in general, and their appearance in the concrete, current debates in the post-war world, and especially in the Hungarian society. In most of these questions, his position was based on the above mentioned views, which were elaborated upon in his works written in the Soviet Union. These ideas, without a doubt, became very essential actors in the Hungarian intellectual life of those days.

Already in the Autumn of 1945, a group of younger philosophers and literary critics could be found around Lukács. The members of this "circle" were József Szigeti, István Király, Imre Lakatos, and György Lázár (alias Leó Lám). It was not an organized circle with regular meetings or any kind of fixed program, but these young people knew each other, and they highly respected Lukács' philosophical work and the personal discussions they had with him. It is supposed that they were frequent visitors at Lukács'. During these – mostly individual – personal visits, it was possible to ask Lukács' opinion on their current works and ideas, and listen to Lukács talk about his current philosophical problems or political views [7]. It is more or less obvious that the relationships between Lukács and the individual members of this group were different in their significance, stability, duration and reciprocity. By about 1947, the "circle" had ceased to exist, Király and Lakatos explicitly criticized Lukács' moderate political opinions, and personal relations between the former members had worsened.

In 1949, in the shadow of the Rajk trial, an attack on Lukács began in the press. As a result of this "Lukács debate," he lost the opportunity to teach regularly at the university, but "won" the opportunity to work on and finish his book on the role of irrationalism in the emergence of fascism. During this period, he was surrounded by a group of students (e.g. István Hermann, Ágnes Heller, Dénes Zoltai), whom he had already recruited from his own students; the development of this group, which lasted for some decades, led to the formation of the Budapest School of philosophy. In 1951–1952, he finished the manuscript of *The Destruction of Reason*, employing his earlier papers and manuscripts, as well. The book appeared in 1954 in Hungarian and in German.

In 1956, Lukács became a member of the first Nagy government for a few days. Because of his political activity, he was put under arrest for some months. After his discharge, he returned to his philosophical work. In this last period of his life, Lukács summarized his aesthetic views, characterizing the specific nature of aesthetics, tried to elaborate upon a Marxist ontology, the ontology of social existence, and emphasized the importance of the further democratization of the socialist political system.

1.2. Biographical remarks on Lakatos

Lakatos was born in 1922, in Debrecen, and died in 1974, in London. He grew up in a Jewish family in Debrecen, the central town of eastern Hungary, in "the Calvinist Rome," as it was sometimes called. He attended the Jewish Gymnasium of the town, where the Hebrew language and Jewish culture were essential parts of the curriculum. He was a very ambitious and excellent student in every field, especially in mathematics. For example, he took part in the traditional mathematical problem solving competition of the journal *Középiskolai Matematikai és Fizikai Lapok* (Mathematical and Physical Journal for Secondary Schools), lasting whole school-year, where the goal was to solve the problems appearing in the journal and send the solutions to the editors from month to month. The journal regularly published the best solutions of the problems, together with a list of the pupils who had sent in a correct solution of the problem in question. In the issues of the journal of 1937–38, we can very frequently find the name Imre Lipsitz (Lakatos' original name); moreover, because of his excellent work in 1937, his photograph – together with those of 40 others – was published.[3] Furthermore, he struggled for (and occasionally won) the position of leader of the "Jewish self-education group" of his school.[4]

It seems to be important that, even at the beginning of his conscious life, he was unavoidably involved with and successfully educated in both ideological, religious and exact, scientific aspects of the culture. It can, obviously, be supposed that his life-long interest in both political and scientific matters, moreover, his vacillation between a political and an academic career, can be associated with these socializing elements of his youth.

At that time, Debrecen had a relatively significant Jewish community with a few thousand members, despite the strong assimilative tendencies of some earlier decades. However, in Lakatos' boyhood – parallel to the spreading of the ideology of national-socialism – an increasing segregation and opposition between the majority and the Jewish minority can be observed [38]. Conflicts resulting from the coexistence of different cultural and religious traditions and values in one town (and even in one person) were an everyday experience for the young Lakatos. How could a young intellectual find or define his identity in this conflict-filled environment?

It seems to have been a very important problem for – not only the young, but also the older – Lakatos to find recipient communities (from family to social class) for himself. He probably desired strongly to accept the entire value system or world-view of receptive communities or that of their prominent masters, because he wanted to become the best member of the society, or the eminent pupil of the great master. In my view, he wanted to select and accept the values of his selected environment; moreover, he wanted to control his own acceptability which was (apparently) determined by his environment. The socio-historical situation made his identity into a problem to be solved, and he did try to solve this problem, but he was educated to find correct, rational solutions to hard (mathematical) problems. Lakatos attempted, as a young boy, to become assimilated into the ruler right wing community, and later, to organize a left wing, secret resistance group.[5] Both solutions were rational, even the second one. The success of such a strategy depends on the quality of the solution, and Lakatos was clever enough to chose the better one.

Of course, such identity problems possess a social dimension, as well. As a genuine member of a socially defined minority and an outlawed community, he must have been sensitive to the similar situation of the oppressed classes of his society. He probably recognized the unsolvable and irrational increase of the social conflicts, which would lead to the coming war and ruination. In this situation, the question naturally emerges: How would it be possible to find a rational solution to (his own and) the difficult social problems? In Marxism, he could find a reasonable treatment of these problems. In my opinion, his early adherence to Marxism was rooted in this situation. It was clear that the Marxian worldview offered an almost scientific description of social systems, with a rational treatment of social conflicts, and also allowed a realm of freedom for truly free development of personality, without any social oppression or the rule of classes. The formation of this new society is full of important an interesting problems; certainly, a very engaging challenge for a young, clever genius. On the other hand, the unselfish, self-conscious, self-sacrificing "soldiers" of the struggle for the new society are incorporated into the new community not as simple members, but as the heroes of the new age, the new society. It can be seen that Lakatos, in his adult life in Hungary, tried to pursue a twofold goal: to actively take part in the political life, and to continue doing science.

Lakatos' "academic life" until 1945

Between 1940 and 1944 he was a student at the University of Debrecen. In the first year he could study only law – in Hungary, the "Jewish-law" of the time prescribed the desired ratio of Jewish students for a given academic field at the university – but in 1941 he was able to move to the Faculty of Humanities, where he began studying mathematics, physics, philosophy and some other humanities.

The University of Debrecen was, at that time, a relatively new and small institution. It was about 20 years old, had no independent Faculty of Sciences, and the mathematical and physical courses of study were organized within the framework of the Faculty of Humanities. The university had relatively few students (about 10 students, for example, graduated together with Lakatos in the field of mathematics and physics). Let us now consider the faculty of the university during the time in which Lakatos studied there.

In mathematics, Lajos Dávid was a lecturer in algebra and analysis. He was sensitive to the historical aspects of mathematics; for example, he published a book on the two Bolyai's (Farkas and János Bolyai). László Rédei was another lecturer in algebra. Ottó Varga and Béla Gyíres taught geometry and differential geometry. This subject probably was presented on a high level – the subject of differential geometry has to this day been taught at an internationally recognized level in Debrecen. For a short period, Lakatos was an assistant of Varga.

Physics was taught by Sándor Szalay, a pioneer of nuclear physics research in Hungary, and János Horváth, who worked in statistical physics. (Lakatos later wrote some short commentaries in Hungarian newspapers and journals on the different aspects of nuclear physics; the atomic bomb,[6] diseases caused by nuclear radiation [26], etc. It could be stated that he was seriously interested in nuclear physics, and that he considered it along with its social context.)

In the humanities, he could listen to the Kantian philosopher Béla Tankó, to Gyula Mitrovics, who worked in aesthetics, psychology, and neo-Kantian philosophy, and Imre Révész, a historian of science and religion. From the later events of Lakatos' life, it can be seen that Sándor Karácsony, who worked on Protestantism and Hungarian studies, had an important influence on the young Lakatos. (I suppose that Lakatos received the intellectual support from Karácsony to find and consolidate his Hungarian identity in the very Protestant environment of Debrecen.)

However, it was first and foremost Árpád Szabó's (1913–2001) lectures and personality that fascinated the young student. Szabó, who was a lecturer in the Department of Classical Philology from 1940, studied the relationship between ancient politics and art, *politics and science* from a common historical perspective. (Later, when he demonstrated the role of philosophical ideas in the emergence of Greek mathematical thinking in his book – entitled *The Beginnings of Greek Mathematics* – he became a world-famous historian of mathematics.) Szabó and Lakatos had a close, friendly relationship, which was further strengthened by their similar political values. (Szabó, for example, supported the progressive student movement of the university, and Lakatos took part in it.) They continued their close relationship until Lakatos' death.

Because of his illegal organizing activity in the beginning of 1944, Lakatos was expelled from the university[7] and completed his studies only after the war.

Lakatos' "political life" until 1945

Lakatos took part in the left wing student movement called *Márciusi Front* (March Front), influenced by the communist Gyula Kállai. (Kállai published during these years in Debrecen some issues of a progressive journal, called *Tovább*. It was soon suspended. When, after the war, this journal was restarted in Budapest, Lakatos worked as its editor.) At least since 1942,

Lakatos had organized an illegal communist group, a small, closed group in which he was the intellectual leader. They had a very weak relationship with the illegal communist party – a few years later, in an official investigation, even the existence of this relationship was refuted. The main activity of the group was studying Marxism: reading and discussing the history of the Bolshevik party, and theoretical and literary works of communist authors. They did not perform any practical actions against the state or its institutions.

When Lakatos had to leave Debrecen, and secretly went to the close by town of Nagyvárad, he changed his name from Lipsitz to Lakatos (in two steps), which served to increase his survival chances during this period which bore the multiple dangers of the holocaust, the anti-Communist state policy, and the war. The activity of the secret group continued under these conditions. Unfortunately, however, they committed a horrible practical offence: the group, acting on Lakatos' suggestion, forced the suicide of a young girl, who belonged to the group. They thought that she would have been dangerous to the conspiracy, had she been arrested by the police. She had to sacrifice herself for the community.[8] This unfortunate action cast the occasional shadow on Lakatos' later political career.

After the war, Lakatos returned to Debrecen, at which point he realized that he had lost most of his family. In 1945 he became a member of the communist party, and soon he went to Budapest to work for the new society.

Lakatos' political and philosophical activity from 1945 to 1950

In Budapest, Lakatos plunged into the center of the lively, post-war reformulation of the country's intellectual and political relationships. He took part in the struggles for communist positions on different fronts of the political fight, but in our view, he never played a central, or at least significant, role in these struggles. Some reports indicate that he seriously aspired to become a prominent politician but, in fact, he occupied only less important official positions, with limited active power. During these years, he had a subordinate position in the Ministry of Religion and Education; he was a member of a special group of the communist party, influencing the communist policy in the fields of education and culture; and for a short time he worked as an editor of a communist journal. However, these charges and their significance changed from time to time. Because of the frequent and quick changes, and uncertain and contradictory recollections, the precise reconstruction of his position-history was impossible – at least for us.

It is, perhaps, indicative of the character of that period, that the Ministry of Religion and Education was essentially re-organized at least *eight times*. In a initial short period Géza Teleki, between November of 1945 and November of 1946 Dezső Keresztury, and after an interregnum, from March of 1947, Gyula Ortutay served as minister of education. The staff rapidly grew, and it was frequently reorganized and changed. (In the first period, many people from Debrecen were appointed to positions – for example, László Kardos, who was Lakatos' teacher at the Jewish Gymnasium.) There are some faint indications that Lakatos was employed around 1945–46, and it seems certain that he was appointed around 1948 to a relatively high position having to do with higher education.

Education was a very important field for the political struggles of that age. After the war, the removal of the relics of fascist ideology and its representatives from the educational system and cultural institutions was begun, and the re-building of a new, progressive Hungarian culture was supported by various political movements. The characteristics of this new culture, however, were open to question. There was a fierce cultural battle between the actors in political life for the dominant role in the cultural and educational field. During this time, the communist party emphasized the importance of the education of the masses, including various forms of public and continuing education; the party supported the cultural development of the ambitious members of the lower classes – instead of the representatives of the cultural élite.

The attack on the 50-year-old *Eötvös Collegium* (Eötvös College) was a part of this policy. Between 1945 and 1947, Lakatos was a member of this college: he occasionally lived there, he worked in its library, he took part actively in its intellectual life. Up to this point, the college had been a special kind of educational institution for the intellectual élite. In 1946, fuelled by leftist educational policy concepts, a struggle began for the radical reconstruction of the college, to transform it into an institution, which would serve the aims of the working class and the new people's democracy. Lakatos took up this struggle; he argued for these goals in the college and wrote a paper, in which he radically criticized the retrograde, liberal bourgeoisie atmosphere of the college and its director (the former minister of education, Dezső Keresztury), and called for its fundamental reorganization for the purpose of educating the new, working class intellectuals [17]. Lakatos had no direct role in the real transformation of the college. Its "transformation" (more precisely, its temporary liquidation) actually saw the college turned into an ordinary dormitory for students; this process dragged on at length [41] and was finished three years later, when Lakatos had already been sent to an internment camp.

The position of his paper is interesting from the vantage point of his later intellectual evolution. After a very critical diagnosis of the intellectual character of the college, he recommended the fundamental revision and reorganization of the *college – public life* and *college – science* relationships. In these respects, he argued against the unchanged continuation of the old college spirit, as in the new social environment it would yield "isolated," "apolitical," "private," "gentry" scientists and science. Instead of the isolated scientific élite, he demanded that there be produced the vanguard of the young intellectuals of the plebeian democracy. He argued for a new Hungarian science, which would break with its bourgeois, gentry-Hungarian character, whereupon a progressive, Marxist scientific character could emerge. It is, perhaps, worth mentioning that in these argumentations Lakatos evidently considered science in its social context; he saw scientific and social developments as parallel, coexisting processes, both of them driven by the progressive aspirations of the masses. Of course, his attitude was more or less accepted and dominant in that intellectual environment. (His paper was criticized, but only its unfavorable diagnosis was rejected – without criticism of his science and scientist concepts.) In these months, Lakatos wrote many other papers considering the problems of society-science-education-politics relationships [18–20, 22, 25]. In these papers, his ideas on these interrelated topics can be seen in more details. It seems to be significant for us, that his position in the political debates and struggles was very conscious, clear and based on serious political and philosophical studies; on the other hand, he also tried to draw the practical conclusions of, and to propagate, his accepted ideas.

Journalism can be identified as another of Lakatos' fields of activity at this time. He wrote on different political problems, and in 1947, he became an editor of the journal *Tovább* (Go Ahead), which was a revival of a journal which had been published about ten years earlier by Gyula Kállai in Debrecen. The political and cultural journal appeared in 26 issues between May and October of 1947, and was considered to be an unofficial weekly paper of the communist party. The editor-in-chief was Géza Losonczy, and it is very probable that Lakatos was the editor of one of its columns, entitled *Technika és tudomány* (Technology and Science). Doing this job he, perhaps, could practically fulfil some of the social tasks of the vanguard of young intellectuals – on which he elaborated in the debate over the college.

Yet what do we know about his studies during this period? When Lakatos came to Budapest, he could continue his "official" studies as a member of Eötvös College. (He signed his papers during this time as "student of philosophy, member of the Eötvös College.") During this time, based on his earlier studies of Marx and Engels, he was already considered to be one of the best young Marxist philosophers. At the college, he met some young people with similar interests, young teachers and ex-members of the college, including István Király (from Debrecen), József Szigeti, Leó Lám and others.

Between autumn of 1945 and the end of 1947, according to the Lukács' biographies, there formed around Lukács the first group of young intellectuals. Lakatos was introduced to Lukács by Szigeti, and presumably visited Lukács on many other occasions, too. (However, only a single postcard from Prague signed by Lakatos – and John Bernal – can be found in the Lukács Archives, as a material sign of their relation.) During these visits, he had opportunities to continue personal discussions with Lukács on current political topics and the most general philosophical problems. According to Szigeti's recollections, at their first meeting Lukács and Lakatos immediately started to discuss a problem of the philosophy of mathematics [42b]. In these days, Lukács' major interests were in aesthetics, including the definition of the progressive Hungarian literary tradition; the characterization of the current crisis of bourgeois philosophy, and the description of the role of irrationalism in the formation of the fascist ideology; and the political problems of the "transition" of the post-war Hungarian society to a popular democracy. In this political question, Lukács' position was not so radical as the official – in some respects rather dogmatic – position of the communist party. He preserved his alignment with a policy of alliance between the progressive classes as a version of the Popular Front policy.

On the other hand, Lakatos obviously attended some of Lukács' lectures and took part in some of his seminars on the above-mentioned topics. Already in 1945, however, when Szigeti first met Lakatos and asked him about Lukács' *History and Class Consciousness,* Lakatos was already very familiar with this topic. (It is unclear how he was able to find a copy of this forbidden and rare book in Hungary.)

In 1947, Lakatos published two small, mainly political articles in the journal *Forum*, which was considered to be the journal of Lukács. In these papers, he critically reviewed the current issues of the journals *Huszadik Század* and *Vigilia.* He criticized the political positions of the Hungarian "civil radicalism" and of progressive Catholicism, and their relationship to communist politics [23, 24]. In this respect he, in principle, could imagine an alliance between the proletariat and the representatives of civil liberalism and Catholicism, but not on the basis of the ideas presented by most of the authors of these journals. (In his criticism, he cited and rejected some political ideas of Michael Polányi, too.) He studied the details of the policy of alliance in another two papers, as well [15, 25].

However, Lakatos had published his first papers in the journal called *Valóság* (Reality), which was a journal devoted to young intellectuals and which was published by the organization of young communists. It was much more than a journal, in fact, because in the summer of 1946, a circle was formed around the journal; this group organized common meetings and debates, and consisted of Árpád Szabó, Zoltán Szabó, István Király, József Szigeti, and Lakatos. In his first paper in *Valóság,* Lakatos reviewed a book, a collection of papers on dialectics written by Erik Molnár [14]. Lakatos conceded the author's merits in the popularization of dialectics and its application in some cases, but this did not suffice. He criticized weak points of the book, chiefly on the basis of Lukács' *History and Class Consciousness* – without mentioning it, however – and Lukács' actual writings in newspapers. He anticipated the struggle between Marxist and Bourgeois science, in which Marxist science was supported by the Marxian method, that of dialectics.

It is important to note that two of Lakatos' further papers in *Valóság* initiated political debates. We have already mentioned, above, his famous paper addressing Eötvös College, but before that, he had published in *Valóság* a lengthy book review of Imre Csécsy's *Világos pillanat* (A Clear Moment). In 1946, Csécsy was the president of the Hungarian Radical Party, and in this book he proposed the harmonization of the value systems of individualism and socialism. Lakatos sharply criticized this position [15], Csécsy replied to his critique [3], and Lakatos then returned to the problem briefly in his paper in *Forum* [23]. According to Lakatos, Csécsy, in his book, described very clearly the awful practical consequences of the "bourgeois" ideology and, confronted with these consequences, he rejected these aspects of "civic" values, but wanted to rescue their positive aspects in the form of an updated "Citoyen" ideology. On the basis of Lukács' study on Thomas Mann [33], Lakatos classified Csécsy as a representative of citoyen consciousness. However, he criticized – in a very sophisticated way – Csécsy's ideas about the present and future role of the intelligentsia, because Csécsy had not accepted clearly enough the leading role of the proletariat in the forthcoming social revolution. In accordance with the communist theory of social revolution, Lakatos argued for the thesis that the proletariat is the only progressive historical actor in the revolution, working in its revolutionary practice together with its allies, the peasantry and the progressive intelligentsia. Lakatos declared – again, implicitly following the ideas of Lukács' *History and Class Consciousness,* and of his Mann study – that in this way, a unity of actions and thinking can be established as a real progressive historical alternative. In his reply, Csécsy pointed out the ideology-ladenness of Lakatos' interpretation of progress. However, Lakatos maintained his position in the later paper in *Forum,* and suggested to the radicalists a more radical self-criticism, in order to reach the real progressive standpoint.

Another front of Lakatos' political struggle was his opposition to the reactionary ideas of Christianity. In this respect, he emphasized the principal possibility of a harmonic relationship between Christian and socialist values, a way in which Christians could be allies of the proletariat, under the guidance of socialist ideology. He criticized the aspirations of the Christian ideologists and scientists to a more dominant social role for their – in this respect, reactionary – ideas [24, 25].

In the *Társadalmi Szemle* (Social Review), which was the most important theoretical journal of the communist party, Lakatos published only two very brief book reviews. These had no more than, perhaps, symbolic significance for him [20, 21].

Alongside this vigorous political activity, he worked on his dissertation, entitled *A természettudományos fogalomalkotás szociológiájáról* (On the Sociology of Concept Formation in the Natural Sciences) [11, 12, 31]. From time to time, he published parts of this work [16, 22]. Comparing these two papers, it becomes clear that they have common parts, which are almost verbally identical. It is very natural to suppose that these common parts were parts of his – later on, lost – dissertation, too. This hypothesis is strongly supported by the passages from the official evaluative essay of Sándor Karácsony, the opponent of his dissertation, from July 8, 1947: "... the dissertation itself was not an idea rushed into the author's mind, it was based on two previous publications in serious places. One of them appeared in the Athenaeum under the title *Criticism of the idealism in physics,* while the other

(*Modern physics – modern society*) appeared in a thick volume for teachers entitled *Továbbképzés és demokrácia*."[9a]

In the identical parts of these works, he considered the problems of modern physics (quantum mechanics, theory of relativity, the physical world-view) in a social context. His paper [16] in the *Athenaeum*, which was the journal of the Hungarian Academy of Sciences and the Hungarian Philosophical Society, begins as a review of Susan Stebbing's book *Philosophy and the Physicists* (London, Pelican, 1943). However, in the review – similarly to his other review papers – he expressed his own ideas, too. He described Stebbing's analysis of Jeans' and Eddington's idealism, but, as he said, he wanted to go further: instead of following the immanent development of science, he tries to find an explanation for their standpoints in their social environment. In this respect, Lakatos analyzed the different historical concepts of nature, and compared their social characteristics and determinant factors. The concept of nature is, in fact, a social category, which expresses historico-social values. In the correct understanding of the distinctions between the Eddingtonian "familiar," "physical" and "real" worlds, the concept of reification has an important significance. From the analysis of the capitalist production processes, Lakatos arrived at the necessary distinctions between the earlier rational, progressive bourgeois attitude and the later irrational, regressive bourgeois attitude toward nature and science. There is no doubt that in these analyses, Lakatos applied the notions and adopted various theses from *History and Class Consciousness* – especially that of its central study, entitled 'Reification and the Consciousness of the Proletariat.'

Lakatos' other "dissertation" paper appeared in a book, which was a collection of papers published by the educational department of the Ministry of Religion and Education. The book was edited by the head of this department, and the studies had many illustrious authors, including, for example, the chemist Tibor Erdey-Grúz, the historian István Hajnal, the minister of education Dezső Keresztury, the sociologist Sándor Szalai, and the classical philologist Árpád Szabó. The aim of the book, entitled *Továbbképzés és demokrácia* (Further Education and Democracy), was to provide a high-level scientific background for school teachers, which was based on the new, more free and democratic ideology. In his contribution [22], Lakatos included most of his earlier review of Stebbing's book, and he filled it out with some further thoughts on the possible – depending on social conditions – positive and/or negative roles of atomic energy. Moreover, he inserted his earlier theses into the broader context of the philosophy and the sociology of science. He depicted the fundamental role of a dialectical view of the development of modern physics. The great achievement of modern physics was, he said, historicism's penetration of the natural sciences. Philosophers have to transcend Hegelianism, because only a Marxian analysis, applying the dialectics of the product-commodity-product relation and the concept of fetishism, can describe the real historical process of science. He emphasized the significance of historical materialism – not dialectical materialism – for these analyses. Of course, these ideas are also borrowed from Lukács' *History and Class Consciousness.* (The English translation of this paper of Lakatos is included in the present volume.)

In 1948, he received his doctorate, *sub laurea almae matris,* from the University of Debrecen. As often mentioned before, all copies of his dissertation have been lost. From the above description of his earlier "dissertation" papers, however, it seems to be almost clear that, in his dissertation, he most likely presented a fundamentally Lukácsian analysis of the history of natural sciences, first and foremost of physics, based on Lukács *History and Class Consciousness.* Perhaps it is also of significance that Lakatos nominated Lukács as the first on the list of twenty-three invited guests at his graduation ceremony.

His interest in physics continued when, in the beginning of 1949, he went to Moscow to study physics. He soon had to break off his "candidature" in Moscow, though, and it seems to us that he interrupted his study of physics for about two decades.

Unfortunately, we know only very little about Lakatos' activity in the party. It is very probable that he had some – informal – influences upon communist policy in some questions; for example, in the democratic reconstruction of the higher education system, and in the policy of science. However, this influence was never a very significant one. In this respect, he was more a tool in the hands of leaders at different levels of the party-hierarchy, than he was a determinant of events. The party was a common field encompassing different struggles – he was a good fighter, but his rivals were, as well. He personally knew some prominent leaders of the party, for example József Révai, who controlled matters of culture in the country. It seems to be unquestionable that Lakatos wanted to build up a career in the party; he did a lot to this end, but he was quite unsuccessful. At the end of 1947, for example, he organized, together with Király and some others, a radical leftist attack on the not-sufficiently-revolutionary Lukács' views – but the party, deeming it to be untimely, stopped it. (Lukács had tried to defend the freedom of the artists against direct political control, citing the idea that "poets are partisans and not soldiers of the party." He was in the minority with his position.) Lakatos also had some connections to the dogmatic philosopher László Rudas, who started an attack against Lukács in 1949.

What Lakatos finally attained was a mid-level position in the Ministry of Education. For by this time, after all, he had dropped out of the closer Lukácsian circle – his earlier wife had become the wife of Szigeti, another member of the circle. The circle had ceased to exist. According to Szigeti, however, it is very probable that Lakatos was in contact with Lukács in 1948 as well, during his work in the Ministry, when he would, from time to time, ask for his advice.

In 1949, Lakatos' official career was interrupted and he was sent to Moscow to do scientific work, probably because of the consequences of a party investigation of his forcing the suicide of a member of their illegal group five years earlier. In Moscow, his behavior was considered to be unpartly-like, which perhaps meant that he had been worsted in the local party fights. A few months later, he suddenly returned home. His unexpected transitions during this time, as well as some other developments of his life, give rise to the suspicion that, perhaps, he was doing work for some kind of secret service. However, this suspicion is not supported by any concrete evidence. His actions and behavior – whether or not they were determined by some secret service – can be explained as an intensified struggle to re-attain his lost, better social position. Returning home, he became a teacher at a famous secondary school in Budapest for a few months. There is a suspicion that he served as an agent in a demonstrative conceptional trial against József Révai [1], but the political situation in the country changed – in 1949, Révai also became a member of the government, as minister of public education – and Lakatos' fate took an unhappy turn.

Lakatos' political and scientific activity from 1950 to 1956

In 1950, Lakatos was imprisoned for three years, for reasons which remain unclear. They might have had to do with the above-mentioned Révai-matter, but a proper denunciation at the proper state or party organs appears a more natural explanation. The injurious trial and the awful conditions of the internment camp did not fundamentally destroy his communist conviction. Together with many others, he came free in 1953; however, he agreed with the political police to prepare secret reports on the activities and views of certain persons, including his closest friends. After his discharge he was practically ostracized from political life, but he tried to follow events as far as he could.

Between 1954 and 1956, he held a position at the Institute of Mathematics of the Hungarian Academy of Sciences, in Budapest. This institute was a very good research center for mathematics, full of world-famous members of the Hungarian mathematical school. Here – in an inspirational environment – he could develop his mathematical and philosophical knowledge. He translated two mathematical books into Hungarian, and obviously could discuss the fundamental problems of mathematical thinking with, among others, Alfréd Rényi, who also published his ideas about this topic in a dialogical form. There are no indications as to whether he met with Lukács in this period frequently or not. However, he – very probably – studied Lukács' *The Destruction of Reason,* which appeared at this time. (According to a letter of Szigeti, Lakatos was present at Lukács when the Hungarian title of the book was found out.[9b])

In 1956, Lakatos gave a talk in a Petőfi-Circle debate on education on 28th of September. These debates of the Petőfi-circle about economics, philosophy, literature, etc., served as a forum for the renewal of the Hungarian intellectual life, following some over-controlled years. It is interesting that

Lakatos actively took part only in the discussion of the problems of education. A few months earlier, there had been a large debate-meeting about philosophy, with the participation of Lukács and chaired by Szigeti. Here, Lukács and some others sharply criticized the current low-level, Stalinist ideology and its representatives, and suggested a return to the original Marxist-Leninist values in the spirit of the 20th Congress of the Soviet Communist Party. Hundreds of intellectuals took part in this meeting. It is hard to imagine that Lakatos was not present at this meeting; however, he did not speak at it. Transcripts of these debate-meetings, including Lakatos' contribution to the discussions, have been published recently [27].

In his Petőfi circle talk, Lakatos concentrated on problems concerning the education of scientists. He emphasized the significance of the legitimation of the fundamental role of *talent* in socialist society. He castigated the earlier sectarian, utopian, and demagogue political and pedagogical practices, which overestimated the role of education at the expense of the role of talent.[10] The talented had been considered as enemies of the socialist system, whereas the "simple," "grey," "disciplined" personality was cultivated. This practice produced societal and personal havoc – a caricature, at best, of the revolutionary, optimistic pedagogical ideals was being realized. Lakatos proposed the re-organization of the Eötvös College and made some other suggestions along these lines. On the other hand, he emphasized the basic right to *original thought,* especially in the cases of scientists and artists. He rejected the earlier, dangerous and meaningless ideological wars against empiricism, curiosity, and original thinking. In his speech, he applied the unusual phrase, "destructors of reason." Sectarian pedagogy condemned some fields of sciences as bourgeois sciences, and inspired fanatism, instead of the real revolutionary Marxian thrust. In scientific education, it is necessary to emphasize respect for the facts, the demands of exact thinking and application of proofs. He demanded the recognition of the right of *dissenting opinion* in politics, and especially, in the sciences; this is reasonable, since only one man can be the first to see the truth.

It is clear that in his talk (what he actually read aloud – as a kind of declaration), Lakatos reflected at the same time upon his own fate, upon the situation of science in society, and upon socialist values in the society at hand. At this time, in other words, he considered himself to be a socialist scientist and a scientific political thinker (at least in the field of education policy). However, his standpoint concerning political values changed significantly with respect to his position in the forties. Now he criticized the impatient sectarian practice which he himself had once subscribed to, mocking for instance, some ideas of his earlier friend and ideological ally, István Király, and suggested that they should return to a quite scientific, searching, doubting, original Marxian ideology of practice in scientific and political life. I suppose this was, at the same time, a personal program for himself, too.

It seems to us that at this time, he was intensely interested in the problems of plurality and of the scientific methods of choice between these plural alternatives – in various fields. For example, he discussed with his acquaintances the possible consequences of the introduction of a pluralist party system instead of the single state-party, and at the same time he translated Pólya's book, *How to Solve It*, which focused on heuristics, and the methodology of reflection and criticism. These problems have "common" parts: what is the best way to select a solution of the actual problems, what kind of epistemology is acceptable for it, what is the role of originality in the solution, etc. It seems to be evident that Lakatos' political ideas and his concept of science changed during these years, but he preserved his way of thinking and considered the scientific and political activities and processes together to be a co-existing, parallel process – which he was able to recognize in his own personality, as well.

In 1956, his revolutionary practice was not so significant. In the end of November 1956, together with his new family, he went to Vienna, and later to England, where he soon became the well-known Lakatos. There were various reasons for his emigration. In Hungary, he was strongly interested both in a political, and in an academic career. Initially, he had good chances, nice perspectives, but for various reasons, he was not very successful in either field. It became clear, in those years when Lukács was arrested for several months, that a serious political position was impossible for him in Hungary. His personal rivals were in good academic positions, so his chances at an academic career were very poor. But perhaps his personal motivations were the most important ones. With his second marriage, he found a family which accepted him. I think his strong desire for total, unconditional acceptance, the unquestionable assimilation into a community, were very essential aspects of his personality. Because the family wanted to emigrate, he had to go with them. He said to his friends that his emigration would be only a temporary circumstance. On the other hand, he escaped, in this way, from the consequences of the unhappy aspects of his political past: the forcing of the suicide in the forties, and his having served as an agent for the political police in the fifties.

2. LUKÁCS' INFLUENCE ON LAKATOS' PHILOSOPHY

On the basis of the above-described biographical and historical details, it is quite obvious that Lakatos' philosophy was strongly influenced by Lukácsian ideas. But sometimes it is not so easy to demonstrate what is obvious. In our case, one source of difficulty was that Lakatos himself usually did not declare that Lukács was his intellectual mentor. In that age, moreover, Marxist ideology was represented in Hungary by many authors, making the identification of the Lukácsian version no simple task. Lakatos, of course, studied other Marxist authors, as well; for example, he cited some works of József Révai, Marx himself, Lenin, Erik Molnár, Stalin, Béla Fogarasi, etc.

In the literature, the Hegelian aspects of Lakatosian ideas are discussed sometimes. It seems to be clear that the young Lakatos got acquainted with and studied the Marxist interpretations of Hegelian thoughts. Later on, perhaps, he was able to separate the interpretation from the original Hegelian ideas – I think that this was unimportant for his general outlook, however. Moreover, the Marxist interpretations of Hegel significantly differed in level, complexity and orientation. For Lakatos, Lukács, the author of the very Hegelian book *History and Class Consciousness,* doubtlessly played the crucial role.

In the recent literature, Val Dusek and John Kadvany have discussed the Lukácsian influence on Lakatos [4, 9, 10a, 10b]. Dusek, in his very interesting and inspiring paper, disclosed two groups of similarities. He noticed in the Lakatosian MSRP the "polemical defence of rationality against irrationalism and the readiness to condemn his opponents as irrationalists of the most extreme kind," which was an echo of the ideas in the late Lukács' *The Destruction of Reason.* Moreover, the central role of 'progress vs. degeneration' in the Lakatosian philosophy of science reminds of the Lukács' book. It seems to us that these relations undoubtedly exist – we will return to them below. Dusek also described another group of relationships: the specifically Lakatosian terminology, the concepts of "hard core" and "protective belt" recall phrases of the Leninist theory of party; and the sharp distinction between formal history and actual history was preferred by both philosophers. In our view, these are important and essential similarities, but they are less specifically Lukácsian ones. Of course, we can find these notions and ideas in his works, but this is the case in many other books and papers of that age. So, this is a direct influence of the political practice and ideology of their age.

John Kadvany stressed another aspect. He emphasized the Hegelian components of Lakatos' philosophy of mathematics, and depicted his *Proofs and Refutations* as a "mathematical *Bildungsroman*" which is modelled after Hegel's *Phenomenology of Spirit*. He also noted the appearance of Hegelian-Marxian dialectics in Lakatos' philosophy of science, which can be associated with his adaptation of the ideas in Lukács' *History and Class Consciousness.* In this respect, Kadvany observed a very interesting conflation of the concepts of "alienation" and "objectification" in Lakatos' *Proofs and Refutations,* which reproduced the "Hegelian mistake" of Lukács' *History and Class Consciousness.* In addition, he mentioned a public debate in England in the seventies, where Lakatos said that his original goal was to be a successor of Lukács.

Most parts of Kadvany's analysis are reasonable; moreover, they can be supported by some elements of the intellectual development of the young Lakatos. As was demonstrated above, Lakatos already knew very well Lukács' *History and Class Consciousness,* including its dialectics, in 1945, before he had met the author. He applied the notions and the basic view of this book in practically all of his early papers. Among others, Hegel's

Phenomenology of Spirit also plays an essential role in Lukács' book; furthermore, Lukács gave lectures on this work of Hegel for a few semesters during the years 1945 and 1946 [45], which Lakatos almost certainly must have attended. The only mention of a Lukácsian view in Lakatos' late works supports this: "the Cunning of Reason, as Georg Lukács, in his more optimistic moments used to say, arrives at the mountain top via a twisting, winding road and not via the steep and direct route."[11] In the thirties and forties, Lukács changed his opinion about the ontological aspects of the Hegelian view; so also did Lakatos have to re-think the positions of the young Lukács, when confronting the ideas of the older one.

The Lakatosian reproduction of Lukács' "mistake," i.e. the conflation of the concepts "alienation" and "objectification," can already be found in another version in his early papers [16, 22], where he stated that "natural things ... as the reificated forms of human relations obtain quantitative features ... the quantities of the reificated nature turn grey both nature and man in an imperialist way, and deprive them from all other features."[12]

Dusek and Kadvany recognized different important Lukácsian influences on Lakatos. It is very important, that these influences on the young Lakatos were preserved, and that they returned in his well-known later works. In the following, we try to summarize our view of the Lukács–Lakatos relationship. Two stages of this relationship will be distinguished: in the first period, Lakatos obtained from Lukács the theory for his political practice, while in the second period he obtained the political practice for his theory.

2.1. *Political practice of rational theory*

First of all, it is important to distinguish between the indirect and direct influence of Lukács on Lakatos. Indirectly, Lukács was a model for Lakatos, a model of the communist thinker, who can serve as politician *and* philosopher, who wants to bring into the practice revolutionary ideas, and who can suggest ideologically well-grounded approaches to the practical problems of the new society. In other words: the main question was how to find the right relation between practice and theory. In this respect, Lukács represented a very attractive alternative regarding the level, as well as the content. The young Lakatos could study this point in the personal life of Lukács, as well as in the activity of his intellectual "heroes" such as Lenin, Goethe, and Thomas Mann. Lukács analyzed and emphasized the theory-practice relation mainly in the political sphere and in art, fields in which he was able to understand, distinguish and represent the progressive and regressive aspects as the rational and irrational alternatives.

In accordance with the dominant aspirations of the age, the young Lakatos also wanted to unify theory and praxis, spirit and action, thought and life. Already in his second printed paper he cited Thomas Mann with whole-hearted enthusiasm: "Act as a thinker and think as an activist!"[13] Sometimes his aspirations met with success and sometimes they did not, but he tried to follow this idea throughout all his life. Born into an irrational world and, in spite of this background, attempting to live a life under the rule of pure rational decisions – this may have been a very important effort for the young Lakatos. He was described many times as a very rational person, without any sign of emotion. This characteristic became visible in his decision to sacrifice the "weakest link" of the secret group, and in his neutral report about it; furthermore, his unscrupulous determination to win in the debates, in his irony, etc. However, a follower of this attitude often, allegedly, already owns the truth, and does not search for it. The source of this type of truth and rationality is some personal, ideological or institutional authority. The young Lakatos accepted a radical, more or less dogmatic communist value system and followed it as the most rational approach to both personal and social problems. He considered these values to be rational, because he was convinced of their progressive character. In this period, Lakatos showed no interest in the search for rationality, but he applied an "abstract version" which, for him, was above question.

In this respect, Lukács' messianic Marxist book, *History and Class Consciousness,* was an important "theoretical background" for the young Lakatos. The above-mentioned early papers demonstrate, moreover, that the influence of other Marxist authors on his philosophical views was practically negligible. To give an interesting illustration, we can find there the name of Engels in only one, unimportant place. It is well-known, that in the *History and Class Consciousness,* Lukács criticized and rejected Engels' views on the dialectics of nature, in contrast to other Marxist authors of the time. Lakatos' implicit adoption of the Lukácsian criticism was obviously a conscious decision.

A further important aspect of the Lukácsian attitude is philosophizing in a political context by putting the problems into a world-historical perspective, and considering any kind of current problems from the point of view of class struggle and the progress of mankind. If the "current" morality confronts with these "eternal" political values, the former can be the loser. This situation can produce a special kind of personality – as somebody remembered: Lakatos was "the most immoral man I had ever met." In our view, the source of this attitude is a rigid and unconditional insistence on the standards of an abstract rationality in all fields of human activity. This preference of an abstract, theoretical rationality can be based on the strong belief in the historical progress of mankind, a belief maintained in spite of an irrational political and everyday practice. This practical and intellectual situation highlights, in an extraordinary way, the problem of how theoretical and practical rationality relate. '*How can we find the true relation of theoretical and practical rationality*' is a very central problem for Lakatos. This is *the* Lakatosian dilemma, a dilemma which is clearly and analogously expressed both in the political and the scientific sphere, in those fields which were of utmost importance to Lakatos. So it is very natural that he tried to interrelate the solutions of problems of the relationship between practical and theoretical rationality in politics and the sciences. In pursuing this line of thought, four periods of Lakatos' intellectual development can be distinguished.

1. *At the beginning of his political career*, between 1945 and 1947, Lakatos' radical political praxis and the ideas of *History and Class Consciousness* of the "young" Lukács were, more or less, harmonized with each other. The sectarian political practice and the utopian, messianic theory were similarly based on the absolute, indubitable possession of practical and theoretical certainty. Lakatos applied these ideas in his early philosophy of science, too. Dusek [4] mentions the similarities between the historicist, subjectivist interpretation of Kuhn, and some views of the *History and Class Consciousness.* Accepting this view a comparison of the philosophy of science of the young Lakatos and that of Kuhn, seems to be a promising task. However, in this period Lakatos had to confront the fact, that the current, the "real" Lukács had shifted to new philosophical and political ideas and, moreover, radically opposed his own earlier views. For example, Lukács sharply criticied the current bourgeois philosophy [34, 35], but he supported the alliance of the proletariat and the citoyens within the framework of a Popular Front policy. In this case, only the proletariat can be identified as the historical agent of the theoretical rationality, but in the practical rationality, i.e. in concrete political practice, the task is shared with its alliances. The early political papers of Lakatos [15, 19, 23–25] reveal that he theoretically accepted and practically propagated these political views. But he did not apply them in his philosophy of science – at least not at this time. This would happen only about 25 years later. In this way, between 1945–47, Lakatos' political philosophy and philosophy of science were essentially influenced by the "earlier" Lukács, but his political views were motivated (and moderated) by the "current" Lukács.

2. *During the next period*, between 1948 and 1950, Lakatos became more radical than Lukács in political practice. So a disharmony emerged; first, between Lakatos' own practice and his Lukácsian political thinking, and later between the two personalities: if Lakatos wanted to preserve the harmony between his practice and his thinking, he had to move away from Lukácsian ideas. During this period he opted for the primacy of political praxis and, from this position, he criticized the Lukácsian political philosophy and ultimately oriented himself towards other, more orthodox communist thinkers, such as Révai and Rudas. He abandoned the actual Lukácsian ideas, because he adopted a different vision of practical progressive alternative of social development and, consequently, of rationality. According to personal recollections, he considered Lukács' standpoint, at the time, to be 'nonsense.' In his criticism of Lukács, however, he preserved the Lukácsian background: the rational point is a decision in favor of progress – only Lakatos' definition of progress was changed and phrased in accordance with the "day-to-day" political rationality. In this period, he did not create a

theory of rationality, did not publish philosophical or scientific papers, but he tried to profit from his abilities in the struggles of party and state bureaucracy. He could study closely, and follow inventively, the logic of the dogmatic political practice of the age. But in the end, he became a loser in these struggles. Hence, he was able to experience both aspects of the workings of power: earlier, he had made decisions against others in the name of an abstract rationality, and later, he became a victim of the same mechanism. During this period, Lakatos maintained his goal of becoming a leading politician and thinker, and did a lot to this end, but his imprisonment made impossible, or at least postponed, the realization of these hopes.

3. *During the next period*, between 1953 and 1956, Lakatos was practically unable to take part in political life. But he tried to rethink the practice-theory relation in all the fields of politics, philosophy and the sciences (especially in mathematics). By this time, he had had enough experience of the political and social practices of fascism and the current socialism, and he obviously wanted to find a theoretical explanation of these developments. To this end, he had to place the concept of rationality into a more historical, individual and pluralistic context. As an employee of the Institute of Mathematics, he studied, besides his mathematical readings, the works of Lukács, Hayek, and Popper.

Lukács' *The Destruction of Reason* [36] appeared in 1954. In this book, the controversial political-practical consequences of abstract philosophical ideas were demonstrated, and the concepts of rationality and irrationality were interpreted in a socio-historical context. Both aspects were very important for Lakatos.

In the 1940s and 1950s, many philosophical analyses confronted the rise of the fascist ideology. Let us mention the works of Adorno and Horkheimer, Popper, and Lukács. In their book, *The Dialectics of Enlightenment,* Adorno and Horkheimer specified de-contextualized *rationality* as an essential source for the fascist value system. On the other hand, both Popper (in his *The Open Society and Its Enemies*) and Lukács (in *The Destruction of Reason*) considered the acceptance of – and adherence to *irrationality* as the basic element of fascist thinking. However, their concepts of irrationality (and rationality) were very different. Popper defended rationality against the followers of dialectics, Hegel, Marx, and against any historical or political contextualization. *Rationalism,* for Popper, is *an attitude of readiness to listen to critical arguments.* The rational and the irrational confront each other. Against the background of irrationalism, he found the domination of emotions and passions. However, the decision for rationality is based on an irrational belief [40, 13].

In Lukács' work there is another, absolutely opposite concept of the rational and the irrational. The problem of rationality and irrationality was a standard element of his worldview, which appeared in very different versions in his works. In 1919, for example, he considered it with regard to the relation of bourgeois vs. proletarian culture; in 1924, in *History and Class Consciousness,* he put the problem into a social context; around 1935, as a part of Popular Front politics, he identified two types of bourgeoises, the reactionary bourgeoisie and the citoyen, who was interested in progress and humanism; in 1943, in his Tashkent manuscript, entitled *How Did Germany Become the Centre of Fascism?* he applied his ideas to German culture; in 1947, in his papers and in the book *The Crisis of the Bourgeois Philosophy*, and finally in 1954, in his *The Destruction of Reason*. The final result of his efforts was a socio-historical contextualization of the problem. According to his final position, *the rational attitude is a decision in favor of progress*, the readiness to choose the progressive alternative, while irrationality can be considered as an inability to choose the progressive alternative. To elaborate: for a rational decision, we must apply the dialectics of Hegel and Marx, must consider the concrete social and political situation, must elaborate a careful and critical analysis of the (world)historical processes, and so on. This decision is based on our understanding and consciousness.

Lakatos knew perfectly these Lukácsian ideas. Aside from in his early studies on *History and Class Consciousness,* it is evident from his above-cited papers that, in the mid-forties, he followed the ideology of Lukácsian Popular Font politics. At the same time, he studied Lukács' works on the crisis of bourgeois philosophy. For example, he reviewed [19] a Lukács' paper that appeared in the journal *Társadalmi Szemle* [34]. (It is perhaps interesting, from the standpoint of the development of the Lakatosian views, that in this paper Lukács analyzed the details of the relationship between intuition and irrationalism.) From his Petőfi-circle talk in 1956, it is clear that Lakatos studied *The Destruction of Reason.* On the one hand, he cited its title in a very special context; on the other hand, his suggestions are in accordance with the tenor of this book. He decided in favor of progress (in the question of the training of scientists), he described this progress and declared the rationality of his proposals. In this talk, we can find thoughts which express some views of Popper and Hayek on empiricism, pluralism and personal freedom, but the Lukácsian view is dominant. This is very natural. In those days, Lakatos considered himself to be a socialist thinker, and the Lukácsian position had, for him, the clear advantage. On this basis, he was able to consider the true, progressive socialist practice to be rational, and to criticize the irrationality of the retrograde features of the earlier regime. In this way, he could again construct a unity of political practice and philosophical theory. In this "solution" of the problem, he continued to accept the identification of the notion of rationality as a "decision in favor of progress," but he studied the *methodology* and emphasized the significance *of the decision-making process.* As its important elements, Lakatos enumerated the individual, critical, and pluralistic aspects of the methodology of the decision process. This methodology is equally useful in the case of problem solving in mathematics, and in the case of the correction of the wrong political practices of socialism.

4. *During his fourth period*, in England between 1956 and 1974, Lakatos did not continue any practical political activity, but he preserved his interest in political matters. It seems to us, however, that his point of view changed in the sixties and, perhaps, the political changes in 1968 (the Czechoslovak events and the student movements in 1968–69 at LSE and other universities) caused a special kind of turn in his attitude.

In his early years in England, he was influenced by Popperian ideas. During this time, he moved away from Lukácsian views and accepted the Popperian attitude. As Feyerabend remembered, "Imre Lakatos ... attacked me in almost identical terms: 'Why did you say X when Popper says Y'."[14] This is probably valid in the case of the concept of rationality, as well. The Popperian concept of rationality – and especially the Popperian methodology, which supports the assertion of rationality – were acceptable to him, but in these years he preferred theory and disregarded political practice. However, later on Lakatos gradually worked out his own position. Feyerabend says, ironically: "Imre was a rationalist of sorts – at least that was how he presented himself, as a crusader for reason, law, and order. He travelled all over the world trying to encourage doubtful rationalists and to recommend his methodology as a nostrum."[15] For us, now, the results of this crusade would be interesting. Finally, his position moved far away from the Popperian conception. Listen again to his anarchist-contemporary: "I always felt that Imre's rationalism was not a matter of personal conviction, but a political instrument that he would use or put aside as the situation demanded."[16] Feyerabend was right. What he did not say, however, was that this was the Lukácsian aspect of Lakatos' position. This is the decision 'in favor of progress' depending on the situation, which has always a political dimension.

In our view, as Lakatos became more sensitive to the current political practice, he became more critical of Popperian ideas, as well, including Popper's concept of rationality. (His estrangement from Popper became significant from the last years of sixties.) Parallel to this, he returned to the earlier, well-oiled Lukácsian conception of rationality, as another representation of the relationship between theory and political practice. This return, of course, did not mean his return to communist ideology. His momentary political standpoint was sometimes outspokenly anti-Communist, but in contrast to these opinions, he accepted this very essential aspect of the political philosophy of Lukács. Moreover, he applied it to his philosophy of science.

2.2. Rational theory of political practice

Lakatos' philosophy of science (first and foremost his methodology of scientific research programs) can be considered as an abstract, theoretical representation of a special kind of political practice. This political practice had two typical forms, and different descriptions in Marxist political thought. Both of them were well-known to Lakatos, partly from his own practice, and partly from Marxist political theories, especially from Lukács' views and writings. In the Marxist political system, concrete *political practice* and abstract, theoretical values form a special kind of unity. It seems to us that for Lakatos, this system was the model of the scientific system, where the unity of *scientific practice* and theoretical rationality could be considered. Some kinds of parallelisms and interrelatedness between political and scientific progress were familiar to Lakatos from the beginning of his intellectual development. He had already faced a version of it, for example, in the works of his friend, Árpád Szabó, in Debrecen. Lukács' *History and Class Consciousness* provided a theoretical framework for these ideas. Lakatos took into account this interrelatedness in his last years, too. As Congdon cites Lakatos' opinion, "in England Lakatos remained convinced that there could be made an 'analogy between political ideologies and scientific theories'" [2]. The analogy is very clearly expressed, between political and scientific treatments of the relationship between theoretical and practical rationality, in the problem which was so essential for Lakatos all during of his life. The methodology of the production of the theory-praxis unity is socio-historically determined, and almost independent from the field of the human activity. The practical realizations of the "abstract," rational political values form the actual political history, and in the same manner the practical realizations of the "abstract," rational scientific values produce the actual history of science. However, "true" historical processes can be reconstructed by considering only the abstract aspects of the processes. So studying the working and development of the political sphere, we can describe some aspects of the scientific sphere.

The political practice in question is the communist practice of the years 1940–50 in Hungary. The organization of the party and the alliance policy of the party represent the two aspects of this practice. The party was organized according to so-called democratic centralism, which entailed a highly hierarchical organization. Under the influence of earlier illegal activity and the strong Stalinist environment, the center of the hierarchy became the representation of every important value (the right knowledge, the ability to make decisions, intellectual and material richness) and the peripheries lost their values, they became the simple operational organs of the center – although, the peripheral organs belonged to *their* respective centers. This organizational structure was multiplied on many levels of the political system, with the party as its central unit. In this way, any kind of criticism or attack against the political system reached at first only the "protective belt" of political power, so that the "hard core" of political power (or of the research program) could forbid the direction of the criticism at itself. On the other hand, this structure could penetrate into one's personality as well, and made it possible to sacrifice oneself (or somebody else) for universal revolutionary values. In these cases, individuals could serve as the "protective belts" of the "hard core" of the ideology.

The alliance policy of the party had a similar function. This policy was embodied by the organizational framework of the Popular Front. The "vanguard" of the proletariat was the party, which governed its struggle for the realization of more perfect social relations. In the organization of the Popular Front, the proletariat was the "vanguard" class and coordinated the struggle of its alliances – the peasantry, the intellectuals, and sometimes the citoyen strata of the bourgeoises – for the common purposes, for the more perfect social relations. The role of these alliances was clear: to defend the hard core of the political system as a protective belt. If everything turns out well, the alliances (the "auxiliary hypotheses") could survive the continuous attacks against the progressive forces and progress further, together with the hard core, but in the worst case they would sacrifice themselves for the progress of the hard core. As a result of the application of the political rules of the alliance policy, progressive, theoretical, abstract ideas could penetrate into the masses and the whole society could emerge to a higher level, and achieve a more rational form.

Lakatos had a clear picture of the structure of the socialist political system. However, there was no real "theory" of this practice. The elements of the theoretical descriptions were written mostly in various party and state rules, or were embodied in laws, traditions and morality. However, Lakatos knew and sometimes applied the different Lukácsian descriptions of some aspects of this system in his early papers, as has been demonstrated above.

On the other hand, Lukács suggested an understanding of progression and regression, which would put the problem into a socio-historical context. He was able to present a theoretical distinction between the progressive and regressive strata of the bourgeoisie, and to explain the emergence of German fascism by applying the concepts of rationality and irrationality. In 1956, on the basis of Lukács' explanation, Lakatos was able to identify the progressive and regressive aspects even of socialist practice. In this way, the Lukácsian theoretical description served as a valid explanation of some kind of social practice. The key element of his philosophical description was his special interpretation of rationality, as a decision in favor of progress. From this standpoint, the necessity of dialectics, and the significance of the social and historical aspects are evident consequences. Lakatos, when constructing his MSRP, applied all of these details of the Lukácsian philosophy.

Let us give a brief outline of Lukácsian elements in Lakatos' MSRP.

a) The socio-historical context: Lakatos considers research programs instead of theories or, as he said, "the problem of appraising historical series of theories, or, rather, of 'research programmes'... ."[17] A research program is a special kind of collection, in which theories are not isolated and independent, but interconnected entities. This view ensures the applicability of a simple social system analogy. The historical perspective was an essential aspect of the Lukácsian analysis of progression; however, it can also be associated with a Hegelian attitude. But a pure Hegelian view would be less useful in the consideration of theoretical and practical aspects of the programs, because of the determinant role therein of abstract reason.

b) The concept of progress: In the Lakatosian MSRP, the concepts of progression and degeneration play a fundamental role.[18] As Dusek mentioned, this can be a sign of his Marxist (Lukácsian) philosophy of history. The new research program will be chosen, if it is better from a theoretical, empirical and heuristic point of view. "One learns not by accepting or rejecting one single theory but by comparing one research program with another for theoretical, empirical and heuristic progress."[19] And "thus progress and learning are marked by instances verifying excess content rather than by falsifying instances."[20] This means that Lakatos applied the universal features of progress, but replaced them with the concrete aspects of scientific progress. It is worth mentioning that he distinguished between the concepts of empirical and theoretical (and heuristic) progress.

c) The concept of rationality: Perhaps this is the most important element of the Lukácsian influence. The Lukácsian concept of rationality (decision in favor of progress) appears in clear form many times in the MSRP, especially in the analyses of crucial experiments,[21] but even in the Lakatosian metacriterion of the methodology, as well: "We reject a rationality theory only for a better one, for one which, in this quasi-empirical sense, represents a *progressive shift*."[22] The Lakatosian progressive shift unifies the theoretical, practical (or empirical) and heuristic aspects of the progression; that is to say, the decision in favor of progress can somehow unify theoretical and practical rationality. In this paper, Lakatos underlined the significance and necessity of a *decision* about the demarcation criterion as a fundamental problem of the philosophy of science, and he sharply criticized Polányi, Kuhn, and Feyerabend because of the dangerous characteristics of their suggestions for this decision. His Toulmin study [30] has similar and, perhaps, much better expounded standpoints.

d) The alliance policy: The "protective belt" around the "hard core" of the program, and their functioning, are the perfect representations of the organization of party *and* Popular Front political praxis. In this respect, the negative heuristic expresses the ruler position of the "vanguard," while the positive heuristic expresses the operational abilities of the centers of power; of a middle-level state official in a ministry, for example, as Lakatos was. This element of the Lakatosian philosophy can be compared with his own political practice and, at the same time, with the Lukácsian Popular Front ideas. In the first case it can be stated, perhaps, that Lakatos' evil forcing of the suicide of the young girl in their illegal group was served with removal on a truly scientific level. However, the fundamental significance of the second relation seems to be more reasonable. Lakatos, in his early political papers on educational policy, had already discussed and expansively applied these political ideas to many different cases, so he had only to return to his own, old thoughts.

e) The historical progress: Applying these ideas, he was able to identify a more meaningful historical process in the case of science, as well; that is to say, "progress in the theory of rationality happens to be marked by historical discoveries or rediscoveries: by the reconstruction of a growing bulk of value-impregnated history as rational."[23] In this process, the progression of the unity of practice and theory can be considered and constructed.

As Congdon mentioned [2], in England Lakatos wanted to elaborate upon his political philosophy. Unfortunately, he could not do that, so we can study his systematic political thinking in its "unhappy" form, in his methodology of scientific research programs.

3. CONCLUSION

Lukács and Lakatos were very different philosophers. However, both of them were confronted with the socio-historical reality of Hungary in the first half of the 20th century. Their hope for the best solution to social problems led both philosophers to become Marxists. This decision was a fundamental constituent of their intellectual and practical life. Accepting the challenges of both theoretical and of practical tasks in the building of a new society, they were interested in political practice, as well as in the philosophical foundations thereof. Their Marxist attitude was eminently expressed in the methodology of understanding the relation between theory and practice. In this view, the (political, scientific, everyday, etc.) practice is primary acts as the source of experiences and as the final criterion for theoretical truths, but it would, in and of itself, be blind. These practices have to be based on and oriented by a theoretical system of (political, scientific, everyday, etc.) values. History (of politics, science, everyday life, etc.) is progress towards the realization of these value systems – mankind is the author *and* the actor of its own drama. In the elaboration of acceptable values, modern thinking prefers reason and consciousness, so there is a very intimate interrelatedness between the relations of theory-practice and rationality-irrationality. Therefore, these relations have to be considered and treated together in a complex system. A characteristic and important aspect of this complex problem is *the relationship between practical and theoretical rationality.*

The Lukácsian suggestion to decide in favor of progress ensures a unified solution to these problems. This progress is meaningful only in a (social, scientific, everyday, etc.) historical context. Choosing the progressive alternative, we decide using theoretical rationality and its practical realization, and see that the rationality of a theoretical value system can verify itself only in its actualizing practice. In other words, the decision in favor of progress is a common criterion of theoretical and practical rationality. On the other hand, progress also has practical and theoretical conditions. Our decision can lead to different practical and theoretical consequences, but we are living in the realm of rationality, under the universal domination of reason. The actual version of the unification of theoretical and practical rationality is a question regarding the concrete historical problem-situation.

Such a treatment of the problem of rationality differs markedly from many other approaches. Because of the essential and special role of practical progress (in his MSRP, Lakatos emphasized "empirically progressive" aspects of the programs) this view can be sharply distinguished from the Hegelian and Popperian views. Of course, the above "Marxian" approach also has many variants besides the Lukácsian one. This is a, more or less, Eastern European solution of the problem. In this underdeveloped region of the world, social practice is far from rational, but in progressive practice, the rational perspective can be seen, and it seems to be practically and theoretically an alternative which can be a choice for local progress. The unconditional necessities of this practice invoked from time to time messianic, realistic, progressive and degenerative theoretical attitudes and movements. During his development, Lukács formulated newer and newer versions of this problem. Unfortunately, there is no room here to follow the details of the analysis of these ideas.

It would be necessary to emphasize, that the Lukácsian "decision with the progress" is primarily a *methodological principle,* and does not directly state anything about the concrete content of progress. (Of course, as Lukács said, any relevant methodological problems have some consequences regarding content as well, but these consequences can be temporarily suspended.)

Lakatos accepted and, in practice, continuously applied this Lukácsian methodological principle in his (early and late) works and in his personal life, as well. As a clever young member of an outlawed community, he became sensible to social injustices and made a decision for practical progress, i.e. for the communist movement. He was able to find a strong theoretical fundament of his practical aspirations in the theory of historical materialism, mediated first of all by the book *History and Class Consciousness.* The concrete content of his political activity, the concrete realization of progress was a question of the concrete situation. Sometimes he was with Lukács, sometimes he was against him, depending on the actual content of their views on progress – but in every case, he remained grounded in the Lukácsian methodology. When he went to England his attitude became weaker for a while, but in the late sixties he returned to this methodology. During this period, the disharmony between his methodology and the contents of his (mainly political) statements had developed a significant and extraordinary character.

Lukács considered the general progressive (and regressive) versions of the relationship between practical and theoretical rationality (and irrationality) in philosophy, and studied their consequences mainly in the political and aesthetic sphere. The latter appeared in his political writings and in his theory of aesthetic realism. Faced with its fundamental significance, Lakatos analyzed their appearance in the fields of politics and sciences.

At the beginning of his development, he accepted the methodology *and* the most of the contents of Lukácsian ideas in the political and scientific spheres. Later, he disregarded the Lukácsian contents of this thinking, but accepted its methodological elements further on. In his following periods, Lakatos tried to involve some elements of the Popperian methodology into the discussion and apply them to his philosophy of science. In the end, he practically returned to the Lukácsian position, at least regarding his main methodological principles, and applied them to his MSRP. However, it was not a return to the contents of the Lukácsian progress concept and his communist value system, and he professed on many occasions his anti-Communist views.

In our view – at least with regard to methodology – Lakatos moved his whole life in a Lukácsian framework. At the beginning of his intellectual life, he accepted the position taken by Lukács in the mid-twenties and, after some twists and turns, finally "returned" to the young Lukács of the years following 1910 in Heidelberg. During that time, Lukács himself said that he tried to combine left wing ethics with a right wing epistemology. As a late recapitulation of this intellectual position Lakatos, in the late sixties and early seventies in England, tried to combine *right wing politics with a left wing epistemology*. Lukács later realized his own mistake.

Acknowledgements

I am very grateful to Alex Bándy, József Szigeti, Dénes Zoltai, László Sziklai, Éva Pap, Gábor Vajda, Márta Fehér, Gábor Palló, and Donald Gillies for the enlightening discussions on the topic. József Szigeti allowed me to read part of the manuscript of his intellectual autobiography, Dénes Zoltai gave me his copy of Magosh's book, and Jancis Long, Lee Congdon, Val Dusek, and John Kadvany enabled me to read their papers before publication. I would like to thank them for their help. Thanks also to Michael Stöltzner and Ladislav Kvasz for their encouragement during my work, and to Matteo Motterlini for his suggestions. Michael Stöltzner, John Kadvany and János Laki recommended many essential improvements to an earlier version of the paper. The work was supported by the Aktion Österreich-Ungarn.

Notes

1. Regarding the purposes of this paper, we found very useful the biographies written by Hermann [8] and Szerdahelyi [42a]. In addition to these books, we took into account Lukács' autobiographical sketch [37] and his works [39, 43, 44].
2. In this respect, we first of all learned (and later, somehow, interpreted) the many important data of Lakatos' life and activity from Jancis Long, Alex Bándy, József Szigeti, Dénes Zoltai,

Éva Pap, Gábor Vajda, Márta Fehér, Donald Gillies, György Litván, László Sziklai, Sándorné Kántor, and the participants of the two Lakatos workshops.

3. See [11], p. 228. Lakatos' photo appeared in the *Középiskolai Matematikai és Fizikai Lapok*, vol. 14, 1938, p. vii.
4. This is mentioned in the memoirs of his classmate in the Jewish Gymnasium, [38] pp. 95 and 108–109.
5. A mention of the first case can be seen in [38], pp. 95. and 147, while some remarks on the second one can be found on pp. 147–151.
6. The [21] reference reviewed very shortly the recent Hungarian publication of a popular book of that age.
7. [38], p. 150.
8. The written and oral accounts of this event were presented and treated in different versions by Long [31], Congdon [1], Freudenthal [6], Kántorné [12] and György Litván.

9a. Karácsony's opinion is cited in [12], p. 280.

9b. A letter from József Szigeti to L.R. (2/10/1999).

10. This view of Lakatos can be considered within different contexts. It probably reflected the general ideological, the concrete personal and, perhaps, even the scientific aspects of the problem as well. (Lakatos' second wife, Éva Pap, studied biology and his father-in-law was a researcher in maize development, where the relationships between "inherited" and "acquired" properties are a central problem. The famous Lysenko affair even emphasized these problems for general public, as well. These problems had pedagogical consequences, which were in accordance with the pedagogical suggestions of Lakatos' talk.)
11. [30], p. 238.
12. [22], p. 361.
13. [15], p. 88. From the context of the citation, it is obvious that Lakatos' views on Mann were based on the [33] Lukács study.
14. [5], p. 109.
15. Ibid., p. 129.
16. Ibid., p. 130.
17. [28], p. 318.
18. See, for example, [29], pp. 33–34.
19. [28], p. 320.
20. Ibid., p. 321.
21. [29], pp. 68–73, 88–93, etc.
22. [28], p. 322.
23. Ibid., p. 323.

REFERENCES

(Items from [14] to [27] represent – to the best of our knowledge – a complete bibliography of Lakatos' early Hungarian papers.)

[1] Lee Congdon, "Possessed: Imre Lakatos' Road to 1956", in: *Contemporary European History*, 6, 3, 1997, pp. 279–294. (I have seen only its Hungarian translation, "Bűn és büntetlenség. Az ismeretlen Lakatos Imre", in: *Replika* 29, 1998, pp. 7–23.)

[2] Lee Congdon, "Lakatos' Political Reawakening", in this volume.

[3] Imre Csécsy, "Szellemi elit és haladó értelmiség. Válasz egy bírálatra" (The Intellectual élite and the progressive intellectuals. Reply to a criticism), in: *Valóság* 2, 10, 1946, pp. 60–63.

[4] Val Dusek, "Brecht and Lukács as Teachers of Feyerabend and Lakatos: The Feyerabend-Lakatos Debate as Scientific Recapitulation of the Brecht-Lukács Debate", in: *History of the Human Sciences* 11, 2, 1998, pp. 25–44.

[5] Paul Feyerabend, *Killing Time*. Chicago: University of Chicago Press 1995.

[6] Gad Freudenthal, "A Note on the Life of Imre Lakatos in Occupied Hungary (1944)", in: Ceci N' EST PAS un festschrift pour Imre Toth (manuscript).

[7] Ágnes Heller, *Bicikliző majom. Élet és korrajz. Kőbányai János interjúregénye* (Bicycling Monkey. Her life and times in an interview-novel by János Kőbányai). Budapest: Múlt és Jövő Lap- és Könyvkiadó 1998.

[8] István Hermann, *Lukács György élete* (Georg Lukács' Life). Budapest: Corvina 1985.

[9] John Kadvany, "The Mathematical Present as History", in: *The Philosophical Forum* 26, 4, 1995, pp. 263–287.

[10a] János Radványi (John Kadvany), "The Extraterritoriality of Imre Lakatos. A Conversation with János Radványi", in: George Marcus (ed.), *Late Editions*. Chicago: University of Chicago Press (forthcoming).

[10b] John Kadvany, *Imre Lakatos and the Guises of Reason*. Durham, NC: Duke University Press, 2001.

[11] Sándorné Kántor, "Utószó" (Postscript), in: Imre Lakatos, *Bizonyítások és cáfolatok*, Budapest: Typotex 1998, pp. 225–235.

[12] Sándorné Kántor (Tünde Varga), "Lakatos Imre, a szintetikus modernizmus megalapozója" (Imre Lakatos, the founding father of synthetic modernism), in: *Debreceni Szemle*, 1998, 2, pp. 273–283.

[13] János Kelemen, "'Az ész trónfosztása' védelmében" (In defence of "The Destruction of Reason"), in: *Világosság* 27, 3, 1986, pp. 137–142.

[14] Imre Lakatos, "Molnár Erik: Dialektika" (Dialectics by Erik Molnár – Book review), in: *Valóság* 2, 3–5, 1946, pp. 75–78.

[15] Imre Lakatos, "Citoyen és munkásosztály" (Citoyen and the proletariat – Book review), in: *Valóság* 2, 6–9, 1946, pp. 77–88.

[16] Imre Lakatos, "A fizikai idealizmus bírálata" (A Criticism of idealism in physics – Book review), in: *Athenaeum* 31–32, 1945–1946, (appeared in 1946) pp. 28–33.

[17] Imre Lakatos, "Eötvös Collegium – Györffy Kollégium" (Eötvös College – Györffy College), in: *Valóság* 3, 2, 1947, pp. 107–124.

[18] Imre Lakatos, "Természettudományos világnézet és demokratikus nevelés (Válasz Csiszár Vilmos cikkére.)" (Scientific worldview and democratic education – A reflection on the Vilmos Csiszár's paper), in: *Embernevelés* 3, 2, 1947, pp. 63–66.

[19] Imre Lakatos (signed with only the monogram, I. L.), "A januári Társadalmi Szemle... – Lapszemle" (The January issue of the journal Társadalmi Szemle ... – A journal review), in: *Embernevelés* 3, 2, 1947, pp. 90–91.

[20] Imre Lakatos (signed with only the monogram, I. L.), "Jeges Károly: Megtanulom a fizikát" (I learn physics, by Károly Jeges – Book review), in: *Társadalmi Szemle* 2, 6, 1947, p. 472.

[21] Imre Lakatos (signed with only the monogram, I. L.), "John Hersey: Hirosima" (Hiroshima, by John Hersey – Book review), in: *Társadalmi Szemle* 2, 7–8, 1947, p. 615.

[22] Imre Lakatos, "Modern fizika, modern társadalom" (Modern physics, modern society), in: Gábor Kemény (Szerk.), *Továbbképzés és demokrácia*. Budapest: Vallás- és Közoktatásügyi Minisztérium nevelésügyi osztálya 1947, pp. 347–369.

[23] Imre Lakatos, "Huszadik Század. Társadalomtudományi és politikai szemle" (Twentieth century. Social scientific and political review – Critical review of the journal *Huszadik Század*), in: *Forum* 2, 1947, pp. 316–320.

[24] Imre Lakatos, "Vigilia" (Critical review of the journal *Vigilia*), in: *Forum* 2, 1947, pp. 733–736.

[25] Imre Lakatos, "'Haladó tudós' a demokráciában" (The "Progressive scientist" in democracy), in: *Tovább* 2, 8, 1947, p. 3.

[26] Imre Lakatos, "Az atomkorszak új betegségei" (New diseases of the atomic age), in: *Tovább* 2, 20, 1947. p. 13.

[27] Imre Lakatos, "Felszólalás a Petőfi kör pedagógusvitáján 1956 szeptember 28-án" (Talk at the pedagogy meeting of the Petőfi-circle on 28th of September 1956), in: András B. Hegedűs and János M. Rainer (Szerk.), *A Petőfi Kör vitái hiteles jegyzőkönyvek alapján VI. Pedagógusvita*. Budapest: Múzsák – 1956-os Intézet 1992, pp. 34–38.

[28] Imre Lakatos, "The Role of Crucial Experiments in Science", in: *Studies in History and Philosophy of Science* 4, 1974, pp. 309–325.

[29] Imre Lakatos, "Falsification and the Methodology of Scientific Research Programmes", in: John Worrall and Gregory Currie (eds.), *The Methodology of Scientific Research Programmes. Philosophical Papers. Vol. I.* Cambridge: Cambridge University Press 1978, pp. 8–101.

[30] Imre Lakatos, "Understanding Toulmin", in: John Worrall and Gregory Currie (eds.), *Mathematics, Science and Epistemology. Philosophical Papers. Vol. II.* Cambridge: Cambridge University Press 1978, pp. 224–243.

[31] Jancis Long, "The Unforgiven. Imre Lakatos' Life in Hungary", in this volume.

[32] Georg Lukács, *Geschichte und Klassenbewusstsein. Studien über Marxistische Dialektik.* Berlin: Malik Verlag 1923. English edition: *History and Class Consciousness: Studies in Marxist Dialectics.* Cambridge: MIT Press 1971.

[33] Georg Lukács, "Auf der Suche nach dem Bürger" (In Search of Bourgeois Man), in: *Internationale Literatur* 6–7, 1945, pp. 58–75. (Reprinted many times in different volumes, for example in his books: *Thomas Mann.* Berlin: Aufbau 1950, 1953, 1957 and *Essays on Thomas Mann.* New York: Grosset and Dunlap 1965.)

[34] György Lukács, "A polgári filozófia válsága" (The Crisis of Bourgeois Philosophy), in: *Társadalmi Szemle* 2, 1, 1947 pp. 25–39.

[35] György Lukács, *A polgári filozófia válsága* (The Crisis of Bourgeois Philosophy). Budapest: Hungária 1947, 1949.

[36] György Lukács, *Az ész trónfosztása.* Budapest: Akadémiai 1954. English edition: *The Destruction of Reason.* London: Merlin 1980.

[37] Georg Lukács, *Record of a Life: An Autobiographical Sketch*, edited by István Eörsi. London: Verso 1983.

[38] György M. Magosh, *Vadfa* (Wild-tree – Autobiography). Budapest: Múlt és Jövő 1994.

[39] George H.R. Parkinson, *Georg Lukács.* London: Routledge 1977.

[40] Karl Popper, *The Open Society and Its Enemies.* Vol. I-II, London: Routledge 1945.

[41] Imre Szász, *Ménesi út. Regény és dokumentum* (Ménesi Street. Novel and Document). Budapest: Magvető 1987.

[42a] István Szerdahelyi, *Lukács György* (Georg Lukács – A Biography). Budapest: Akadémiai 1988.

[42b] József Szigeti, *Intellektuális önéletrajzom* (Intellectual autobiography). Budapest: Eötvös József Könyvkiadó 2000.

[43] Victor Zitta, *Georg Lukács' Marxism. Alienation, Dialectics, Revolution.* The Hague: Nijhoff 1964.

[44] Dénes Zoltai, *Egy írástudó visszatér. Lukács György 1945 utáni munkásságáról* (Return of a Clerk. On Georg Lukács' activity following 1945). Budapest: Kossuth 1985.

[45] Dénes Zoltai, "Lukács – a nevelő" (Lukács as a pedagogue), in: *Doxa* 8, 1986, pp. 7–26.

Department of History and Philosophy of Science
Eötvös University
H – 1518 Budapest, Pf. 32.
Hungary

Lakatos with Tim Smiley, Dean Acheson, Nina Leather and others
(By kind permission of Éva Pap)

LEE CONGDON

LAKATOS' POLITICAL REAWAKENING

For approximately his first ten years in England, Imre Lakatos held aloof from direct involvement in politics, though he remained convinced that there was an "analogy between political ideologies and scientific theories."[1] A valid scientific theory might therefore pave the way for a responsible political ideology. But he was no hurry to explore that possible connection. After his shameful – or to his mind, misguided – activities as a communist in Hungary,[2] he was quite willing to limit himself to scientific work and the pursuit of women. Indeed, for Lakatos the two activities went hand in hand; in 1958, when his marriage to Éva Pap was failing, he confided to his father that he couldn't work without love.[3]

Lakatos searched for the affection he craved in affairs with students and, in due course, with Gillian Page, a devoted woman who was Director for Journals at Cambridge University Press. Eventually he bought a home in Hampstead and moved in with Page, but not before an extended period of doubt and hesitation. He was only half joking when, in 1968, he told his father about his romantic uncertainties. "I am very busy and have practically no time for my private affairs but I think I shall discard both Gillian and Éva [who wanted him back]. The trouble is that refutation of a [scientific] theory is only convincing when one has a new, better one; and in this respect women are like theories."[4]

Despite his professional and romantic preoccupations, however, Lakatos did enjoy some lighthearted political banter with his Hampstead neighbor George Lichtheim, a German Jew who learned English when, as a boy, he attended school in London for a period of two years. After Hitler came to power, Lichtheim returned to the British capital but soon moved on to Palestine, where he worked as an editor for the *Palestine Post*. Still drawn to Britain, however, he settled in London in 1946.[5]

A superb English stylist, Lichtheim soon made a name for himself as a contributor to intellectual journals on both sides of the Atlantic. He enjoyed commenting on international affairs, but as a Hegelian Marxist of an unorthodox sort he devoted ever more of his time to the study of socialist ideas. In 1961 he published what was to remain his finest work, *Marxism: An Historical and Critical Study*. The book's critical success brought him academic offers and during the 1960s he did agree to lecture at Stanford and conduct research at Columbia. By then he had met Lakatos, with whom he

G. Kampis, L. Kvasz and M. Stöltzner (eds.),
Appraising Lakatos: Mathematics, Methodology and the Man, 339–349.

shared not only a profound knowledge of Marxism but also a mischievous sense of humor. The tone of their correspondence may be suggested by an excerpt from a letter Lichtheim posted to Lakatos from California on September 8, 1965:

> Judging from the British papers, your Prime Minister Mr. [Harold] Wilson is not a success. This is due to lack of Marxist training. Why don't you join [Thomas] Balogh and the others, and do something about it? I am sending my dollar savings home, but it is really suicidal, the way things are going. It is the fault of [Nicholas] Kaldor and the other Hungarians, who should never have been admitted to England.[6]

In short, while the two Central Europeans forged a close friendship, they do not seem to have discussed politics in any very serious way. In fact, Lakatos scoffed at the idea that he and Lichtheim had formed a Marx study group.[7] It was therefore not until the student revolt, the so-called "Troubles," at the London School of Economics, that Lakatos discovered that he was "still excitable politically."[8]

The Troubles commenced during the summer of 1966, when a Selection Committee comprised of members of the LSE Court of Governors and Academic Board chose Walter Adams to succeed Sydney Caine as School Director. Adams had previously served as Secretary of the LSE and as Principal of the new interracial University College of Rhodesia and Nyasaland. There he had pursued an anti-discriminatory policy with respect to admissions while permitting residential segregation. He was, that is, an incremental reformer, hardly the sort of man to please the LSE Socialist Society (Soc Soc), the members of which were already disgruntled as a result of the Wilson government's insufficient militancy. The Society's magazine, *The Agitator*, quickly issued a report in which it charged the incoming Director with being soft on racial discrimination and on Southern Rhodesia's Unilateral Declaration of Independence (1965).[9]

On the day the *Agitator* report appeared, David Adelstein, President of the Students' Union, wrote to the Chairman of the Governors, Lord Bridges, to ask how much of the information contained in it was accurate. Bridges refused to enter into discussion or to violate the confidentiality of the appointment process, a refusal he repeated in a letter to *The Times*. When Adelstein requested permission to respond, in his official capacity, in the same forum, Director Caine refused it. But the youth went ahead with his plan anyway, and escaped punishment at a Board of Discipline trial.

Enter Marshall Bloom, an American graduate student who had had experience as a civil rights worker in Alabama. His ideas concerning "teach-ins" and "sit-ins" led to the calling of a "STOP ADAMS" meeting in the School's Old Theatre on January 31, 1967. But because the flyer announcing the meeting threatened "direct action," Caine refused the organizers permission to use School facilities. The student protesters then circulated a new leaflet announcing their intention to ignore the Director's decision. At the appointed time, Caine went to the Old Theatre to explain his position but the crowd of students quickly transformed itself into a mob and forced its way in. In the chaos that ensued, one of the School's porters collapsed and died.

The tough and intelligent Secretary of the LSE, Harry Kidd, instituted disciplinary action against Adelstein, Bloom, and four other students, but once again the Board of Discipline opted for leniency. It found the four lesser figures not guilty of violating School regulations and it suspended Adelstein and Bloom until the end of the summer term, in other words for about three months. And even that penalty was suspended after the two young men signed a statement signaling their intent, in future, to respect constitutional procedures.

Soon thereafter, Adams assumed his new responsibilities as Director of the LSE. The fall (Michaelmas) term passed without incident, but proved to be merely the lull before the storm. In February 1968, the Machinery of Government Committee, formed in 1966, released its Majority and Minority Reports. The former called for modest student representation on the Court of Governors and on the newly proposed "Council" (the financial authority) and "Senate" (the academic authority); it was accepted, with reservations, by two of the five student members of the Committee. The other three, one of whom was Adelstein, issued a long Minority Report in order "to challenge the assumptions and limitations of the report, to raise deeper problems, ones relating to educational aims."[10]

For Lakatos, those Reports were the last straw. In a letter to an old friend written a few months later, he confessed that he could no longer resist his "political instincts."[11] Indeed, as soon as he read the Reports he penned an open "Letter to the Director of the London School of Economics."[12] In it he took strong issue with the recommendation that students help formulate academic policy. Such an innovation would, he argued, violate the principle of academic autonomy, according to which academic policy was exclusively the responsibility of professors who had attained seniority.

> I came from a part of the world where this principle has never been completely implemented and where during the last 30–40 years it has been tragically eroded, first under Nazi and then under Stalinist pressure. As an undergraduate I witnessed the demands of Nazi students at my University to suppress "Jewish-liberal-marxist influence" expressed in the syllabuses. I saw how they, in concord with outside political forces, tried for many years – not without some success – to influence appointments and have teachers sacked who resisted their bandwagon. Later I was a graduate student at Moscow University when resolutions of the Central Committee of the Communist Party determined syllabuses in genetics and sent dissenters to death.

Lakatos insisted upon the principle of academic autonomy for the same reason that he focused on the internal history of mathematics and science, namely that it freed reason from external, and hence irrational, pressures and made possible the growth of knowledge. Well-trained and experienced men and women, in their capacity as scholars, embodied reason and pledged to search for truth in the way determined purely by science's inner life.[13] Those who stood outside the community of scholars and who wished scholarship to be subordinated to some goal – usually political – external to it introduced irrationality into science and in so doing forced it to deviate from the path of knowledge and truth. The Lysenko case in the Soviet Union was only the most egregious example of what Lakatos had in mind.

This argument brought Lakatos very close – closer than he seemed to recognize – to his countryman Michael Polanyi, an eloquent defender of pure scientific research. At the same time it distanced him even further from his close friend, Paul Feyerabend, who was capable of writing the following: "It often happens that parts of science become hardened and intolerant so that proliferation must be enforced from the outside, and by political means. Of course, success cannot be guaranteed – see the Lysenko affair. But this does not remove the need for non-scientific controls on science."[14] Based upon his experience of Nazi and Soviet rule, Lakatos knew where Feyerabend's argument was likely to lead.

Similarly, he knew that the authors of the Majority Report were far too sanguine about radical student aims. In a *Times* piece of March 18, 1968, Adelstein himself had proclaimed that student representation on the School's governing bodies was merely a first step. "The *next thing*," the former Students' Union President declared, "is for students to begin to run their own courses, initially through their own societies, and then to demand that they should run a particular part of a course: its content, how it is taught, and who teaches it."[15] With his own, communist-inspired, destruction of Eötvös College (Budapest) clearly in mind,[16] Lakatos warned that radicals like Adelstein, once they became members of the Council or Senate, would employ what Hungary's Stalinist dictator Mátyás Rákosi had called "salami tactics" – one slice of academic autonomy at a time until it disappeared entirely.

What the radicals really wanted, Lakatos concluded from personal experience in Hungary, was not the right to criticize, a right that they should and did enjoy, but the winning of power that would enable them first to destroy universities as centers of learning and then, in their place, to create centers for the dissemination of social and political propaganda. The Director and the Academic Board would be well advised, the Hungarian concluded, to reject both the Majority and the Minority Reports.

Initially, Lakatos was not very optimistic about the cause he defended. "They yield to Students' Power in a cowardly and unreasonable way," he wrote to his father, "and I am trying to turn the tide. I am very sceptical about the success." But only two months later, in June 1968, he was able to report that his "Letter" was circulating in some 1500 copies and had nearly become official policy.[17] He could claim some credit, therefore, when, in the fall, the LSE Academic Board passed a motion that "the responsibility on behalf of the School for the determination of matters involving general academic standards must rest, and be seen to rest, entirely with the academic staff of the School."[18]

By that time, however, the LSE had entered a new and more perilous time of Trouble. It began when student radicals demanded that School buildings be open to protesters who would be in London on Sunday, October 27 to demonstrate against the war in Vietnam. Fearing violence, the Court of Governors authorized Director Adams to post notice that School buildings would remain closed that day. The ball was then in the court of the radicals, led by two Americans, Vic Schoenbach and Paul Hoch. Hoch was actually enrolled at Bedford College, and as Professor David Wiggins has recalled, he was never able to state clearly what it was that he wanted.[19] Whatever it was, he and Schoenbach were not satisfied when students and protesters defied Adams' order without suffering any consequences.

A month after the weekend "occupation" of School buildings, Lakatos wrote to his father. "The students," he said, "are not as good, I am afraid, as you think. Frankly, I became rather conservative and although I think that the world is bad as it is but the students could make it even worse."[20] Shortly thereafter he informed a Hungarian friend that "the London School of Economics (LSE for short) is ceaselessly on the front page of the daily newspapers, because university infighting has a sharper edge here than anywhere else in England; and I have not been able to resist drawing upon my old [communist] party experiences."[21] He was, that is, prepared to go for the jugular, particularly after the events of January 1969.

On January 17, Lord Robbins, the distinguished economist who had replaced Lord Bridges as Chairman of the Court of Governors, addressed the LSE Students' Union. He did not meet with a friendly reception. Radical students shouted hostile questions concerning Adams' past and the investment of School funds in companies that traded in South Africa. Robbins attempted to answer but the radicals were not interested; as Hoch and Schoenbach later boasted, they could smell blood.[22]

Lakatos' colleague, John Watkins, remembered what happened next. "When Robbins left the building, nearly everyone else did too. The next item on the agenda was student accommodation; there was no mention of the 'gates' (that had been installed at strategic locations in School buildings in an effort to prevent occupations). However the Student Union's constitution allowed emergency resolutions to be introduced by a two thirds majority. And the now poorly attended Union meeting was manipulated by Soc Soc members into passing (a motion demanding that the gates be removed within seven days)."[23] Failing that, the students threatened to take matters into their own hands.

Seeking a compromise, Adams agreed to remove some of the gates, but that only emboldened the students, who knew irresolution when they saw it. On January 24, the Union voted for "direct action" and members set out to destroy the gates. Some faculty members, Watkins included, attempted to block their way, but they were pushed aside so that the sledgehammers and pickaxes could accomplish their purpose. Too late, the police arrived, the vandals scattered, and Adams declared the School closed indefinitely.

On February 2, Lakatos brought his father up to date. "This is only a short note," he began, "to say that the LSE has become the centre of new vandalism in Britain. It is now closed down. The Parliament discussed for hours the LSE and for two weeks it has been in the headlines each day. It is all very depressing and we do not know whether the University will survive or not. This is a Fascist movement with Communist slogans and their first aim is to destroy the Universities as centres of learning and turn them into strongholds of a social revolution."[24]

Determined to save the LSE from a fate similar to that he had imposed upon Eötvös College, Lakatos had, three days earlier, placed a motion before the School's Academic Board. It read as follows:

> This Board recognizes that it is the confessed aim of certain revolutionary groups to transform the L.S.E., if necessary by force, from an institution of learning into a centre of social revolution; that they have made shrewd use of any pretext to disrupt scholarly and educational work and to destroy the confidence between staff and students; that the damage already inflicted on the L.S.E. is catastrophic; and that the very survival of the L.S.E. is now at stake. Therefore this Board calls upon its constitutional authorities to use any means open to them under the constitution of the School and the law of the land to restore L.S.E. by sustained and determined efforts to its former status.[25]

The motion was seconded by the conservative political philosopher Michael Oakeshott.

The LSE remained closed for 25 days, but Lakatos had every right to believe that he had helped to stiffen the authorities' backbone. In the end they adopted strong measures, including the issuing of injunctions against 13 student radicals. By November Lakatos could report to his father that the student revolt was over, but he added darkly that "the government now tries to do to the universities all the harm that the students had intended: namely to turn universities into government policy making agencies and consultative organs in their day-to-day problems and destroy them as centres of independent research. So this is the next political battle."[26] All along, in fact, that had been his principal fear. In his open "Letter" to Walter Adams he had warned that "there are no arguments for Student Power that would not be arguments also for Government Power."[27] And government power, he knew from experience, would be far more difficult to defend against.

Thus, despite the subsiding of the student revolt, Lakatos remained pessimistic about the future of English education – and of English politics. In letters to his father he complained often and bitterly about the leftist direction of England's policy. "Here the Conservative Government [of Edward Heath] is as a matter of fact Social Democrat," he wrote in November 1973. "The Labour party swung to the ultra left." A month later he charged that the trade unions were dominated by communists "whose only aim is to destroy the country and the trade unions dominate the Labour party. The Conservatives are weak-kneed, lack good leadership and God only knows where we end up. The immediate question is how to survive 1974."[28]

Having been led by the student revolt to turn his attention back to politics after a self-imposed retreat that had lasted several years, Lakatos did not confine his interest to domestic affairs. He was naturally fascinated, for example, by Alexander Dubček's attempt to give Czechoslovak socialism a "human face." To his colleague at the LSE, the Czech-born Ernest Gellner, he wrote that "the question is whether the Russians succeed in preventing the Czech attempt to refute the thesis that socialism leads to serfdom ... My congratulations for being a Czech these days."[29]

The answer to Lakatos' question came during the night of August 20–21, when Soviet forces, joined by their Warsaw Pact allies – Romania excepted – invaded Czechoslovakia. The day following Lakatos wrote to Karl Popper. "The third invasion of Czechoslovakia, I am afraid, may be no less ominous than its invasion in 1939, and the [Communist] putsch in 1948. I find it difficult now to concentrate on induction."[30] And so it was. Indeed, the Soviet invasion of Czechoslovakia led him to exaggerate the geopolitical threat posed by the U.S.S.R. His letters from the early 1970s return again and again to that threat and to the central role the United States had to play in confronting it.

To his considerable irritation, most of his friends were far more ready to denounce U.S. actions in Vietnam than they were Soviet repression in Eastern Europe. To Paul Meehl, a psychology professor at the University of Minnesota, he tried to clarify his position: "I regard Anglo-Saxon military power, even if partially evil, as the main and only safeguard of the world from some new Middle Ages." Then he added a word of explanatory self-analysis. "You must remember that most of my family and childhood friends died in gas chambers and too many of my later friends were hanged, tortured and/or ruined by the *Communists*."[31]

When, in 1972, George McGovern won the Democratic nomination for president, Lakatos began to have apocalyptic visions. "Like you," he wrote to his father, "I am worried about politics. I am of course not very enthusiastic about Nixon but at least he is not a disaster like McGovern would be. I am convinced that if McGovern becomes president the Russians will turn Europe into Vietnam. Unfortunately people forget how Russia occupied Eastern Europe in the 1940s and think that Western Europe is different from Eastern Europe. What happened there in the 1940s could happen here in the 1970s. It is frightening that if one says this here people believe that one is mad." More reasonably–and more presciently–he added that "if one could win a couple of decades then one may escape the Soviet danger because, after all, sooner or later if they can't make war and can't expand, they will have some major internal trouble."[32]

When Nixon won a landslide victory in November, Lakatos heaved a sigh of relief. And yet he brooded over America's ability to lead the Western world, not least because he had gained first-hand knowledge of the country as a result of visiting professorships at the University of California, San Diego (1964–65) and Boston University (1973). Indeed, in the fall of 1973, he witnessed a nation that seemed to be coming unraveled as a result of Watergate. From Princeton he wrote to his father concerning the "mad Decline of the American Empire" and characterized the unrelenting attack on Nixon as a "witchhunt."[33]

For the beleaguered president's enemies, Lakatos displayed nothing but contempt. "The American press," the Hungarian master of invective told his father, "is a witchhunting crowd of crooks and liars. I know Archibald Cox from Boston, he is a common crook. Sam Erwin is senile and corrupt. [John] Sirica is regarded as a legal idiot and he is the judge whose sentences are most frequently reversed in higher courts. All these liberal heroes are common gangsters. Unfortunately Nixon, while his policies are good, is exactly the type who is bound to become the victim of a witchhunt."[34]

That was the kind of angry attack that one could make only in a private letter, and Lakatos knew it. By the time he wrote it, he had already begun to make plans to write a more carefully-reasoned political philosophy, a subject he urged his departmental colleagues to revive at the LSE. In a 1970 letter to Kay (Mrs. Thomas) Kuhn, for example, he expressed his wish to discuss with her scholar-husband "problems not only in the history/philosophy of science but problems of politics and Weltanschauung" as well. And in a 1972 letter to his friend and Boston University colleague Marx Wartofsky, he wrote that he had some work to complete in the philosophy of mathematics and the philosophy of science. "And then I still hope to have enough energy left for a spell of political philosophy."[35]

Lakatos died suddenly on February 2, 1974 without having written his philosophy of politics. There is no doubt that he would have rejected any philosophy originating on the far left, but that does not mean that he would have embraced the right; "unfortunately," he wrote to a friend in 1968, "many people regard me as one of the pillars of the right wing."[36] For the Hungarian, the quintessential conservative political thinker was his LSE colleague Michael Oakeshott. He knew that Oakeshott's views on politics consciously paralleled Polanyi's on science;[37] and the latter, in his view, made of science an irrational enterprise. Both Polanyi and Oakeshott maintained that methodological rules could not be formally and exhaustively stated, that they were tacitly known and embodied in a tradition of practice. Hence in the case of politics, only those steeped in a tradition of political practice – only an elite – could be counted on to govern well. Those, Oakeshott argued, who insisted upon a "rational" politics invariably attempted to reconstruct society on the basis of allegedly clear and universal principles. Just as certainly the attempt led to the suppression of freedom.

Lakatos entertained a good bit of respect for Oakeshott's philosophy but he could not, in the end, overcome his objection to the idea that an elite possessed a privileged knowledge that could not be criticized on the basis of abstract principles. "Only a privileged élite," he wrote in 1973, "has the craft of politics. All élitists lay great stress on the inarticulable, on the "tacit dimension" of science.[38] But if the 'tacit dimension' plays a role in normative appraisal, the layman obviously cannot be a judge. For the tacit dimension is shared and understood (*verstanden*) only by the élite. Only they can judge their own work. Thus in this tradition we have a combination of élitism and of a cult of the unarticulated and, indeed, of the inarticulable."[39]

Lakatos was too wedded to the idea that an *impersonal* critical rationalism was the motor of progress to accept Oakeshott's conservatism. In formulating his own political philosophy, he intended to adapt the methodology of research programs that he had championed in the philosophy of science. He would have eschewed a "degenerating" program such as Marxism, because it had never successfully predicted a novel fact. "Never!," he wrote a year before his death. "It has some famous unsuccessful predictions. It predicted the absolute impoverishment of the working class. It predicted that the first socialist revolution would take place in the industrially most developed society. It predicted that socialist societies would be free of revolutions. It predicted that there will be no conflict of interest between socialist countries."[40]

Almost certainly, Lakatos would have selected an alternative Enlightenment research program, one that he believed was mercifully free of dogma and hence capable of predicting novel facts and of solving political problems by piecemeal engineering: the slightly left of center "Open Society" liberalism of his quondam friend Karl Popper. Ernest Gellner recognized this when he referred to Lakatos as "the Zhdanov of the Open Society."[41] Open Society liberalism stood in sharp contrast to the dogmatic and closed Marxism of his youth. Political ideologies of the latter sort, he wrote in a revealing note to *Proofs and Refutations*, "which first may be debated (and perhaps accepted only under pressure) may turn into unquestioned background knowledge even in a single generation: the critics are forgotten (and perhaps executed) until a revolution vindicates their objections."[42] He had in mind, of course, the Hungarian Revolution of 1956.

NOTES

I am grateful to Dr Mary Lincoln, Director of Northern Illinois University Press, for allowing me to republish material that first appeared in my book, *Seeing Red: Hungarian Intellectuals in Exile and the Challenge of Communism* (2001).

1. Imre Lakatos, *Proofs and Refutations: The Logic of Mathematical Discovery*, eds. John Worrall and Elie Zahar (Cambridge: Cambridge University Press, 1976), p. 49n.
2. On Lakatos' years in Hungary see my article: "Possessed: Imre Lakatos' Road to 1956", *Contemporary European History*, VI, 3 (1997), 279–94, and the article by Jancis Long in this volume.
3. See Jacob Lipsitz's letter to Imre Lakatos, March 12, 1958, Imre Lakatos Papers, File 13/265, British Library of Political and Economic Science (hereafter BLPES), London, England. I am grateful to Professor John Worrall for permission to quote from the Lakatos Papers.
4. Imre Lakatos' letter to Jacob Lipsitz, May 18, 1968, Imre Lakatos Papers, File 13/266, BLPES.

5. For details of Lichtheim's life I have relied upon Walter Laqueur, "George Lichtheim, 1912–1973", *Commentary*, LVI, 2 (1973), 45–52.
6. George Lichtheim's letter to Imre Lakatos, September 8, 1965, Imre Lakatos Papers, File 13/545, BLPES.
7. See Eugene Kamenka's letter to Imre Lakatos, November 1, 1961, and Imre Lakatos' letter to Eugene Kamenka, January 8, 1962, Imre Lakatos Papers, File 12.1, items 2 and 3, BLPES.
8. Imre Lakatos' letter to Jacob Lipsitz, October 23, 1968, Imre Lakatos Papers, File 13/266, BLPES.
9. For this account of the Troubles I have relied heavily upon Harry Kidd, *The Trouble at L.S.E., 1966–1967* (London: Oxford University Press, 1969) and Ralf Dahrendorf, *LSE: A History of the London School of Economics and Political Science, 1895–1995* (Oxford: Oxford University Press, 1995), pp. 443–75.
10. Cited in Dahrendorf, *LSE*, p. 437.
11. Imre Lakatos' letter to Vilma (Mrs. Tibor) Balázs, November 8, 1968. Personal possession of Mrs. Balázs. I am most grateful to Mrs. Balázs for permission to quote from this and other letters to her from Lakatos.
12. Imre Lakatos, *Mathematics, Science and Epistemology. Philosophical Papers*, II, eds. John Worrall and Gregory Currie (Cambridge: Cambridge University Press, 1978), pp. 247–53.
13. Ibid., p. 258.
14. Paul Feyerabend, *Against Method: Outline of an Anarchistic Theory of Knowledge* (London: Verso, 1978 (1975)), p. 51.
15. Adelstein, cited in Lakatos, *Mathematics, Science and Epistemology*, p. 248.
16. On this matter see my "Possessed", 284–87.
17. Imre Lakatos' letters to Jacob Lipsitz, April 6, 1968 and June 27, 1968, Imre Lakatos Papers, File 13/266, BLPES.
18. Cited by John Worrall and Gregory Currie in Lakatos, *Mathematics, Science and Epistemology*, p. 253n.
19. Personal conversation with Professor David Wiggins, London, October 1995.
20. Imre Lakatos' letter to Jacob Lipsitz, November 27, 1968, Imre Lakatos Papers, File 13/266, BLPES.
21. Imre Lakatos' letter to Vilma (Mrs. Tibor) Balázs, December 18, 1968.
22. Dahrendorf, *LSE*, pp. 464–65.
23. Personal communication from Professor John Watkins, November 5, 1997.
24. Imre Lakatos' letter to Jacob Lipsitz, February 2, 1969, Imre Lakatos Papers, File 13/266, BLPES.
25. Imre Lakatos Papers, File 11/19, BLPES.
26. Imre Lakatos' letter to Jacob Lipsitz, November 1, 1969, Imre Lakatos Papers, File 13/267, BLPES.
27. Lakatos, *Mathematics, Science and Epistemology*, pp. 249–50.
28. Imre Lakatos' letters to Jacob Lipsitz, November 11, 1973 and December 23, 1973 (the latter was misdated 1974), Imre Lakatos Papers, File 13/268, BLPES.
29. Imre Lakatos' letter to Ernest Gellner, July 17, 1968, Imre Lakatos Papers, File 13/314, BLPES.
30. Imre Lakatos' letter to Karl Popper, August 22, 1968, Imre Lakatos Papers, File 13/736b, item 131, BLPES.
31. Imre Lakatos' letter to Paul Meehl, July 13, 1971, Imre Lakatos Papers, File 12.1, item 11, BLPES.
32. Imre Lakatos' letter to Jacob Lipsitz, June 30, 1972, Imre Lakatos Papers, File 13/268, BLPES.
33. Imre Lakatos' letter to Jacob Lipsitz, October 28, 1973, Imre Lakatos Papers, File 13/268, BLPES.
34. Imre Lakatos' letter to Jacob Lipsitz, November 5, 1973, Imre Lakatos Papers, File 13/268, BLPES.
35. Imre Lakatos' letters to Kay Kuhn, February 13, 1970, Imre Lakatos Papers, File 12.5, item 124, and to Marx Wartofsky, March 30, 1972, File 12.1, item 12, BLPES.
36. Imre Lakatos' letter to Vilma (Mrs. Tibor) Balázs, November 8, 1968.

37. See, for example, Michael Oakeshott, *Rationalism in Politics and Other Essays* (Indianapolis: Liberty Press, 1991), p. 13n.
38. See Michael Polanyi, *The Tacit Dimension* (Garden City, New York: Anchor Books, 1967).
39. Lakatos, *Mathematics, Science and Epistemology*, p. 227.
40. Imre Lakatos, *The Methodology of Scientific Research Programmes. Philosophical Papers*, I, eds. John Worrall and Gregory Currie (Cambridge: Cambridge University Press, 1978), pp. 5–6.
41. Ernest Gellner, "The Last Marxists", *Times Literary Supplement*, September 23, 1994, p. 4. Andrei Zhdanov was Stalin's cultural enforcer in the years immediately following World War II.
42. Lakatos, *Proofs and Refutations*, p. 49n.

Department of History
James Madison University
Harrisonburg, Virginia 22807
USA

IV.

Imre Lakatos' Hungarian Dissertation

A Documentation Arranged by Gábor Kutrovátz

Imre Lakatos' Hungarian Dissertation

Imre Lakatos' philosophy of science is rooted in a number of different fields, not all of them purely scientific. During his years of education, he was influenced by mathematics and natural sciences as well as by philosophy, but the role of political ideologies also cannot be denied. His basic philosophical ideas – such as the rationality of science, the continual growth of knowledge, the social determinism of scientific activities, and the indispensable role of historical attitudes in the philosophy of science – are definitely in accordance with his early devotion to Marxism (and Lukács' philosophy), both in theory and in practice.

One can easily find clear evidence that Lakatos saw basic connections between the theoretical sciences he studied and the practical principles he followed in politics. This is demonstrated by the early papers he published in various journals, and it must have played an important role in the doctoral dissertation he wrote in 1947. Unfortunately, not a single copy of this dissertation can be found now. There are several theories as to when and why the paper disappeared, but most probably Lakatos himself "stole" it some time before leaving Hungary in 1956. Later on, he hinted several times that he was rather unsatisfied with it, regarded it as "immature," and he also said that he would not have minded if nobody had ever seen it. After many failed attempts at finding it, we believe that the dissertation has been lost for ever.

Fortunately, we are not left without traces of the contents of this work, because important parts of it were already published while it was being written. In this documentary section we publish the first English translation of a paper that was an important part of the lost dissertation and provide an overview of the archival records found at Debrecen University.

Imre Lakatos took his doctoral examination on July 1st, 1947. According to the records at the Archive of the University of Debrecen, he answered the following questions:

Philosophy:
Scientific conceptualisation of pre-Socratic philosophers.
Modern times: the intrusion of social and historical concepts into scientific concepts.
Descartes system of scientific concepts.
Optimism in the rising – and pessimism in the declining – capitalist literature.

G. Kampis, L. Kvasz and M. Stöltzner (eds.),
Appraising Lakatos: Mathematics, Methodology and the Man, 353–374.

Physics:
Conservation of mass. Materialization, annihilation.
The experimental basis of relativity theory.
The mechanistic world view and causality.
Quantum theory and natural philosophy. Elements of Schrödinger's theory.

Mathematics:
Different (and still identical) definitions of complex functions and their analyticity.
Isolated singularities and related theorems. The principle of analytic continuation.
Measure theory of point sets according to Jordan: Lebesque-type generalization.
The notion of measurable functions. Blaire-type function classes.

He received the grade of "excellent" for every answer.

His doctoral dissertation was entitled *A természettudományos fogalomalkotás szociológiájáról* [On the Sociology of Concept-building in the Natural Sciences]. The dissertation had two opponents: *Sándor Karácsony*, who was a very popular philosophy professor at Debrecen University at the time; he was also the most influential instructor of Lakatos (besides the later world-famous *Árpád Szabó*). The other opponent was *Ottó Varga*, professor of mathematics.

We have two evaluations from Karácsony and one from Varga which are provided in the appendix. They give some clue on the content of the dissertation as a whole, but do not suffice to reconstruct the argument. In Karácsony's second evaluation, however, we find the following remark:

> His dissertation is not based on a sudden idea; it matured through two previous publications, both published in very serious journals: the first in *Athenaeum*, under the title "Physical Idealism," and the second in a thick volume written for educators, *Further Education and Democracy*, entitled "Modern physics, modern society."

Now, if we compare the two papers mentioned by Karácsony as forerunners of the dissertation, we come to see that the essential body of the earlier one (*Criticism of Physical Idealism*) is almost literally identical to a great part of the longer paper (*Modern Physics, Modern Society*). The small differences are either stylistic or explanatory, since the journal *Further Education and Democracy*, an ideological collection of writings for teachers (published by the Ministry of Religion and Education), served more popular purposes than the rather scientific *Athenaeum*, the journal of the Hungarian Academy of Sciences and the Hungarian Philosophical Society. Hence, it is most likely that this text contains most of young Lakatos' essential thoughts and ideas concerning the position, the development and the function of science.

Criticism of Physical Idealism is a critical essay discussing *Susan Stebbing*'s book, *Philosophy and the Physicists* (London, Pelican, 1943). Lakatos, however, criticises not only Stebbing's analyses of Eddington's and Jeans' idealism, but he also adds his own criticism of the two scientists' world views, which he considers to be typical examples of "bourgeois" science. Instead of focusing on the immanent development of science, he decides to look for explanations outside of science. He emphasises the indispensable role of sociological and economic influences on scientific concept building, and he concludes that the world view of a given scientific age or community is nothing more than a historical category. The whole argumentation appears again in *Modern Physics, Modern Society*, supplemented by some further ideas and more loose associations; the context becomes broader and the investigations more fundamental. Here, we are given a deeper (Marxist and Lukacsian contra Hegelian) analysis of the "dialectical structure" of the modern scientific view, determined by social relations and motions. And if we imagine going further in this direction, then we will surely come very close to the text of the lost dissertation.

EDITORIAL NOTE

A complete translation of the paper, *Modern Physics, Modern Society* is given here. Footnotes marked with numerals are those of the author; comments and specifications of the translator are put in square brackets. Further references not given by Lakatos are specified in footnotes marked with asterisks. We also indicate the beginning and the end of the part of the text which is common to the two papers mentioned above. All footnotes in the three evaluations of Lakatos' thesis were added by the editor.

I would like to thank Dezső Gurka for hunting down the original documents concerning the dissertation, László Ropolyi for helping to provide the information contained in the notes, and Michael Stöltzner for a critical reading of the whole section.

IMRE LAKATOS

MODERN PHYSICS, MODERN SOCIETY

When examining the significance of modern physics in the history of sciences and of humankind in general, we must first and foremost emphasise three aspects:

1. the exploration of nuclear energy as a productive force;
2. the considerable development of newer physical terminology in terms of epistemology and logic;
3. the preparation and the establishment of large scale working-forms of natural sciences, and of their planned co-operation.

Let us examine these issues in more detail.

1. If heat and electric energy multiplied man's power over nature, created an artificial environment on the ever smaller surface of Earth, and tore out a sizeable piece of free time from the time needed for the constant direct war against nature, then, with proper technical utilisation, these results can be increased incredibly by the deliverance of nuclear energy: It disconnects the remaining "natural" elements from the picture of the Earth's surface (hence, it finally eliminates the dependence of industrial plants on coal source areas, waterfalls, etc.), it makes our planet a narrow homeland (leaving the Earth becomes possible for us), and it soon eliminates almost entirely the direct war against nature (the required labour time dropping to a few minutes a day).

But "by thus acting on the external world and changing it, he [man] at the same time changes his own nature. He develops his slumbering powers and compels them to act in obedience to his sway."[1] Hence, human nature is changed *primarily* not by "spiritual renewals," nor by "re-education," but by transformation of the mode of production. Therefore, there is no doubt that the whole habit of human personality will be changed in the nuclear era.

Heat and electric energy played an indispensable role in the shaping of the human personality's present historical phase. They gave immense strength to capitalism to destroy once and for all the feudal constraints on individuals, the dependence on other persons; this unfolded in man an incredibly rich and varied mass of abilities inconceivable before, partly by reducing labour time, partly by – with the help of instruments using heat and electric energy in transportation, telecommunications and the lighting industry (that brought the village closer to the town, brought together the continents, and lengthened the day) – widening the borders of free time so much that more and more populous segments of humankind could be reached by the collection of the cultural treasures of previous centuries.

Just as, however, capitalism started this splendid progress on the basis of heat and electric energy; just as, through the extension of human labour, it extended man himself, who is formed by labour, it simultaneously initiated another progress, as well, which turned this process of labour upside down, deformed it, and transformed it from the extensor of man into the spoiler, and then the remover of him. Because the product taken out of man's hands – dead labour – becomes alive in the commodity production, it gains a form of independent existence not ruled by, but ruling over, man, determining him, either when – like capital – it throws itself as an evil vampire onto living labour in order to suck over-labour with insatiable appetite, or when – like the simple product realising its value only on the market (or not realising it at all) – it deprives average men of the result of their labour, or forces them into crises and wars not to be influenced by human will. And not only the product of labour is separated from the worker, opposing him as an enemy power in this enchanted world, but the instrument of labour that once was his personal property, and the elongation of his hands becomes alien to him, even an enemy, something to be destroyed. If the guildsman's tool is the elongation of his fingers, then the modern worker at the production line is the mere elongation of his tool. The process of labour is understood only by the engineer who works in his laboratory outside the factory. Physical labour and mental labour are differentiated, they begin to fight, and in the meanwhile men sink to the condition of being partial workers: these are the characteristics of this strange distorted motion. Man disappears, and there remain turners and engineers, Germans and Frenchmen, townsmen and villagers, capitalists and workers. Labour is not forming, but deforming man. Men can only be men outside labour hours: what a contradiction! Heat and electric energy have brought people close to one another, but at the same time – under the supervision of capital – they squeezed them into blind character masks. Bringing them together was still a historical achievement, and it was achieved by the bourgeoisie through the vast extension of productive forces.

But what can the bourgeoisie do with nuclear energy? A society where production is not regulated by the law of value, where man is the master of the objectual aspects of his labour, could definitely use it; with it, the differences between physical labour and mental labour, village and town, nation and nation could be abolished, and a new historical form of labour could be shaped, of which the *Entfremdung* would no more be a constitutive characteristic – as Hegel believed it to be, always postulating the passing category of civil labour in the *"sich entfremdete Geist"* – which would no longer be measurable in hours; and of which the "tasteless story" of Menenius Agrippa will no more be true, for hands and brain would not be projected into different persons. Division of labour, quantitatively measurable labour, the irrationality of the product motion of labour: these can be left behind.

If, however, the productive forces that emerge in classical capitalism launch a *contradictory* evolution, then declining capitalism, with its last great productive force, nuclear energy, is able to do nothing *but* destroy. The brilliant source of humankind's future wealth has been distorted into a terrible weapon of the world bourgeoisie. The remarkable difference between the introduction of electric and nuclear energies is not a mere accident.

And the physicist? The physicist doesn't want to destroy. But the atomic bomb, the product of his labour, withdrawn from his hands, becomes his master in the same way as the shoe becomes the master of the worker when it is taken from his hands. *The physicist created the conditions of one of the most important leaps in humankind's development, the conditions of a society without classes; and the product of his labour has become a weapon in the war against the classless society, a weapon in the pompous fortress of capitalism.*

2. It cannot be surprising that such a discovery, with its unmatched significance in the history of humankind, causes crucial changes in human thinking as well. As a result of the exploration of the atomic structure, the alliance of metaphysical materialism and formal logic collapsed, though it had been ruling over natural sciences for hundreds of years, and it had connected the very concept of matter to the notion of material structure of the 18th and 19th centuries: the notion of rushing billiard balls in empty space with only measurable properties. As Du Bois-Reymond (since then classically) put it in 1872, "cognition of nature – that is, acquiring scientific knowledge of the world of bodies with the help of and in the terms of theoretical natural sciences – lies in reducing changes in the world of bodies to motions of atoms, motions caused by *time-independent* central forces. Hence cognition of nature is the dissolving of natural processes in the mechanics of atoms." Since then, it has turned out that this picture is an approximation valid only in earthly dimensions, and it is as useless in the million light year regions of the universe as in the tiny cosmos of the atomic nucleus. The rapid, eruptive development of science has made deeper and deeper demands on human thinking, which was used to intermediate dimensions when building scientific concepts: the exploding qualitative transformation of live and inanimate nature – the unfolding of matter's history full of revolutions – on the one hand, and the constant changing of the picture we conceive of nature on the other, have proved clearly that science is a reflection of an ever changing material world on an ever changing human mind, where the coincidence of image and object will never follow. The structure of matter is constantly changing – we can observe on a small scale how cosmic radiation, the mere oscillation of the electromagnetic field, transforms into masses bearing negative and positive charges[2] – and we are beginning to realise a vast tide (a tide in which the subsequent states are

connected not only by the thin chain of causation but by relations much broader, much more complicated, more alive) of which our earthly world is only a provincial corner in space and time with its generally peacefully balanced atomic configurations,[3] and with its slow particles with relatively constant masses.[4] In this idyll, however, everything is a mere historical category – even the carbon atom which, with its peculiar character, has launched this wonderful and maybe momentary awakening of matter to self consciousness. In the immense density chaos of distant stars[5] we would look in vain for this strange symbiosis of six neutrons and six protons,[6] not to mention the shell of six electrons which, with its fellows, is able to join in the magnificent round dance, the basis of the organic world,[7] but we would probably search in vain for the dialectics of protons and neutrons: matter is a dense ancestral mass of neutrons,[8] which hardly knows anything about the tension of electric charges dissolved in the dynamic balance of the atom, nor about the electric charge itself; here the opposites of matter are, as of yet, in a sleep of germs, ripening the emergence of diverse forms of their motion. Chemical laws are not valid yet, since there are no atoms that could fuse as molecules and then part again; and what meaning could Coulomb's Law have when electron and proton haven't parted yet, when positrons and electrons in a beam of gamma light are waiting for the proper historical moment[9] to show their corpuscular aspects? And if we go further, peeping into dim shapes with groping imagination, towards the final, undifferentiated unity of mass and energy, of mass and space, where, perhaps, existence is boiling without quantitative law, then it is not difficult to convince ourselves that the present state cannot be the final one (which is implied gloomily by the red shift in the spectra of stars, without knowing whether it suggests the explosion of our universe). Therefore, as the only thesis of materialism, there remains an infinitely broad Leninian axiom postulating an existence *independent of our mind*, of which our mind reflects constantly more and more, without ever exhausting it.

This picture differs qualitatively from the world view of classical physics, which described a seemingly ever moving universe – for example, it revealed the firm iron bars of bridges to be busy swarms of molecules which can, with certain probability, disperse at any moment – but this motion was governed by eternal laws. Even Hegel himself, who recognised the revolutionary dialectics of the history of society, thought of nature as the limit of his method. In his opinion, historical change, as opposed to nature where change is only circular, that is, the repetition of the same, takes place not just on the surface but in the notion. The notion itself – the law – is what changes.

The great achievement of modern physics is planting historicism into natural science (of course not in the narrow idiographic, Rickertian sense!), thus eliminating all types of nomothetic-idiographic dualism.[10] And here we can think not only of the results of astrophysics, which deprive the earthly forms of matter from their absolute and exclusive character, because we can think of those contradictory structures so frequent in modern physics: the opposites of mass and energy, of light and particle, of particle and wave, contrasts that exclude each other conceptually but are still inseparable, transforming into each other. (According to the classical scientific ideal, which could deal with problems only in which the identity or opposition of only two things could be examined at once, a scientific result, which postulates the identity and the opposition of two poles at once, is viewed as false: contradiction, therefore not true. The Hegelian dialectics for the case of the spirit, and its firm version, the Marxist economy for the case of the society, show that such contradictions do not point to the incorrectness of our opinions, but belong to the self-development of the spirit, to the essence of the motions of society. Modern physics has resolved the temporary splitting of natural and social realities that came this way.)

These dialectic structures are naturally not yet processed by logic at all today; but it is certain that the mere conceptual list of dialectical structures is absolutely unimportant and *épatez le bourgeois*-like as far as they are emphasised only as exotic (the more they upset our present intuition, the more effective they are); but they immediately become precious heuristic principles when we start looking for the historical dialectics lying behind the contingent, formal elements of conceptual dialectics. The opposition of the exchange-value and the use-value within commodity was already recognised by Ricardo. Marx's revolutionary deed was to show that the commodity, as a unity of these opposites, is a historical category, the antithetic phase in the dialectic line of product-commodity-product.[11] The problem is not simple. Since, after all, Marxist dialectics literally – conceptually as much as historically – is properly supported by examples only in *Capital*,[12] it is difficult to deal simply in analogies. One thing is for certain: conceptual dialectics is useful only in cases where it is backed by historical dialectics. *The appearance of conceptual dialectics in modern physics means the necessity of the historical method.*

We can see now that the world of eternal, universal ideas borrowed from mathematical conceptualisation failed not only in the field of the history of ideas, but in the natural sciences, as well. But if the old ideal in the mathematical sciences held that the cognitive subject must withdraw totally from the result of cognition (so that the image can really be objective), and that knowledge becomes truly scientific when all the subjective characteristics of the cognitive subject have vanished from it (*"Subjekt überhaupt"*), then historicism – the ideological intensification of class antagonisms following the wavering of the ideological supremacy of the bourgeoisie – denies even the possibility of such cognition, but naturally only in the field of social existence. But the legacy of the resulting dualism has met serious doubts owing to the Heisenberg uncertainty relation, which determines the smallest change one can cause in nature when touching it with the finest "*finger*," a photon or an electron. (Note that with humankind, biological evolution has virtually come to an end; humans evolve in their instruments of labour, as if these were elongated organs.) It is, of course, conceivable that the

Heisenberg relation is just a temporary, approximate result of science. One can assume that this minimal uncertainty can decrease or transform structurally. But it will have no influence on the fact that, while as a condition of scientific cognition, it has been held until now that the observer must not disturb the observed object, today this disturbance is acknowledged as a condition of scientific cognition, without which no effect would get to us at all. (The sociological condition of the emergence, and then the disintegration, of a world view from the primitive man's demonic view of nature, a world view in which nature is independent of man and is ruled by strict superhuman laws, upon which it is *possible* to reflect as an outer, passive "observer"; this condition can easily be explained by the theory of commodity fetishism.[13])

Atomic physics therefore has a huge importance concerning the general world view and human thought. *The rigidly separated – in subject as well as in method – disciplines of metaphysical materialism and historicism are taken over by the uniform cosmos of historical materialism.*

***

But the bourgeoisie is as incapable of exploiting the results of modern physics in philosophy, as it is in production. And like the atomic bomb, the practical result of the atomic physicist's work, having left the laboratory, becomes independent of his will, and then turns against him; so does the theoretical result of his labour, independent of or against his will, become a servant of the demands of the bourgeois class ideology, the authenticator of all kinds of fideist-idealist chaos in the ideological front of the struggle against socialism. This is how the questions of free will and of moral responsibility, the problems of materiality of existence, of God and the world, could move into the centre of the philosophical arguments concerning modern physics; such complexes, the examination of which did not at all help towards the specific developments of natural and human sciences. This is obviously necessary in the case of ideologists belonging to a class, whose existential uncertainty leads to an obsessive search for the elixir of life in every field, and which thus loses all objective connections to reality.

Two prototypes of these ideologists are *Jeans* and *Eddington.* In Jeans' case, the main question is the material, versus the spiritual character of existence. Susan Stebbing, in her book[14] on modern physical idealism, shows how Jeans[#] makes the picture of the universe subjective when he slips "senseless" distances and "lonely" stars into it, in order to get to a hostile, cold cosmos where we live our momentary lives as miserable nothings, where our lives are not at all purposeful or essential . This picture does not require any specific element from modern physics; Kant could have associated the same to the vast distances of the sky, instead of the experience of the moral world order. (Poetry of course reflects society more quickly and more sensitively than physical or philosophical world views: 50 years before Kant, Pope writes: "A mighty maze, but not without a plan," while 50 years before Jeans, Matthew Arnold says: "And we are here as on a darkling plain, swept with confused alarms of struggle and flight, where ignorant armies clash by night.")[15]

Out of this hopelessness leads, then, the Jeansian metaphysics: the cosmos is only a thought in God's mind. For if there is no "matter," there are no "real" physical relations, then the insignificance of man and the terrible emptiness of the world have ceased.

Hence it is not surprising that Jeans' philosophy stands on such a subjective, we could say, emotional base, and that his epistemological concepts are incredibly confused and contradictory. It is true, though, that exact, distinct thinking, and the existential "anxiety"[16] which makes all objective relations to the outer world impossible, are incompatible notions as well. It is no wonder, therefore, that Jeans, who thinks of nature as a subjective category, holds that Heisenberg showed: "nature has more horrors of exactness and distinctness than of anything else."

Jeans' nature actually is a social category; in it, the capitalist society is extended to be a cosmic vision. In capitalist society, the position of individuals is characterised – as an upside-down appearance of a social constraint stronger than in any previous social form – by the loneliness in the hostile, alien world depicted by Jeans, where existence is senseless, where man is nothing, where the mood of death depresses the souls. (The *realistically* disadvantageous cosmic perspectives do not seem to make a special impression on men: they did not at all take notice of the heat death of the universe made plausible by the entropy theorem!) But all these are not eternal, unchangeable natural laws, but a human network of relations that can pretend to be a natural law only in the extreme phase of the reification of society.

From Jeans' conception there follows unavoidably what Stebbing calls "the escape of Jeans": the fideist construction mentioned above by which he becomes a trivial apologist for the anti-human capitalism (the hopeless and senseless world is a mere illusion: drive matter out and the nightmare disappears). If "the world is purely thought" then the spirit is no longer an "accidental trespasser in the realm of matter" but something "creative and controlling." With this mental construction, however, Jeans could not put an end to the painful reality of capitalism reflected by his cosmos so faithfully; this – somewhat indirect – ideology can put at ease only those for whom the dehumanising effect of capitalist production is only momentary discomfort, "*Unbehagen*."[17]

The fact that in Jeans' last book[18] (published in 1942), we can find, instead of his well-known brave and certain style, only careful ideas that mention his old theories only as possibilities, is closely connected with the wavering of the feeling of security of the English bourgeoisie to an extent that cannot be recovered by *such* an ideological weaponry.

The basis of *Eddington's* philosophy is the three-fold structure of reality: there is a "familiar" world, the world of colours, sounds, fragrances; there is a "physical" world, the world of oscillations, potentials, Heisenberg-matrices; and there is an unknowable, "real" world, of which the physical world is only an empty shadow or frame. This "real" world used to be the realm of theologians, but now the roles have changed: if physicists used to refer to theologians as to the possessors of "higher" knowledge, then today theologians refer to idealist physicists: "science itself tells us so." Modern physics, of course, does not postulate such an unknowable world any more than classical physics did. The Eddingtonian three-fold splitting of the world is still not Eddington's arbitrary mental construction but the historical result of an ideological process, necessary in its main lines. In the following, we attempt to outline this process.

This was not the first example in history of such an attack against "the physical world." Think of *Goethe*. He, however, started his attack from the "familiar" world, and not from an unknowable world, as did Eddington. "Goethe's theory of colours must be viewed as an attempt, which is meant to save the reality of sense impressions from the attack of science" – writes *Helmholtz*.[*] Indeed, Goethe claims not to "seek after something behind the phenomena"[**] – he voices "the superiority of ordinary human reason" over "the torture chamber of the concepts."[***]

But how did concepts become torture chambers ? How did this mental (yet two-fold) splitting of uniform reality happen, characterised by Heisenberg as analogous to "Newton's and Goethe's theories of colour dealing with two different layers of reality"?[19] According to *Whitehead*[20] who illustrates the same process in the case of Wordsworth and Shelley – this originates from the discord[****] of man's aesthetic intuition and the mechanism of science. This discord, however, is not an eternal category at all. The same quantitative and mechanistic structures that Wordsworth felt to be too narrow, and which were "torture chambers" for Goethe, had mediated the harmony of the spheres for *Kepler*.

The explanation for the phenomenon is simple, on the whole. When products became commodities (things deprived of every perceptible feature, excepting exchange-value), and when the concrete, directly human form of production was transformed into a rationalised mechanism, torn into partial processes measurable with a watch – then the whole field of human knowledge, from geometry through physics and biology to ethics, was flooded by a new set of abstractions (the notion of the quantitative in the first place). This new group of concepts, despite its being *objective, discovering the essence of nature*, and therefore independent of its manifestation in the motion of commodities, has unrecognisably fused with the spiritless number-world of the motion of commodities, with mechanical nature prevailing over man in capitalist production. Therefore, in a society where class rule appears as a "feature" of machines to rule over men, where social and natural relations – as a result of social relations becoming natural ones – seem to transform into one another, there a realistic illusion arises by necessity that quantitative concept-building, and hence the concept of "physical world," is "responsible" for the dehumanisation of capitalism. This dehumanisation, however, does not unfold itself in the era lasting for centuries when the capitalist production is yet an island in the feudal social order: in this time the capitalist is the citizen, the representative of humanity, the hopeful deposit of its infinite extension, just like quantitative concept-building seems to open up infinite perspectives for science – and for the social success of science. But bourgeois reality soon changes these perspectives into illusions.

In the "physical world" natural things become mere coat-racks for numbers – which reflect their objective essence! –; but at the same time they, as reificated forms of human relations, obtain quantitative features: the land will be characterised, for example besides by its potassium content, by its annuity, or gold will be described not only by its density and atomic mass, but also by its interest. And just as we have characterised the Sun by giving the numerical value of its energy, then the energy of man appears in joules as well.

But while the characteristic mathematical data of the "physical world" supersede neither the beauty of the Earth nor the brilliance of gold, the quantities of reificated nature turn grey both man and nature in an imperialist way, and deprive them of all other features.

In reality, therefore, we are indeed faced with a double world: the unity of the non-excluding opposites of the physical world and the familiar world which reflects the *natural* existence of things and, as an opposition to it, the fetish world reflecting the *social* existence of things. But the coincidence that the dialectical structure of the motion of commodities and of factory production turned out to be isomorphic with the structure of the essence of nature, and the fact that man was led by the economic unfolding of his own historical movement to, so to say, resound with this isomorphism, strengthen to a vast extent the objectual skin of social relations that was necessarily deposited in commodity production, and the objective splitting is forced to appear in the seemingly rigid opposite of familiar and physical worlds; the physical world (actually, through the development of productive forces, the extensor of humanity!) will seem to drown humanity, and the familiar world will seem the *only* object of direct human relations to the world. In reality, natural nature is obviously restricted to the familiar world only in the eyes of the men of feudalism. Wordsworth and Goethe stay imprisoned in this restricted narrow perspective when they start their magnificent attacks on dehumanised nature.

But if this is the war of humanism against capitalism (admittedly, in a restricted perspective), then the situation is absolutely different in the case of the present criticism of science in the name of humanism. The problem is new even in its form: up to now, we have observed the splitting of our picture of nature into two, but in the new world view, we can speak of a three-fold split.

There arises the question of why criticism of the physical world cannot appear from the side of the familiar world in the age of imperialism, as well. This is so because even the memory and the possibility of all realistic forms of non-reificated man-nature relations have vanished from the intuition fascinated by the fetishes of capitalist production. But does this mean that, with the vanishing of our critical standpoint, the physical world, as the new relation of man and nature, can be established without contradictions? No; the criticism remains there, even as a formless, momentary *Unbehagen,* in order to proceed soon from a new direction: from the direction of the *unknowable world.*

In Goethe's criticism of Newton, a crucial role was played by the fact that, in Goethe's time, the physical world had not yet penetrated sufficiently into technology, it had not measured itself in the vast practical field that it created later in the 19th century. The Goetheian standpoint, according to which man's power is greater in the familiar world, can arise only in a society where the productive forces have not yet been transformed and suddenly increased by physics. Goethe is answered from capitalism by *Heisenberg*, who holds that for our scientific world view – which establishes our domination over nature – "we had to pay a great price," we had to give up our direct relation to nature.[21] Thus the capitalist development splits progressive humanism into the Heisenbergian "progressio," neglecting humanity and the Goetheian "reactionary" humanism. But now we want to emphasise that the vanishing of the familiar world as a possible relationship between man and nature, and the further evolution of reification, are made possible, even necessary, by progress, by the (specifically capitalist) development of the productive forces.

The argumentation of Heisenberg, however, does not prove adequate at all when examined in more detail. In the age of world-wars, the development of physics does not further increase man's domination over nature, and thus the physical world seemingly becomes a bloodless, abstracted (from reality) empty frame in the negative sense, for which it is not worth making any Heisenbergian "sacrifice" at all.

This is the way in which all concrete forms of relation to nature become senseless for man. Thus the desire is increased to get to *real* nature in another way, which is different from all previous ways, be it the Bergsonian "épouser l'essence des choses," or any other form of direct cognition without concepts. What is blind for the progressive citizens, like *Anschauung ohne Begriff,* becomes the only source of light for the declining bourgeoisie! And so – in the hopeless deadlock of these experiments – the unknowable nature is born. Nevertheless, the unknowable world signifies the failure of bourgeois rationality.[22] As Goethe could not separate the physical world from the reificated world, so does modern physical idealism confuse the failure of the bourgeois reason towards nature, with the failure of reason itself towards nature. Bergson and Eddington are victims of the same ideological shift; both of their theories show the bourgeois ideology becoming illusion, its purposes becoming unreachable, its rationality turning to irrationality.

How incapable the bourgeois ideologists are of evaluating objectively the important logical results of modern physics, is especially well shown by the example of the *principle of causality.* Modern physics has undoubtedly modified our notion of causality. It showed that the universe is a machine which is not perfectly fixed, it has a certain allowed "play," determined by the

Planck constant. Only the smallness of the Planck constant, compared to the quantities of our world, makes it possible that the statistical laws of microphysics can take the form of causal laws that are valid, by strong necessity, in the macro-world.

But this is not the first time that physics has modified the principle of causality. Causality – as the general but concrete law of the objective relations of the material world – will undoubtedly appear to be a historical category, and its mental reflection has already proven to be a historical category. We must not forget that one of the most basic laws of the macro-world, the Newtonian law of motion, finds a cause for the change in velocity, but not for the change in location. (This was not understood by Kepler, and this is why he could not interpret his kinematic laws in terms of dynamic ones. Since he – standing on the ground of a blind notion of causality – was looking for the cause of *change in location* in forces effective along the tangent of the *orbits* which, naturally, were not found.) In this sense, we can say that just as modern physics holds that the familiar principle of causality is made possible by the minuteness of the Planck constant, so does classical physics hold that this principle is based on the smallness of n in the law

$$m\frac{d^n x}{dt^n} = K$$

(n is not greater than 2); were it not valid, motions would have a much greater – in a sense different from the one used in quantum mechanics – *"acausal"* play.

It is now obvious that the mental reflection of the objective respect of causality, the law of causality, has reached a more advanced state of development. The only remaining problem can be the one of whether this development has any influence on the question of free will. The answer is: no, because the brain processes initiating human actions are macroscopic, and the present phase of the theory of causation offers, as of yet, no guideline to deal with the problem of free will.

But such an objective evaluation of scientific problems is not possible in a society which expects from physics, "guidance too, a world view that guarantees the utmost earthly good, the inner peace of the soul,"[23] but without the criticism of the capitalist relations of production.

And so it happened that newer developments in causality divided physicists into two opposing groups: the "jubilants" and the "mourners."[24]

Capitalist society in its imperialist era puts more and more shackles on the individual, surrounding him with its drowning, seemingly objectual net in which he cannot move. Limitation and freedom, of course, are generally not mutually excluding concepts, but presuppose one another. In the capitalist society these split into – realistically – rigid opposites. I cannot go now into the details of this process, which is overwhelmingly peculiar to the capitalist order. I can only hint that while Goethe, for instance, could still say that "in der Beschränkung zeigt sich erst der Meister, und das Gesetz nur kann uns Freiheit geben"[*****], this conception is totally alien to the art of modern civil society; epigonal classicism on the one hand and the anarchy of -isms on the other have taken over the place of the great realism. Only two standpoints are possible within the capitalist society: the acceptance of limits, "trust in, and total devotion to, the higher power ruling our whole lives,"[25] or the absolute rejection of limits in the *action gratuite*.[26] The dissolution of the unity of limit and freedom is therefore not only a subjective appearance, not only an "error" in the ideology of bourgeois thinkers, but a very real appearance, the real reflection of a real historical process. The only way towards the tertium datur is socialism, which absolutely transcends the capitalist society.

Our two poles are clearly demonstrated in the ideological movements swarming around physics: the *action gratuite*, this miserable and distorted copy of freedom, is recognised and seen verified by the "jubilants" in the unpredictable, playful electron of quantum mechanics. But the "mourners" turn against this groundless anti-capitalist rebellion, distorted to be reactionary in its result, and they, starting from the incompatibility of the *action gratuite* and human dignity, from this seemingly right argument which essentially idealises the capitalist reality (for where in capitalist society is any human dignity?), reach the conclusion that the domination of causality must be restored in quantum mechanics, and that only causality (in its old form!) as an eternal category makes scientific thought possible.[##] Hence, the bourgeois ideology today leaves only two paths open: the paths of irrationalism and of the restriction of rational theory.

We can see, therefore, how declining capitalism turns back all the advancements of humankind; how the most brilliant, most revolutionary era of natural sciences is reflected, in the appearance of the seemingly insoluble crisis, by a scientific public opinion fascinated by the sight of the capitalist crisis. Naturally it is not a pure coincidence that, while the productive kinds of capitalist energy produced an atheist and rational atmosphere, the mass murderer "atomic energy" hypocritically arouses religious and irrational ideologies everywhere. This is how an idealist monism can be built from the blocks of a uniform materialist world view.

***

3. As capitalism has, from the individual producer, moved by way of the concentration of the instruments of production to institutionally organised production – which is hence systematic in its cells, but chaotic across the whole range – so has it, in science, shifted from the isolated scientist, by way of the growing concentration of the scientific apparatus, to institutionally organised scientific work – which is hence systematic in its cells, but chaotic across the whole range. Moreover, in the crisis of the bourgeois world power which by today has become permanent, when scientific equipment worth millions is concentrated in one laboratory that works strictly according to a systematic plan, then not only is it impossible to speak about planned scientific work in a world-wide range, but the internationality of scientific

research also vanishes; it falls into hermetically isolated Anglo-Saxon, French, Russian sciences. As the exchange of the products of industrial production, world trade, has ceased, the exchange of scientific results, international science, has ceased in the same way.

And, as in the capitalist development of production, a parallel process has started, of which the essence is to deprive and to alienate workers from their instruments of production, so the development of science has led, through the abolition of private libraries and of private laboratories, to scientific factories owned by the bourgeois state or a certain tycoon.

***

The purpose of this brief discourse has been to show how closely the natural sciences are tied to society, how strongly their development is determined by the economic foundation. We saw that, as in capitalist society, the development of productive forces involves the seeds of their limitation, the general theoretical advancement launched by the natural sciences is distorted into the narrow canon of class ideology, and scientific organisation cannot exceed the state of development of the organising forms of production. But does the dialectics of conceptual development not remain (delayed or accelerated, perhaps changed in its motives by the speed of social progress) untouched? Will physics, therefore, not be a continually more and more exact reflection of nature? In this way, through the development of conceptual reflection, we can discover the pure, immanent sphere of science, and only its radiation, the material limits of its motion, can be touched by the distortion of a society struggling with controversies!

But what should we mean by "more and more" exact reflection? Conceptual development is not a quantitative development that renders ever larger concentric circles of exactness. Nor it is "logical," for conceptual development changes logic as well. So what guarantees the "more and more" exact system of concepts?

Does the sociological sphere not penetrate into scientific concept-building itself?

Dealing with this question on this occasion falls outside the limits of this paper. But it is at any rate certain that the examination of the subjective, sociologically determined aspects of scientific concept-building can only support us in the belief that the objective aspects play an ever larger role in the development of science; nature, which is independent of the human mind, is reflected more and more completely, in an ever larger domain of human practice, in our "physical world."

Notes

1. *Marx: Das Kapital.* [*Capital*, Vol. I, Part III, Ch. 5, Sect. 1 (*The Labour Process*), second paragraph].
2. Materialisation of gamma rays.
3. Protons, neutrons and electrons, the stable forms of this dynamic balance, are rather rare phenomena in the universe (the word "generally" refers to unstable systems, the radioactive elements).
4. According to the theory of relativity, the mass of a particle travelling at a considerable velocity compared to the speed of light is increased compared to its rest mass.
5. The latest findings of astrophysics tell us that in some stars, matter is condensed at extremely high density.
6. The carbon nucleus consists of six protons and six neutrons, surrounded by an electron cloud of six electrons.
7. We think of the benzene ring here.
8. See Landau's paper in Nature, 1938, p. 333 ("neutronic state of matter").
9. It must be near a heavy nucleus, in order for the impulse law to be satisfied.
10. In Rickert's opinion, reality becomes nature if we look at it with regard to the universal, and becomes history if we look at it with regard to the particular and the individual. Having this in mind, Rickert – following Windelband – opposes the universalising, nomothetic procedure of natural sciences to the individualising, idiographic procedure of historical sciences.
11. *Molnár, Erik:* Dialektika, p. 116.
12. On the logic of *Capial* see: Molnár: Dialektika, pp. 88–117, and Fogarasi, Béla, Marxizmus és logika. [*Marxism and Logic*].
13. Lukács, G. Geschichte und Klassenbewusstsein. [*History and Class Consciousness*] Berlin, 1923, p. 109.
14. *Stebbing:* Philosophy and the Physicists, Penguin edition, 1937.

\# Beginning of the part which is greatly identical to the body of the paper *Criticism of Physical Idealism.*

15. *Whitehead:* Science and the Modern World, pp. 99–101. [Lakatos quoted the original English text in the footnote. The cited poems are: Alexander Pope: *Essay on Man*, and: Matthew Arnold: *Dover Beach*].
16. Its philosophical representation is the Heideggerian *Angst*.
17. *Freud:* Das Unbehagen in der Kultur.
18. *Jeans:* Physics and Philosophy, 1942. Published in Hungary in 1945: *Új fizikai világkép.*

* In: On Goethe's Scientific Researches, p. 359. A lecture from 1859.

** In: Goethe: Scientific Studies, Suhrkamp Edition, Vol. 12, p. 307.

*** Loc. cit.

19. Heisenberg's talk in Budapest on Goethe's theory of colours. In: *Matematikai és Fizikai Lapok*, 1943 (Translated by Faragó Péter).
20. *Whitehead:* Science and Modern World, 1926. [See chapter: *The Romantic reaction*].

**** Lakatos uses the term "diszkordia."

21. *Heisenberg:* Wandlungen in den Grundlagen der Naturwissenschaft, Leipzig, 1935 [in Hungarian: *Változások a természettudomány alapjaiban*, Egyetemi nyomda, 1946].
22. This cannot be applied mechanically to the Kantian epistemology, because the latter is based on Wolffian dogmatism – and is therefore directed *mainly* against feudal ideology.
23. *Planck:* Sinn und Grenzen der exakten Wissenschaft [*Die Naturwissenschaften* 30 (1942), pp. 125–133].
24. Eddington's terminology.

***** A master only first reveals himself in the face of limitation, and law alone can give us freedom.

25. Planck, loc. cit. p. 131.
26. "Free action" that falls outside every causal chain.

\#\# End of the part common to both publications.

SÁNDOR KARÁCSONY'S FIRST EVALUATION

[Archive file number: XXVI. 3/b. 832/1946–47.]

Imre Lakatos' dissertation brings together three originally distinct chapters. The core of the dissertation is the third chapter, which was probably written first, and its problems called for writing the first two chapters as preliminaries. Lakatos argues that scientific concept-building is not objective, not the alphabet of nature, but stand in relation to and is a function of the prevailing social structure. Therefore, natural science, during the course of its progress, occasionally transcends its proper system of concepts, as long as socialist society has not yet reached its static state of rest, at which time it will force scientific concept-building into a form more independent of time and facts.

When dealing with this smartly formulated problem, Imre Lakatos approaches the very same (and unsolved) question I raised in my paper "Reform of our Scientific Life."[1] It is a serious defect of our scientific world view that our relation to that other entity [*valami*] is not objective enough – what we know is not nature itself but the method by which we try to come to know nature.

In the second part of his dissertation, Imre Lakatos deals with the sociology of the sociology of knowledge [sic!], and he concludes that our relation to that other entity can never, and in no way, be objective. He is led by Scheler and Mannheim along this line, but I think they lead him to a dead end. According to Lakatos, synopsis [*egybelátás*] is an irrational and aristocratic operation, and it represents the standpoint of the "intellectual elite" floating above classes. The intelligentsia, glorified as *Subjekt überhaupt*, is faced not with the object itself but with polarised judgements which are determined, in the first place, by social existence. This is the opinion of the bourgeoisie that won over feudalism, and it can be held only as long as it remains humankind's general opinion too, that is, as long as the word of the proletariat is not yet voiced independently.

We must not forget, however, that in our author's view, physics and society, as well as the slow process of nature's cognition and history, are conjoined. This is most obvious from the first part of his dissertation, where he connects modern physics with modern society. He emphasises three aspects: 1) the exploration of nuclear energy as a productive force, 2) the development of newer physical terminology in terms of epistemology and logic, 3) the preparation and the establishment of large scale forms of work in the natural sciences, and of their planned co-operation. First he repeats the well-known Marxist theses (human nature is changed not by "spiritual renewals," nor by "re-education," but by transformation of the mode of production,[2] etc.). Then he goes on to argue that the bourgeoisie can do nothing else with nuclear energy but destroy. In the second section he shows that, in philosophy, capitalism cannot handle the newer findings of physics. He continues Susan Stebbing's line of thought, who recognises the peculiarities in Jeans' and Eddington's works, but does not care about their causes.

The reason for Jeans' "senseless" distances and "lonely" stars is his capitalist, civil way of thinking, as opposed to facts that have already grown out of this context. The same applies to the three-fold structure of Eddington's reality, but this time the physical world is attacked not from the "familiar world," but from the "unknowable world."

Lakatos clearly recognises the process of atomisation in scientific life, and he dwells on this point in the third section. Today we cannot speak about international research works, but only about hermetically isolated Anglo-Saxon, French, or Russian ones. We fear that physics will not be the more and more exact reflection of nature, and the sociological sphere penetrates deeper and deeper into scientific concept building.

Imre Lakatos' thinking is not quite systematic yet, and there is a very lively intuition. (Why is he so much afraid of it!) But his examples are interesting, and the way he raises problems is fruitful. It provides a way to deal with scientific problems in general. It can show the way to all the problem situations in scientific life. I warm-heartedly accept his dissertation as summa cum laude basis for the oral examinations.

Debrecen, 30 June 1947
Sándor Karácsony

SÁNDOR KARÁCSONY'S SECOND EVALUATION

[Archives of Debrecen University Library, file XXVI. – 3/b, 217/1947–48.]

Extract

from the Record of the 15th Session of the Faculty of Humanities, Debrecen University, held on 8 July 1947.

No. 182

Dr. Sándor Karácsony, full university professor, presents the following proposal:

Honoured Faculty! At our university, Imre Lakatos passed the doctoral examination on July 1st of the present year, with philosophy as the main subject and mathematics and theoretical physics as secondary subjects"[3] with the distinction, "summa cum laude."[4] The topic of his dissertation is: "The socio-historical aspects of scientific concept building."[5] Both the dissertation and the oral examination proved that the candidate's knowledge is rich, of high quality, and above average.

I became interested in the previous scientific activities of this young man, partly because I read most of them soon after they were published, partly because my fellow opponent[6] formed the opinion at the oral exam that, as it was wished and asked by the candidate, he should be awarded the sub laurea Almae Matris[7] doctorate. At this point, I can see all of Imre Lakatos' work as a whole, and I am of the opinion that it meets the standard. His dissertation is not based on a sudden idea; it matured through two previous publications, both published in very serious journals: the first in *Athenaeum*,[8] under the title "Physical Idealism,"[9] and the second in a thick volume written for educators, *Further Education and Democracy*,[10] entitled "Modern Physics, Modern Society."[11] He published most of his papers in *Valóság*,[12] in the following temporal order: "Molnár Erik: Dialectics," "Citoyen and Working Class."[13] The latter is a response to Imre Csécsey's book, the "Enlightened Moment."[14] Both essays are much more than simple book reviews. In the first, one can read corrections of this kind: "It is not a mere coincidence that Molnár fights against the physical vulgar-materialism, at a time when the fight against the Jeansian physical idealism is so timely now." – he intends, therefore, to repeat in its original form the proletariat's message, which censorship distorts into esoteric jargon. The other essay is more than a review even by virtue of its form. It is a polemic divided into four chapters between the radical citoyens and the proletariat, in which the author takes the side of the latter. He is answered by Imre Csécsey, who respects his adversary as someone of his own level. The situation is the same in their second polemic. In the journal *Forum,* Lakatos examines the role played by *Huszadik század*[15] in the past, and he criticises its role in the present. Csécsey, the editor of the magazine[16] reflects upon this article in the same magazine, and it is evident from his style that he holds the young author in high respect. *Valóság* also published another paper by Imre Lakatos, entitled "Eötvös College – Györffy College."[17] This essay has provoked a lively debate among many former and present students of Eötvös College (*Valóság*, *Március Tizenötödike*, and twice the *Válasz*, also twice *Szabad Nép*, *Tiszatáj*, *Köznevelés*, *Ember-nevelés*);[18] but this debate has been wholesome and fertile, and even when they expressed strong opposition to the motivating essay, they did not despise or revile it. This article is a criticism of past and present education, though it is a bit biased (meaning that it is somewhat dogmatic, class-struggle minded, and it requires the past to meet today's standards), but at the same time it is consistent, high-quality and moderate.

The journal *Embernevelés*[19] also published a paper by Lakatos, which bore the title: "Democratic education and scientific world view."[20] Its most essential statement is that democratic education teaches humbleness towards the facts, and that it imbues the student with the desire to face reality instead of mere views. The original democracy of natural sciences is to be emphasised: their facts and theories can be controlled by anyone, and this control drives them forward.

The foregoing scientific works of Imre Lakatos are based on dialectic Marxism, but in its modern and not in its orthodox form.[21] And it is only a basis, since he himself has original and particular things to say, even more so now than earlier. His originality is increasing. The philosophy behind all his opinions is consistent and systematic. Considering all that has been said, I recommend him warm-heartedly for the sub laurae Almae Matris doctoral degree.

The Faculty accepts this proposal and, as Imre Lakatos satisfies all the demands and conditions prescribed by the related guidelines and by the Faculty itself, the doctoral appointment of Imre Lakatos is hereby submitted to the Council.

This record is authentic:
Debrecen, 12 July 1947

OTTÓ VARGA'S EVALUATION

[Archive file number: XXVI – 3/b. 832/1946–47.]

In the history of human thought, rationalism and irrationalist romanticism have been fighting since the 18th century. The latter does not find the categorical forms of reason's thought adequate for the interpretation of life and the world, and it turns toward the irrational aspects. Thus, an organic view is created. Laplace saw the world as a reasonable system of elements, and sought after a formula from which the future state of the world could be derived; but after him the basic notion of totality has been given an increasing importance, according to which the parts of which the universe is composed are connected to one another not mechanistically but teleologically. Therefore, the 20th century advocates the primacy of the spirit, it prefers spiritualism to materialism. Each science is a different apparent form of the very same essence and should be handled with this principle in mind. In philosophy, in social sciences or in economics, the consequence of this view is Spann's system: universalism.

In his dissertation, our candidate points out these changes and contradictions in 19th and 20th century thought. He especially examines the view of Du Bois-Reymond, who was a typical representative of rationalism in the natural sciences and stated that cognition of nature follows from the mechanics of atoms. This view is confronted with the present ideas of quantum mechanics concerning the structure of atoms which follow from Heisenberg's uncertainty principle. This latter view states that atoms and their building elements, the various corpuscles, have a certain individuality from a physical point of view. Lakatos argues that this individuality means irrationalism as opposed to rational physics. He thinks that the dissolution of contradictions is made possible only when we view any scientific statement as a creation of the society which created that science – that is, truth is always a function of history. As a consequence of this opinion, he emphasises the indispensable unity of the natural and the moral sciences. Of course, he sees this unity not in the spirit, and his world view is not a continuation of Hegel's irrealist system – Spannism – but, starting from Marx's historical materialism, he seeks after, and thinks to find, this unity in the natural sciences, in the comparative analysis of classical and modern physics.

Imre Lakatos starts to examine this problem with a profound analysis of natural scientific concepts, but his solution is essentially based on the method of dialectical materialism. As I am only the second opponent of this dissertation, my task here is not more than the evaluation of its scientific aspects, and thus I conclude that, from this point of view, it shows a surpassing knowledge of the subject, and shows a reliable judgement. Therefore, I accept this dissertation as summa cum laude basis for the oral examinations.

Debrecen, 26 June 1947
Ottó Varga

NOTES

1. "A tudományos életünk reformja", *Forum* 1946/4 (December), pp. 285–293. This was the introductory essay of that number, and it was motivated by Lukács.
2. Note that in the paper "Modern Physics, Modern Society," Lakatos writes that "human nature is changed *primarily* not by ..." (italics in the original).
3. That is, the candidate received the best possible grade in every subject.
4. As a rule, every philosophy student had to choose aesthetics and pedagogy as secondary subjects for the doctoral examination. Ottó Varga, however, filed a petition with the Faculty (on the 30th of May, 1947) that Lakatos be allowed to take mathematics and physics, because "the subject of his dissertation belongs to natural philosophy, which is a border field between physics and mathematics."
5. As we know, the full title was "On the sociology of scientific concept building" [In Hungarian: *A természettudományos fogalomalkotás szociológiájához*].
6. Ottó Varga.
7. *sub laurea Almae Matris*: it means that both the exam and the dissertation were *summa cum laude*.
8. *Athenaeum*: Philosophical journal of the Hungarian Philosophical Society and the Hungarian Academy of Sciences.
9. The full title is "Criticism of physical idealism" [*A fizikai idealizmus bírálata*].
10. [*Továbbképzés és demokrácia*] This volume consisted of a collection of scientific and philosophical papers written for schoolteachers.
11. [*Modern fizika – modern társadalom*].
12. "Reality": A journal for social sciences.
13. [*Molnár Erik: Dialektika* and *Citoyen és munkásosztály*].
14. [*A világos pillanat*] Imre Csécsey was the president of the Hungarian Radical Party.
15. "Twentieth Century": An old journal for social sciences.
16. This is a mistake: Imre Csécsey was never an editor of *Forum*.
17. [*Eötvös kollégium – Györffy kollégium*] These colleges represented two very different kinds of educational institutes in Hungary. The Eötvös College was (and is) an exclusive and elitist educational institute, while Györffy College presented a high quality popular education. Lakatos suggested that Eötvös College be transformed following the model of Györffy College, because Hungary needed a new kind of popular elite.
18. The English titles of these journals are: "Reality," "Fifteenth of March," "Reply," "Free People," "Tisza-country" "Common Education," "Human Education."
19. "Human Education": A pedagogical journal.
20. [*Demokratikus nevelés és természettudományos világnézet*].
21. By the "modern" form of Marxism, Karácsony probably means Lukács' philosophy.

LAKATOS' HUNGARIAN PAPERS BETWEEN 1946–1956

This bibliography arranged by László Ropolyi contains, to the best of our knowledge, all of Lakatos' scientific or philosophical papers which appeared in Hungarian in the years between 1946–1956. We have taken into account the bibliography provided in Lakatos' *Philosophical Papers* (*Vol. I–II.* Cambridge: Cambridge University Press 1978) and those papers published in recent years mainly in Hungary [14]. We corrected some wrong data or misprints where necessary. The bibliography does not include newspaper articles presumably written by Lakatos. Papers [3, 9] represent the only available sources to reconstruct the content of Lakatos' lost dissertation.

[1] Imre Lakatos, "Molnár Erik: Dialektika" (Dialectics by Erik Molnár – Book review), in: *Valóság* 2, 3–5, 1946, pp. 75–78.

[2] Imre Lakatos, "Citoyen és munkásosztály" (Citoyen and proletariat – Book review), in: *Valóság* 2, 6–9, 1946, pp. 77–88.

[3] Imre Lakatos, "A fizikai idealizmus bírálata" (Criticism of idealism in physics – Book review), in: *Athenaeum* 31–32, 1945–1946, (appeared in 1946) pp. 28–33.

[4] Imre Lakatos, "Eötvös Collegium – Györffy Kollégium" (Eötvös College – Györffy College), in: *Valóság* 3, 2, 1947, pp. 107–124.

[5] Imre Lakatos, "Természettudományos világnézet és demokratikus nevelés (Válasz Csiszár Vilmos cikkére)" (Scientific worldview and democratic education – A reflection on Vilmos Csiszár's paper), in: *Embernevelés* 3, 2, 1947, pp. 63–66.

[6] Imre Lakatos (signed by the monogram I. L. only), "A januári Társadalmi Szemle... – Lapszemle" (The January issue of the journal Társadalmi Szemle ... – A journal review), in: *Embernevelés* 3, 2, 1947, pp. 90–91.

[7] Imre Lakatos (signed by the monogram I. L. only), "Jeges Károly: Megtanulom a fizikát" (I learn physics by Károly Jeges – Book review), in: *Társadalmi Szemle* 2, 6, 1947, p. 472.

[8] Imre Lakatos (signed by the monogram I. L. only), "John Hersey: Hirosima" (Hiroshima by John Hersey – Book review), in: *Társadalmi Szemle* 2, 7–8, 1947, p. 615.

[9] Imre Lakatos, "Modern fizika, modern társadalom" (Modern physics, modern society), in: Gábor Kemény (Szerk.), *Továbbképzés és demokrácia.* Budapest: Vallás- és Közoktatásügyi Minisztérium nevelésügyi osztálya 1947, pp. 347–369.

[10] Imre Lakatos, "Huszadik Század. Társadalomtudományi és politikai szemle" (Twentieth century. Social scientific and political review – Critical review of the journal Huszadik Század), in: *Forum* 2, 1947, pp. 316–320.

[11] Imre Lakatos, "Vigilia" (Critical review of the journal Vigilia), in: *Forum* 2, 1947, pp. 733–736.

[12] Imre Lakatos, "'Haladó tudós' a demokráciában" ("Progressive scientist" in the democracy), in: *Tovább* 2, 8, 1947, p. 3.

[13] Imre Lakatos, "Az atomkorszak új betegségei" (New diseases of the atomic age), in: *Tovább* 2, 20, 1947, p. 13.

G. Kampis, L. Kvasz and M. Stöltzner (eds.), Appraising Lakatos: Mathematics, Methodology and the Man, 375–376.

[14] Imre Lakatos, "Felszólalás a Petőfi kör pedagógusvitáján 1956 szeptember 28-án" (Talk on the pedagogy meeting of the Petőfi-circle on 28th September 1956), in: András B. Hegedűs and János M. Rainer (Szerk.), *A Petőfi Kör vitái hiteles jegyzőkönyvek alapján VI. Pedagógusvita*. Budapest: Múzsák – 1956-os Intézet 1992, pp. 34–38.

Index of Names

G. Kampis, L. Kvasz and M. Stöltzner (eds.),
Appraising Lakatos: Mathematics, Methodology and the Man, 377–382.